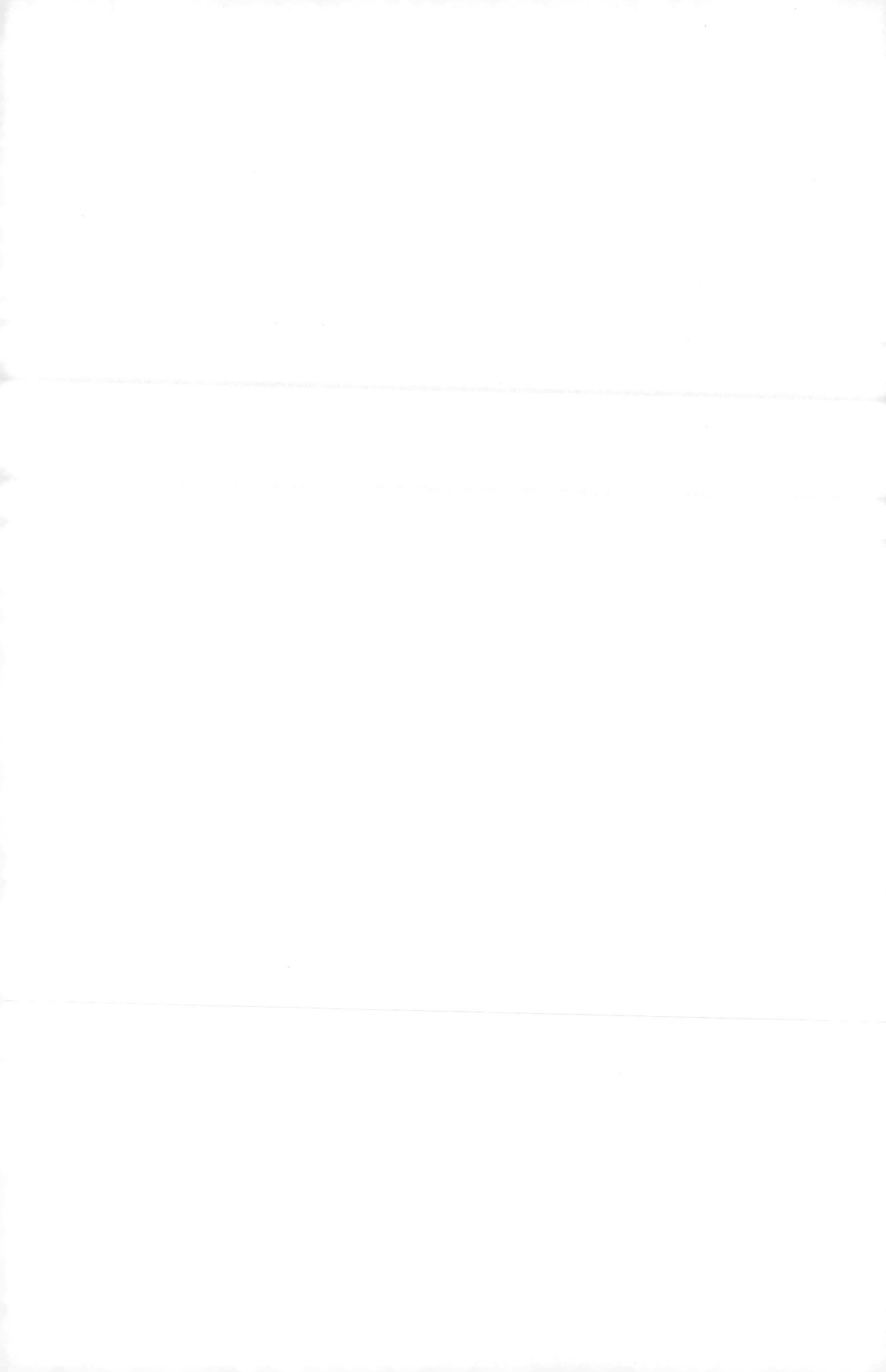

Fast NMR Data Acquisition
Beyond the Fourier Transform

New Developments in NMR

Editor-in-chief:
William S. Price, *University of Western Sydney, Australia*

Series editors:
Sharon Ashbrook, *University of St Andrews, UK*
Bruce Balcom, *University of New Brunswick, Canada*
István Furó, *Industrial NMR Centre at KTH, Sweden*
Masatsune Kainosho, *Tokyo Metropolitan University, Japan*
Maili Liu, *Chinese Academy of Sciences, Wuhan, China*

Titles in the series:

1: Contemporary Computer-Assisted Approaches to Molecular Structure Elucidation
2: New Applications of NMR in Drug Discovery and Development
3: Advances in Biological Solid-State NMR
4: Hyperpolarized Xenon-129 Magnetic Resonance: Concepts, Production, Techniques and Applications
5: Mobile NMR and MRI: Developments and Applications
6: Gas Phase NMR
7: Magnetic Resonance Technology: Hardware and System Component Design
8: Biophysics and Biochemistry of Cartilage by NMR and MRI
9: Diffusion NMR of Confined Systems: Fluid Transport in Porous Solids and Heterogeneous Materials
10: NMR in Glycoscience and Glycotechnology
11: Fast NMR Data Acquisition: Beyond the Fourier Transform

How to obtain future titles on publication:
A standing order plan is available for this series. A standing order will bring delivery of each new volume immediately on publication.

For further information please contact:
Book Sales Department, Royal Society of Chemistry, Thomas Graham House, Science Park, Milton Road, Cambridge, CB4 0WF, UK
Telephone: +44 (0)1223 420066, Fax: +44 (0)1223 420247
Email: booksales@rsc.org
Visit our website at www.rsc.org/books

Fast NMR Data Acquisition
Beyond the Fourier Transform

Edited by

Mehdi Mobli
The University of Queensland, Brisbane, Australia
Email: m.mobli@uq.edu.au

and

Jeffrey C. Hoch
UConn Health, Farmington, CT, USA
Email: hoch@uchc.edu

THE QUEEN'S AWARDS
FOR ENTERPRISE:
INTERNATIONAL TRADE
2013

New Developments in NMR No. 11

Print ISBN: 978-1-84973-619-0
PDF eISBN: 978-1-78262-836-1
EPUB eISBN: 978-1-78801-135-8
ISSN: 2044-253X

A catalogue record for this book is available from the British Library

The Royal Society of Chemistry is a charity, registered in England and Wales, Number 207890, and a company incorporated in England by Royal Charter (Registered No. RC000524), registered office: Burlington House, Piccadilly, London W1J 0BA, UK, Telephone: +44 (0) 207 4378 6556.

For further information see our web site at www.rsc.org

Printed in the United Kingdom by CPI Group (UK) Ltd, Croydon, CR0 4YY, UK

Foreword

Few branches of spectroscopy match the versatility, applicability and implications of Magnetic Resonance. In its molecular analysis mode, NMR, it provides structural and dynamic information in the widest range of situations: solids, organics, pharmaceuticals, proteins and nucleic acids, cells, and metabolism in living organisms. In its imaging mode, MRI, it provides one of the most widely used forms for understanding biological function and for non-invasive diagnosis of disease. A common denominator of nearly all contemporary NMR and MRI experiments relates to their need to unravel complex, overlapping information. This challenge is solved *via* one of magnetic resonance's most insightful propositions: the multidimensional NMR/MRI experiment. By spreading and correlating information onto several dimensions, multidimensional NMR/MRI stands as one of the intellectual jewels of modern spectroscopy. While originally proposed by Jeener as a tool to assign *J*-coupled peaks in a spectrum, Ernst and others rapidly realized the value of multidimensional magnetic resonance to obtain images of opaque objects, to detect invisible coherence states, to provide the resolution needed to elucidate complex chemical systems, and to determine the spatial structure of biological machines under near physiological conditions. Multidimensional approaches have since been adopted by other branches of spectroscopy—electron paramagnetic resonance, mass spectrometry, IR and visible optics—and thereby taken an additional number of unique roles in chemistry and biochemistry. But in no area of scientific research have multidimensional experiments retained such central roles as in NMR and MRI. Just to give an idea of the breadth of these applications, suffice it to mention that 2D-mediated observations of radiation-less multiple-quantum transitions is essential to understand the structure of complex materials, that 2D correlations between distant nuclei in small molecules often serve as the "eyes" with which organic and pharmaceutical chemists identify their

New Developments in NMR No. 11
Fast NMR Data Acquisition: Beyond the Fourier Transform
Edited by Mehdi Mobli and Jeffrey C. Hoch
© The Royal Society of Chemistry 2017
Published by the Royal Society of Chemistry, www.rsc.org

products, that correlations of low-γ evolutions with ^{1}H spin detection have been essential to endow NMR with the sensitivity needed by the structural biologist seeking to understand biochemical function *in situ*, that tens of millions of yearly 3D MRI scans are at the core of radiological exams preventing and treating the widest range of maladies, and that neither biology's nor psychology's contemporary understanding of living bodies and minds would stand where they do today without multidimensional functional MRI correlations.

Despite these invaluable and extraordinarily diverse roles of one and the same experiment, a grand challenge stands in the road of these MR implementations: the additional time that multidimensional experiments demand *vis-à-vis* their 1D counterparts. This is a demand that was "built-in" and accepted from the genesis of these methods onwards, but which is often onerous and far from inconsequential. Indeed, extended acquisitions have a penalty that goes far beyond the "time is money" concept: by increasing their duration in a manner that grows exponentially with the number of dimensions involved, high-dimensional experiments on the complex systems on which they are most essential rapidly become incompatible with their practical realization. Complex systems tend to have a dynamics of their own, and can rarely withstand extremely long examinations in their natural conditions. In few instances did this become as apparent as in the medical applications of MR, where it was clear that often infirm patients could not be subject to high-definition three- or four-dimensional acquisitions lasting for hours on end. This triggered a slow but steady departure from the discrete Fourier transform principles that dominated the nD MRI acquisition over its first two decades. To this end, phycisists joined efforts with computer scientists, leading eventually to the kind of sparse sampling techniques that nowadays enable the delivery of 256^3 or 512^3 3D images in a matter of minutes. These principles are finding an increased translation into NMR experiments, suffering as they do from the additional sensitivity penalties associated with lower spin concentrations and to mixing processes that, active in-between the various dimensions, tax this kind of acquisition even further. The results of these efforts within the field of NMR, particularly as they have shaped over the last decade, are summarized in the pages of this monograph. These include the use of fast-switching gradients to unravel indirect spectral dimensions, the introduction of regularization procedures in order to bypass the otherwise overtly strict sampling demands of the fast Fourier transform algorithm, the joint sampling of multiple dimensions in a "back-projected" fashion, and the design of metrics to assess the reliability of all these techniques. Coming to the aid of the much lower sensitivities characterizing NMR *vis-à-vis* MRI are relaxation-enhanced methods, which over recent years have become an indispensible tool in multidimensional biomolecular NMR.

While it is clear that accelerated nD NMR acquisitions are rapidly become a mature topic, I would like to challenge the reader by venturing to say that their final form is far from settled. Additional improvements and combinations of new spin physics and data processing will surely keep enhancing the performance of high-dimensional NMR, including perhaps spectroscopic-oriented analogues of common MRI modalities, such as multiband excitations and parallel receiving, which so far have not received all the NMR attention they might deserve. Furthermore, it is unlikely that one single approach will fit best the hundreds of multidimensional experiments normally used in solid and solution phase NMR—a diversity that in both dimensions and interactions is much higher than that occupying our MR imaging colleagues. I therefore conclude by thanking the authors and editors of this volume for offering its material as timely "food for thought", while encouraging all of us to read these pages with a critical, open mind. Chances are that the ultimate treatise on fast multidimensional NMR still remain to be written...

Lucio Frydman
Rehovot

Preface

NMR spectroscopy is ubiquitous in structural elucidation of synthetic compounds, metabolites, natural products and materials in chemistry, as well as structural and functional characterisation of biomolecules and macromolecular complexes. The versatility of NMR spectroscopy derives from multiple-pulse experiments, where nuclear correlations are encoded in multidimensional spectra. However, the direct result of an NMR experiment is not a spectrum, but a time series. The NMR spectrum is generated from the time response of the pulsed experiment through the application of a method for *spectrum analysis*, which constructs a frequency-domain spectrum from, or consistent with, the time-domain empirical data. Signal processing and pulsed NMR therefore go hand-in-hand in modern NMR spectroscopy. The inherently weak NMR signal has made signal processing a vital step in the varied applications of NMR. Basic understanding of signal processing is therefore a pre-requisite for the modern NMR spectroscopist.

In recent years we have witnessed an explosion in the variety of methods for spectrum analysis employed in NMR, motivated by limitations of the discrete Fourier transform (DFT) that was seminal in the development of modern pulsed NMR experiments. Prime among these limitations is the difficulty (using the DFT) of obtaining high-resolution spectra from short data records. An inherent limitation of the DFT is the requirement that data be collected at uniform time intervals; many modern methods of spectrum analysis circumvent this requirement to enable much more efficient sampling approaches. Other modern methods of spectrum analysis obtain high-resolution spectra by implicitly or explicitly modelling the NMR signals. Alternatively, we have witnessed the development of approaches that collect multidimensional data *via* multiplexing in space—exploiting the physical dimensions of the sample—rather than *via* sampling a series of indirect time dimensions, or approaches that tailor the pulse sequence in ways that enable

New Developments in NMR No. 11
Fast NMR Data Acquisition: Beyond the Fourier Transform
Edited by Mehdi Mobli and Jeffrey C. Hoch
© The Royal Society of Chemistry 2017
Published by the Royal Society of Chemistry, www.rsc.org

drastically faster sampling in time. Together, methods based on nonuniform sampling in time or sampling in space enable a class of experiments described as Fast NMR Data Acquisition. The methods that increase the speed of data acquisition through non-conventional pulse sequence design include SOFAST-NMR and Single Scan NMR, covered in the first two chapters of this book. Methods based on modelling the signal to obtain high resolution spectra from short data records or those that support nonuniform sampling are described in Chapters 3–4 and Chapters 5–10, respectively. The latter are further categorized by those that sample in a deterministic matter, *i.e.* uniformly along radial or concentric patterns (Chapters 5 and 6) or those that seek incoherence in the distribution of the sampling times (Chapters 7–10).

In this book we have brought together contributions from leading scientists in the development of Fast NMR Data Acquisition to provide a comprehensive reference text on this rapidly growing field. The popularity and rapid expansion of fast acquisition methods is evident in the literature. For example, a search for non-uniform sampling (NUS) terms (non-uniform, non-linear, projection, radial, *etc.*) and NMR revealed 185 publications since 2000 (Scopus). 13 of these were published between 2000 and 2005, when projection reconstruction and reduced dimensionality experiments were being developed. In 2005–2010 the impact of these experiments and their relationship to data sampling led to 44 publications, and in 2010–2015 the field further expanded with 105 publications, with the introduction of various "compressed sensing" techniques and the elucidation of their relationship to established methods. Similarly, citations of these articles have risen from 237 citations in 2010 to ~800 citations in 2015. These numbers, although crude, nevertheless show how interest in fast acquisition techniques has exploded over the past two decades, moving from relative obscurity to the mainstream. The widespread adoption of Fast Acquisition Methods is perhaps most evident in the rapid adaptation of modern spectrometers to these methods. In 2005, purpose-written pulse sequences had to be used to perform NUS, whilst today it is treated as simply another standard acquisition parameter during experimental setup by most commercial NMR instruments.

There is no doubt that fast acquisition methods are now firmly established as a part of modern NMR spectroscopy and we hope that this text book will serve to orient the spectroscopist in this new era, providing improved understanding of the many methods on offer and enabling informed decisions on how to make the most of the faint nuclear signals to resolve complex chemical and biological problems.

We are deeply indebted to our colleagues who contributed to this volume, and we express special thanks to Professor Lucio Frydman for contributing his perspective in the Foreword. We are also grateful to the editorial staff of the Royal Society of Chemistry for their enthusiasm for this project and their tireless efforts during editing and production. Finally, MM wishes to acknowledge support from the Australian Research Council in establishing fast acquisition methods towards the automation of protein structure determination by NMR (FTl10100925). JCH wishes to acknowledge the generous support of the US National Institutes of Health *via* the grant P41GM111135, which enabled the establishment of **NMRbox.org**: National Center for Biomolecular NMR Data Processing and Analysis. All of the authors of computer codes for non-Fourier methods represented in this book have generously consented to distributing their software *via* NMRbox.org.

Mehdi Mobli and Jeffrey C. Hoch

Contents

Chapter 1	**Polarization-enhanced Fast-pulsing Techniques**	**1**
	Bernhard Brutscher and Zsofia Solyom	

1.1	Introduction	1
	1.1.1 Some Basic Considerations on NMR Sensitivity and Experimental Time	2
	1.1.2 Inter-scan Delay, Longitudinal Relaxation, and Experimental Sensitivity	3
1.2	Proton Longitudinal Relaxation Enhancement	6
	1.2.1 Theoretical Background: Solomon and Bloch–McConnell Equations	6
	1.2.2 Proton LRE Using Paramagnetic Relaxation Agents	7
	1.2.3 Proton LRE from Selective Spin Manipulation	8
	1.2.4 Amide Proton LRE: What Can We Get?	11
	1.2.5 LRE for Protons Other Than Amides	14
1.3	BEST: Increased Sensitivity in Reduced Experimental Time	14
	1.3.1 Properties of Band-selective Pulse Shapes	14
	1.3.2 BEST-HSQC *versus* BEST-TROSY	18
	1.3.3 BEST-optimized ^{13}C-detected Experiments	22
1.4	SOFAST-HMQC: Fast and Sensitive 2D NMR	24
	1.4.1 Ernst-angle Excitation	24

New Developments in NMR No. 11
Fast NMR Data Acquisition: Beyond the Fourier Transform
Edited by Mehdi Mobli and Jeffrey C. Hoch
© The Royal Society of Chemistry 2017
Published by the Royal Society of Chemistry, www.rsc.org

        1.4.2   SOFAST-HMQC: Different Implementations
               of the Same Experiment   26
        1.4.3   UltraSOFAST-HMQC   28
   1.5   Conclusions   29
   References   30

Chapter 2   **Principles of Ultrafast NMR Spectroscopy**   **33**
               Maayan Gal

   2.1   Introduction   33
        2.1.1   One- and Two-dimensional FT NMR   34
   2.2   Principles of UF NMR Spectroscopy   35
        2.2.1   Magnetic Field Gradients   35
        2.2.2   Generic Scheme of UF 2D NMR
               Spectroscopy   38
        2.2.3   Spatial Encoding   39
        2.2.4   Decoding the Indirect Domain Information   40
        2.2.5   The Direct-domain Acquisition   42
   2.3   Processing UF 2D NMR Experiments   43
        2.3.1   Basic Procedure   43
        2.3.2   SNR Considerations in UF 2D NMR   45
   2.4   Discussion   46
   Acknowledgements   47
   References   47

Chapter 3   **Linear Prediction Extrapolation**   **49**
               Jeffrey C. Hoch, Alan S. Stern and Mark W. Maciejewski

   3.1   Introduction   49
   3.2   History of LP Extrapolation in NMR   50
        3.2.1   Broader History of LP   51
   3.3   Determining the LP Coefficients   51
   3.4   Parametric LP and the Stability Requirement   52
   3.5   Mirror-image LP for Signals of Known Phase   53
   3.6   Application   54
   3.7   Best Practices   57
   Acknowledgements   58
   References   58

Chapter 4   **The Filter Diagonalization Method**   **60**
               A. J. Shaka and Vladimir A. Mandelshtam

   4.1   Introduction   60

4.2 Theory 64
    4.2.1 Solving the Harmonic Inversion Problem:
          1D FDM 64
    4.2.2 The Spectral Estimation Problem and
          Regularized Resolvent Transform 68
    4.2.3 Hybrid FDM 70
    4.2.4 Multi-D Spectral Estimation and Harmonic
          Inversion Problems 71
    4.2.5 Spectral Estimation by Multi-D FDM 72
    4.2.6 Regularization of the Multi-D FDM 76
4.3 Examples 78
    4.3.1 1D NMR 78
    4.3.2 2D NMR 82
    4.3.3 3D NMR 86
    4.3.4 4D NMR 91
4.4 Conclusions 93
Acknowledgements 93
References 94

**Chapter 5 Acquisition and Post-processing of Reduced
Dimensionality NMR Experiments 96**
Hamid R. Eghbalnia and John L. Markley

5.1 Introduction 96
5.2 Data Acquisition Approaches 98
5.3 Post-processing and Interpretation 100
5.4 HIFI-NMR 102
5.5 Brief Primer on Statistical Post-processing 102
5.6 HIFI-NMR Algorithm 103
5.7 Automated Projection Spectroscopy 107
5.8 Fast Maximum Likelihood Method 110
5.9 Mixture Models 111
5.10 FMLR Algorithm 112
5.11 Conclusions and Outlook 114
References 114

Chapter 6 Backprojection and Related Methods 119
Brian E. Coggins and Pei Zhou

6.1 Introduction 119
6.2 Radial Sampling and Projections 120
    6.2.1 Measuring Projections: The Projection-slice
          Theorem 120
    6.2.2 Quadrature Detection and Projections 123

6.3   Reconstruction from Projections: Theory   125
      6.3.1   A Simple Approach: The Lattice of Possible
              Peak Positions   125
      6.3.2   Limitations of the Lattice Analysis and
              Related Reconstruction Methods   128
      6.3.3   The Radon Transform and Its Inverse   130
      6.3.4   The Polar Fourier Transform and the Inverse
              Radon Transform   134
      6.3.5   Reconstruction of Higher-dimensional
              Spectra   135
      6.3.6   The Point Response Function for Radial
              Sampling   137
      6.3.7   The Information Content and Ambiguity
              of Radially Sampled Data   149
6.4   Reconstruction from Projections: Practice   151
      6.4.1   The Lower-value Algorithm   151
      6.4.2   Backprojection Without Filtering   153
      6.4.3   The Hybrid Backprojection/Lower-value
              Method   154
      6.4.4   Filtered Backprojection   156
      6.4.5   Other Proposed Approaches to
              Reconstruction   157
6.5   Applications of Projection–Reconstruction
      to Protein NMR   158
6.6   From Radial to Random   161
6.7   Conclusions   166
Acknowledgements   167
References   167

Chapter 7   **CLEAN**   **169**
            Brian E. Coggins and Pei Zhou

7.1   Introduction   169
7.2   Historical Background: The Origins of CLEAN
      in Radioastronomy   170
7.3   The CLEAN Method   171
      7.3.1   Notation   171
      7.3.2   The Problem to be Solved   174
      7.3.3   CLEAN Deconvolves Sampling Artifacts *via*
              Decomposition   176
      7.3.4   Obtaining the Decomposition into
              Components   177

	7.3.5	The Role of the Gain Parameter	179
	7.3.6	Reconstructing the Clean Spectrum	180
7.4		Mathematical Analysis of CLEAN	181
	7.4.1	CLEAN and the NUS Inverse Problem	181
	7.4.2	CLEAN as an Iterative Method for Solving a System of Linear Equations	185
	7.4.3	CLEAN and Compressed Sensing	193
7.5		Implementations of CLEAN in NMR	200
	7.5.1	Early Uses of CLEAN in NMR	200
	7.5.2	Projection–reconstruction NMR	203
	7.5.3	CLEAN and Randomized Sparse Nonuniform Sampling	203
7.6		Using CLEAN in Biomolecular NMR: Examples of Applications	207
7.7		Conclusions	217
		Acknowledgements	218
		References	218

Chapter 8   Covariance NMR   220

*Kirill Blinov, Gary Martin, David A. Snyder and
Antony J. Williams*

8.1		Introduction	220
8.2		Direct Covariance NMR	221
8.3		Indirect Covariance NMR	226
	8.3.1	Principle	226
	8.3.2	Unsymmetrical Indirect Covariance (UIC) and Generalized Indirect Covariance (GIC) NMR	226
	8.3.3	Signal/Noise Ratio in Covariance Spectra	228
	8.3.4	Artifact Detection	231
	8.3.5	Applications of Indirect Covariance NMR	236
	8.3.6	Optimizing Spectra for Best Application to Covariance	239
	8.3.7	Applications of Covariance Processing in Structure Elucidation Problems	242
8.4		Related Methods	245
8.5		Conclusions and Further Directions	246
8.6		Computer-assisted Structure Elucidation (CASE) and the Potential Influence of Covariance Processing	247
		References	249

Chapter 9 **Maximum Entropy Reconstruction** 252
Mehdi Mobli, Alan S. Stern and Jeffrey C. Hoch

9.1 Introduction 252
9.2 Theory 253
9.3 Parameter Selection 257
9.4 Linearity of MaxEnt Reconstruction 258
9.5 Non-uniform Sampling 259
9.6 Random Phase Sampling 260
9.7 MaxEnt Reconstruction and Deconvolution 262
    9.7.1 *J*-coupling 262
    9.7.2 Linewidths 263
9.8 Perspective and Future Applications 263
References 265

Chapter 10 **Compressed Sensing ℓ_1-Norm Minimisation in Multidimensional NMR Spectroscopy** 267
Mark J. Bostock, Daniel J. Holland and Daniel Nietlispach

10.1 Introduction 267
10.2 Theory 269
10.3 Algorithms 271
    10.3.1 Greedy Pursuit 273
    10.3.2 Convex Relaxation Methods 275
    10.3.3 Non-convex Minimisation 277
    10.3.4 Other Approaches 278
10.4 Implementation and Choice of Stopping Criteria 278
10.5 Terminology 282
10.6 Current Applications 283
10.7 Applications to Higher Dimensional Spectroscopy 291
10.8 Future Perspectives 299
10.9 Conclusion 300
References 300

Subject Index 304

CHAPTER 1

Polarization-enhanced Fast-pulsing Techniques

BERNHARD BRUTSCHER*[a,b,c] AND ZSOFIA SOLYOM[a,b,c]

[a] Institut de Biologie Structurale, Université Grenoble 1, 71 avenue des Martyrs, 38044 Grenoble Cedex 9, France; [b] Commissariat à l'Energie Atomique et aux Energies Alternatives (CEA), Grenoble, France; [c] Centre National de Recherche Scientifique (CNRS), Grenoble, France
*Email: bernhard.brutscher@ibs.fr

1.1 Introduction

The concept of multidimensional NMR spectroscopy[1,2] lies at the basis of an amazingly large number of pulse sequence experiments that have made NMR spectroscopy a versatile tool for almost all branches of analytical sciences. Increasing the dimensionality of the experiment provides the required spectral resolution to distinguish individual sites in complex molecules such as proteins and nucleic acids. It also allows encoding information about inter-nuclear spin interactions as contributions to resonance position, line width or intensity, providing valuable information on the molecular structure and dynamics. However, major drawbacks of multidimensional NMR remain its inherent low sensitivity—a direct consequence of the weak magnetic spin interactions—as well as the long experimental times associated with repeating the basic pulse scheme a large number of times. Therefore, efforts are being made by scientists and NMR spectrometer manufacturers to improve the sensitivity and speed of NMR data acquisition. In this chapter we will describe and discuss a class of recently developed NMR experiments, so-called fast-pulsing techniques,

New Developments in NMR No. 11
Fast NMR Data Acquisition: Beyond the Fourier Transform
Edited by Mehdi Mobli and Jeffrey C. Hoch
© The Royal Society of Chemistry 2017
Published by the Royal Society of Chemistry, www.rsc.org

which provide increased sensitivity in shorter overall experimental times. These techniques have been developed in the context of biomolecular NMR studies of proteins and nucleic acids, but the same ideas can also be applied to other macromolecular systems.

1.1.1 Some Basic Considerations on NMR Sensitivity and Experimental Time

Before discussing the particular features of fast-pulsing techniques and their experimental implementation, we will provide a brief reminder of the main factors that determine the experimental sensitivity and the required minimal data acquisition time as these are of prime importance for designing new experiments and for choosing the most appropriate pulse sequence for a given application.

The signal-to-noise ratio (SNR) obtained for a particular NMR experiment on a given NMR spectrometer depends on a number of factors that can be summarized in the following way:

$$\text{SNR} \propto N_{\text{exc}} \gamma_{\text{exc}} (\gamma_{\text{det}})^{3/2} (B_0)^{3/2} f_{\text{probe}} \, P_z^{ss} f_{\text{seq}} \sqrt{N_{\text{scan}}} \Big/ \sqrt{2}^{n-1} \qquad (1.1)$$

Obviously, the SNR is proportional to the number of NMR-active spins that are available for excitation (N_{exc}). Increasing the sample concentration is therefore a valid method to improve SNR. However, for biological molecules sample concentration is often limited by molecular aggregation. SNR also depends on the gyromagnetic ratio of the excited (γ_{exc}) as well as the detected spin species (γ_{det}). Note that they may be different in heteronuclear correlation experiments. High-γ nuclei, such as protons, are therefore preferentially used as starting spin polarization sources and for NMR signal detection. This explains why most NMR biomolecular NMR experiments rely on proton excitation and detection. The NMR instrumentation contributes to the achievable SNR *via* the magnetic field strength (B_0) and the quality of the probe used for signal detection (f_{probe}). This has lead to the development of stronger magnets, currently reaching up to 23.5 T (1 GHz ^{1}H frequency), and cryogenically cooled high-Q probes that yield a drastic reduction in the electronic noise. Finally, SNR also depends on the NMR experiment to be performed *via* a number of parameters: the steady state polarization P_z^{SS} of the excited spins, as will be discussed in more detail in the next section; an attenuation factor f_{seq} taking into account signal loss during the pulse sequence owing to spin relaxation, pulse imperfections, and limited coherence and magnetization transfer efficiencies; the number of experimental repetitions N_{scan}; and finally the dimensionality of the experiment leading to a $\sqrt{2}$ reduction in SNR for each indirect dimension owing to the requirement of phase-sensitive quadrature detection. Note that the additional $\sqrt{2}$ signal loss can be avoided by so-called sensitivity-enhanced quadrature detection schemes.[3] However, this is typically only possible in one of the indirect dimensions, and also leads to additional relaxation-induced signal loss.

The minimal experimental time for recording an n-dimensional (nD) data set, with indirect time domains $t_1, t_2 \cdots t_{n-1}$, is given by the duration of the basic pulse scheme T_{scan}, multiplied by the number of repetitions required for phase cycling and time domain sampling.

$$(T_{\text{exp}})_{\min} = N_{\text{scan}}^{\text{sampling}} T_{\text{scan}} = \left(n_{PC}(n_1 n_2 \cdots n_{n-1}) 2^{n-1} \right) T_{\text{scan}} \qquad (1.2)$$

Here, n_{PC} is the number of repetitions required to complete a given phase cycling scheme that is used for artifact suppression and coherence transfer pathway selection. Nowadays, with the availability of pulsed field gradients that perform a similar task to phase cycling, n_{PC} can often be limited to a small number, typically 2 or 4. n_k is the number of time points acquired in the k-th time domain. For optimal performance of the experiment (in terms of resolution and sensitivity) n_k should be set to a value close to $n_k = SW_k/\Delta\nu_k$, with $\Delta\nu_k$ the natural line width of the frequency edited spin species and SW_k the corresponding spectral width, resulting in typical values of $n_k = 100$–200. The additional factor 2^{n-1} arises from the quadrature detection scheme in each of the $n-1$ indirect dimensions. Consequently, using conventional data acquisition schemes (uniform linear sampling of the grid), each additional dimension increases the experimental time by about two orders of magnitude.

Neglecting, for the sake of simplicity, relaxation-induced signal loss during the incremented time delays t_k, the SNR increases with $\sqrt{N_{\text{scan}}}$. For each experimental setup (sample, NMR spectrometer, NMR experiment) there exists a minimal number of scans, $N_{\text{scan}}^{\text{SNR}}$, required to achieve sufficient SNR to distinguish the NMR signals from the noise. We can thus distinguish two different scenarios: if $N_{\text{scan}}^{\text{SNR}} \gg N_{\text{scan}}^{\text{sampling}}$, we call this experimental situation *sensitivity-limited*, while the case $N_{\text{scan}}^{\text{SNR}} \ll N_{\text{scan}}^{\text{sampling}}$ is referred to as *sampling-limited*. In the sensitivity-limited regime, the basic pulse scheme has to be repeated several times, in order to further improve SNR. Fast multidimensional NMR methods, which are the topic of this book, become of interest either for experiments in the sampling-limited regime or if the overall experimental time does become prohibitively long.

1.1.2 Inter-scan Delay, Longitudinal Relaxation, and Experimental Sensitivity

A schematic drawing of an NMR experiment is shown in Figure 1.1a. It consists of the pulse sequence (suite of pulses and delays) of duration t_{seq}, a signal detection period (t_{det}), and an inter-scan delay (t_{rec}) that is also called the recycle or recovery delay. This latter delay, t_{rec}, is required to allow the spin system, which has been perturbed by the radio frequency pulses, to relax back toward thermodynamic equilibrium before repeating the pulse sequence. The effective recycle delay, during which longitudinal spin relaxation takes place, is $T_{\text{rec}} = t_{\text{det}} + t_{\text{rec}}$, and the total scan time, as introduced in eqn (1.2), is given by $T_{\text{seq}} = t_{\text{det}} + t_{\text{rec}}$.

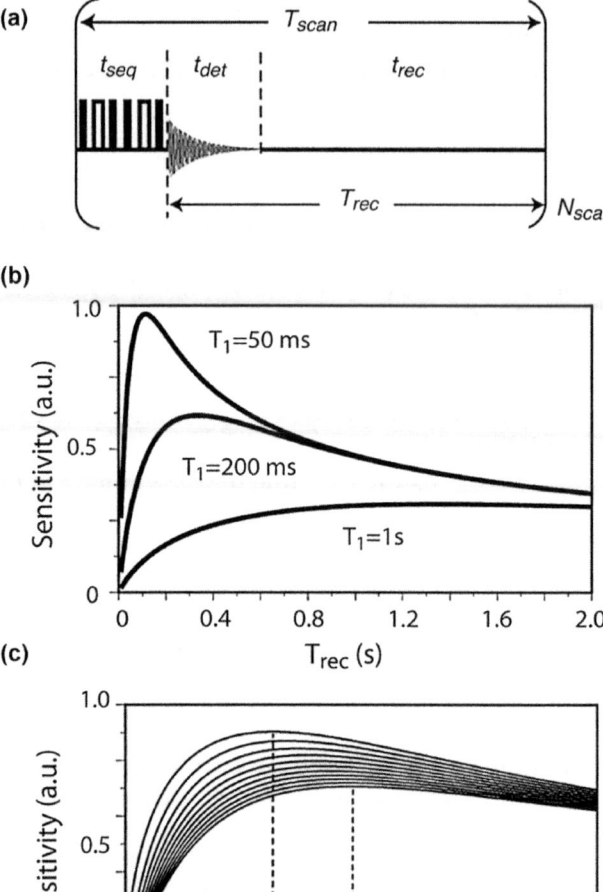

Figure 1.1   (a) Schematic representation of an NMR pulse sequence experiment, consisting of a series of pulses and delays (t_{seq}), a data detection period (t_{det}), and a recycle delay (t_{rec}). This basic scheme needs to be repeated N_{scan} times for enhancing experimental sensitivity, phase-cycling purposes, and time incrementation in indirect dimensions for multidimensional data acquisition. (b) Sensitivity curves, calculated according to eqn (1.3), and plotted as a function of the recycle delay T_{rec} for different longitudinal relaxation times T_1. The duration of the pulse sequence was set to $t_{seq} = 100$ ms for the calculation. (c) Dependence of the experimental sensitivity on the duration of the pulse sequence $t_{seq} = \lambda T_1$. The plotted curves have been computed for $0 \leq \lambda \leq 1$ in steps of 0.1 (from the top to the bottom). The sensitivity maximum shifts from $T_{rec}/T_1 = 1.25$ for $\lambda = 0$ to $T_{rec}/T_1 = 1.9$ for $\lambda = 1$.

In order to evaluate the effect of these pulse sequence parameters, and in particular the recycle delay T_{rec}, on the performance of the experiment in terms of SNR, we can calculate the experimental sensitivity, defined as the signal-to-noise ratio obtained for a fixed amount of time T_{exp}:

$$(\text{SNR})_{T_{exp}} = f(T_{rec}) \propto \frac{P_z^{ss}\sqrt{N_{scan}}}{\sqrt{T_{exp}}} = \frac{1 - \exp(-T_{rec}/T_1)}{\sqrt{T_{scan}}} \qquad (1.3)$$

In eqn (1.3), we have assumed that spin relaxation is mono-exponential, and can thus be described by a single characteristic time constant T_1. We would like to emphasize here that the parameter T_1 describes the time evolution of proton polarization recovery, but can not be directly associated to a constant with a physical meaning.

Throughout this chapter, we will refer to the dependence of the SNR on the recycle delay as a *sensitivity curve*. Examples of such theoretical sensitivity curves computed for different T_1 time constants are plotted in Figure 1.1b. The sensitivity curves show a maximum as the result of two counteracting effects: on one hand, the steady-state spin polarization, and thus the detected NMR signal, increases for longer T_{rec}. On the other hand, when increasing T_{rec}, the number of repetitions that can be performed in a given experimental time decreases, thus reducing SNR. For shorter relaxation time constants T_1, higher overall sensitivity is achieved with the maximum shifted toward shorter recycle delays. As long as the sequence duration t_{seq} is negligible with respect to T_1, the optimal recycle delay and the maximal sensitivity are given by:

$$T_{rec}^{opt} \cong 1.25\ T_1 \quad \text{for } t_{seq} \ll T_1 \qquad (1.4a)$$

$$(\text{SNR})_{T_{exp}}^{max} \propto \frac{1}{\sqrt{T_{rec}^{opt}}} \propto \frac{1}{\sqrt{T_1}} \qquad (1.4b)$$

Otherwise, if t_{seq} becomes comparable to T_1, a longer T_{rec} has to be chosen in order to reach the highest SNR and the maximal sensitivity slightly decreases. Sensitivity curves computed for $0 < t_{seq} < T_1$ are plotted in Figure 1.1c.

Eqn (1.4) implies that under conditions of optimal sensitivity, the required overall data acquisition time is proportional to T_1, while the achievable sensitivity is proportional to $1/\sqrt{T_1}$. Therefore, reducing T_1 provides a convenient way of increasing experimental sensitivity, while at the same time reducing the minimal required experimental time. This has motivated the development of longitudinal relaxation enhanced (LRE) NMR techniques that allow speeding up NMR data acquisition, as will be presented in the following sections.

1.2 Proton Longitudinal Relaxation Enhancement

In this section, we will briefly discuss the spin interactions that govern longitudinal proton spin relaxation in slowly tumbling macromolecules, such as proteins, oligonucleotides, or polysaccharides, before discussing in more detail the experimental LRE schemes that have been developed over the last decade.

1.2.1 Theoretical Background: Solomon and Bloch–McConnell Equations

The main mechanisms responsible for proton polarization buildup are the numerous ^{1}H–^{1}H dipolar interactions that are present in proton-rich molecules, and that are responsible for polarization transfer from one proton to another *via* cross-relaxation effects. Proteins, for example, contain on average about eight protons per residue that are generally tightly packed together within a globular fold. Formally, the time evolution of the polarization of each proton spin in the molecule is given by the *Solomon equations*, a set of coupled first-order differential equations:

$$-\frac{d}{dt}\begin{pmatrix} H_{1z} \\ H_{2z} \\ \vdots \\ H_{nz} \end{pmatrix} = \begin{pmatrix} \sum_j \rho_{1j} & \sigma_{12} & \cdots & \sigma_{1n} \\ \sigma_{21} & \sum_j \rho_{2j} & \cdots & \sigma_{2n} \\ \vdots & \vdots & \ddots & \vdots \\ \sigma_{n1} & \sigma_{n2} & \cdots & \sum_j \rho_{nj} \end{pmatrix} \begin{pmatrix} H_{1z} & - & H_{1z}^0 \\ H_{2z} & & H_{2z}^0 \\ \vdots & & \vdots \\ H_{nz} & - & H_{nz}^0 \end{pmatrix} \tag{1.5}$$

where H_{iz} denotes the z-component of the polarization of proton i and H_{iz}^0 is its thermal equilibrium value that, for the sake of simplicity, will be assumed to be equal to 1 throughout this chapter. The different ρ and σ terms stand for auto- and cross-relaxation rate constants, respectively, with values depending on the distance separating the two protons involved as well as the global and local dynamics of the protein experienced at the sites of the interacting protons. The auto-relaxation rate constants that are responsible for energy exchange with the lattice (molecular motions) contain contributions from dipolar interactions with all neighboring protons, and also with neighboring hetero-atoms such as ^{13}C and ^{15}N in isotopic enriched molecules. If the rotational tumbling of the molecule is slow compared to the proton Larmor frequency ($\tau_c > 1/\omega_H \approx 10^{-9}$ s at high magnetic field strengths), cross-relaxation rates become negative ($\sigma_{ij} < 0$), and as a consequence lead to spin diffusion within the dipolar-coupled spin network. It is worth mentioning here that the cross-relaxation rates (absolute value) increase with the effective rotational correlation time, making spin diffusion more efficient for larger molecules, or molecules studied at lower temperature.

For chemically labile proton sites, *e.g.* amide and hydroxyl protons, hydrogen exchange processes with the bulk water protons provide an additional relaxation source that under certain sample conditions (pH and temperature) may even become predominant, as we will see later on. In order to account for this exchange effect theoretically, the set of first-order differential equations (eqn (1.5)) has to be expanded to include exchange with the water ^{1}H polarization, resulting in the so-called *Bloch–McConnell* equations:

$$-\frac{d}{dt}\begin{pmatrix} W_z \\ H_{1z} \\ H_{2z} \\ \vdots \\ H_{nz} \end{pmatrix} = \begin{pmatrix} \rho_w & 0 & 0 & 0 & 0 \\ 0 & \sum_j \rho_{1j} & \sigma_{12} & \cdots & \sigma_{1n} \\ 0 & \sigma_{21} & \sum_j \rho_{2j} & \cdots & \sigma_{2n} \\ 0 & \vdots & \vdots & \ddots & \vdots \\ 0 & \sigma_{n1} & \sigma_{n2} & \cdots & \sum_j \rho_{nj} \end{pmatrix} \begin{pmatrix} W_z - W_z^0 \\ H_{1z} - H_{1z}^0 \\ H_{2z} - H_{2z}^0 \\ \vdots \\ H_{nz} - H_{nz}^0 \end{pmatrix}$$

$$+ \begin{pmatrix} 0 & 0 & 0 & 0 & 0 \\ -k_{ex,1} & k_{ex,1} & 0 & 0 & 0 \\ -k_{ex,2} & 0 & k_{ex,2} & 0 & 0 \\ \vdots & 0 & 0 & \ddots & \vdots \\ -k_{ex,n} & 0 & 0 & \cdots & k_{ex,n} \end{pmatrix} \begin{pmatrix} W_z \\ H_{1z} \\ H_{2z} \\ \vdots \\ H_{nz} \end{pmatrix} \qquad (1.6)$$

with $k_{ex,i}$ the exchange rate between proton i and water. In eqn (1.6) we have assumed that the bulk water proton polarization W_z is not changed by hydrogen exchange with the protein, which is a reasonable hypothesis in view of the $\sim 10^5$ times higher concentration of water in the NMR sample tube. The chemical exchange rates k_{ex}^{chem} depend on the local chemical environment of the labile proton, as well as the pH and temperature. Roughly, k_{ex}^{chem} doubles when increasing the sample temperature by 7 °C or the pH by 0.3 units. The apparent exchange rates k_{ex} are further modulated by the solvent accessibility of the labile proton that formally is described by a protection factor $P = k_{ex}^{chem}/k_{ex}$, varying from $P = 1$ (no protection) to very large values, $P > 1000$ (highly protected), in the core of stable globular proteins

1.2.2 Proton LRE Using Paramagnetic Relaxation Agents

Eqn (1.6) provides the theoretical basis for the development of longitudinal-relaxation enhancement schemes. A first strategy for accelerating longitudinal proton relaxation consists of the introduction of an additional auto-relaxation mechanism, for example by adding a paramagnetic relaxation agent to the sample. Dipolar interactions between the proton spins and the large dipolar moment of the unpaired electron(s) in the paramagnetic molecule enhance proton spin relaxation by adding a contribution to the rate constants ρ. A list of possible paramagnetic compounds can be found in a recent review by Hocking *et al.*[4] We can distinguish two different types of paramagnetic relaxation agents: (i) those acting directly on the protein nuclear spins and (ii) those acting primarily on the water protons. An example of a paramagnetic chelate belonging to the first category is Ni^{2+}-DO2A.[5,6] At 10 mM concentration, Ni-DO2A has been shown to reduce T_1 relaxation time constants of amide protons in proteins to about 50–200 ms.[6] Of course, the LRE effect strongly depends on the solvent accessibility of the nuclear spins, which translates into a relatively large enhancement for surface residues, while the effect of the paramagnetic relaxation agent is less pronounced for residues in the interior of a protein. The use of paramagnetic relaxation agents is therefore particularly well adapted to the NMR study of highly flexible molecules where most of the residues are solvent exposed.[6] Another drawback of such relaxation agents is that they may cause significant line broadening, especially in the case of specific interaction with the protein. Ni^{2+} has a short electronic relaxation time of $T_{1e} \approx 10^{-12}$ s, which limits the paramagnetic induced line broadening. Another class of paramagnetic compounds selectively enhances the ^1H T_1 of the bulk water. A number of such water relaxing paramagnetic chelates have been developed as contrast agents for magnetic resonance imaging (MRI). An example is gadodiamide (Figure 1.2a), Gd^{3+}(DTPA-BMA), which has been commercialized as clinical contrast agent under the name "Omniscan". Adding gadodiamide at 0.5 mM concentration to an aqueous solution reduces the water T_1 from a few seconds to about 400 ms (Figure 1.2a). Gd^{3+} complexes have been success-fully used for relaxation enhancement in small globular proteins and IDPs,[7] as well as large perdeuterated proteins.[8,9]

1.2.3   Proton LRE from Selective Spin Manipulation

An alternative spectroscopic method that does not require any chemical modification of the sample is to exploit the fact that the relaxation of an individual proton spin depends on the spin state of all other protons in the protein and the bulk water. In particular, solving eqn (1.6) shows that in cases where only a subset of the proton spins is of interest, a *selective* excitation of these spins results in a much faster recovery than a *non-selective* excitation.[10,11] This is of significant practical interest to a large number of biomolecular NMR experiments that excite and detect only a subset of proton spins, *e.g.* amide, aromatic, and methyl ^1H in proteins, or imino, base, and sugar ^1H in nucleic acids.

In order to achieve selective excitation of only a subset of the proton spins in a molecule, two different experimental approaches have been proposed.

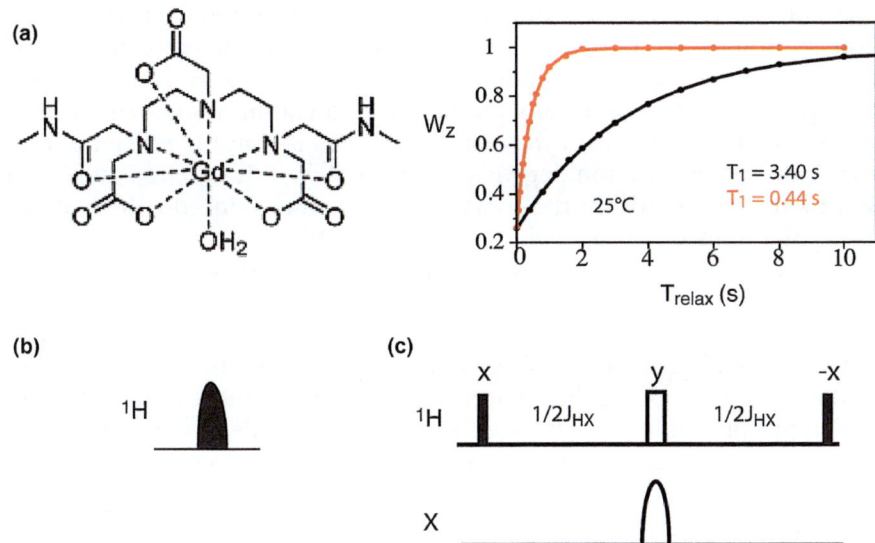

Figure 1.2   Options for ^{1}H longitudinal relaxation enhancement (LRE) techniques: (a) a paramagnetic relaxation agent is added to the protein sample. As an example the chemical structure of the Gd^{3+} chelate gadodiamide, Gd^{3+}(DTPA-BMA), is shown on the left. The effect of 0.5 mM gadodiamide on the water ^{1}H longitudinal relaxation has been measured and is shown in the right panel. The apparent T_1 is reduced from 3.4 to 0.44 s at 25 °C. LRE pulse schemes based on selective ^{1}H excitation: (b) band-selective shaped pulses, and (c) scalar-coupling-based flip-back scheme.

A first method (Figure 1.2b), that we will refer to as *"frequency-selective"*, uses band-selective shaped pulses to manipulate only the spins of interest, while leaving all others unperturbed, or at least little affected by the pulse sequence. The obvious disadvantage of this approach is that it is limited to proton species that are sufficiently different chemically to resonate in distinct spectral regions. Furthermore, in highly structured molecules, such as proteins, ring-current effects may shift a particular resonance far away from the typical chemical shift region, and as a consequence such protons may not become excited by the selective pulses. A second "coupling-selective" method (Figure 1.2c) exploits the presence or absence of a scalar coupled heteronuclear spin X (typically ^{15}N or ^{13}C) to achieve selective proton excitation. For this purpose, broadband ^{1}H pulses are used to excite (and refocus) all proton spins together. Scalar-coupling evolution during a time interval equal to $1/J_{HX}$ then creates a 180° phase shift of protons coupled to X, with respect to all others. A final 90° (flip-back) pulse restores the ^{1}H polarization for the uncoupled spins. In addition, the use of a band-selective 180° X pulse allows even more selectivity by restricting the excitation, for example, to ^{13}C$^\alpha$-bound protons, while protons bound to other aliphatic or aromatic ^{13}C are flipped back at the end of the sequence. This scheme is especially attractive for heteronuclear correlation experiments performed on protein samples with partial or no isotope enrichment. It is also a good choice for performing ${}^1\text{H}^\alpha$ excitation while leaving the water protons that resonate in the same spectral window, close to equilibrium. A major draw-back of such *coupling-based flip-back* techniques is that the efficiency of re-storing the ${}^1\text{H}$ polarization depends on the quality of the applied pulses, and even more importantly on the transverse-relaxation induced coherence loss during the spin evolution delay $1/J_{HX}$. This explains why, in practice, the frequency-selective pulse excitation scheme outperforms the coupling-based flip-back scheme whenever both approaches are feasible.

In the following, we will focus on amide ${}^1\text{H}$ selective experiments of proteins. The large majority of amide protons resonate in a narrow spectral window ($\sim$6 to 10 ppm) that is well separated from most other (aliphatic) protein protons and the bulk water (Figure 1.3a). Amide protons can thus be selectively manipulated by means of appropriate shaped pulses, as discussed in more detail later on. The effect of selective *versus* non-selective proton

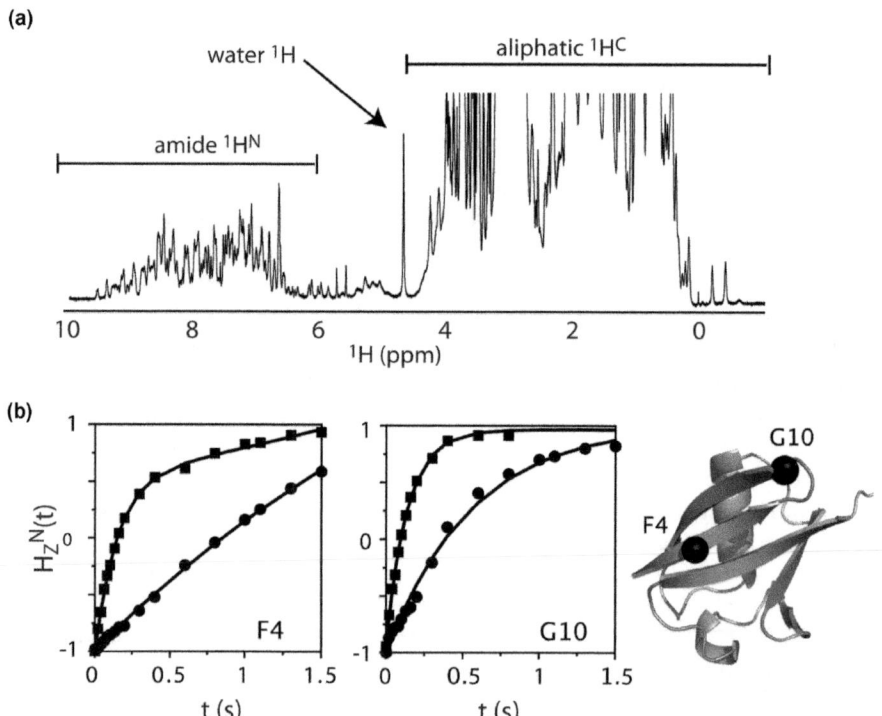

Figure 1.3   (a) ${}^1\text{H}$ spectrum of ubiquitin. The spectral regions for amide, aliphatic, and water protons are highlighted. (b) Experimental ${}^1\text{H}$ polarization recovery curves for selected amide sites in ubiquitin,[11] highlighted by spheres on the ubiquitin structure (right panel). The solid lines are bi-exponential fits to the data measured for selective (filled squares) and non-selective spin inversion (filled circles).

spin inversion on the longitudinal relaxation behavior of amide protons is illustrated in Figure 1.3b for two amide proton sites in the small protein ubiquitin, located in a well-structured β-sheet (F4), and a highly dynamic loop region (G10). For both amide protons, selective spin inversion leads to significant longitudinal relaxation enhancement, while of course the exact relaxation properties also depend on the local structure (average distance to other protons, solvent accessibility) and dynamics (local effective tumbling correlation time), resulting in significant differences in the apparent relaxation curves. Another observation is that the polarization recovery after non-selective spin inversion is well described by a mono-exponential curve requiring a single relaxation time constant ($T_1^{\text{non-sel}}$). This is generally not the case after selective spin inversion that is best described by a bi-exponential behavior with a fast relaxation component (T_1^{fast}) and a slow relaxation contribution (T_1^{slow}). The magnitude of the latter is to a good approximation equal to the relaxation time constant in the non-selective case ($T_1^{\text{slow}} \approx T_1^{\text{non-sel}}$). Although this is not a rigorous treatment, in the remainder of this chapter we will characterize the polarization recovery under selective conditions by a single time constant $T_1 = T_1^{\text{fast}}$ that is obtained either by fitting the first part of the inversion-recovery curve to a mono-exponential function, or from the zero-crossing time t^0 as $T_1 = t^0/\ln 2 = 1.44t^0$.

1.2.4 Amide Proton LRE: What Can We Get?

In order to quantify the achievable LRE effects for amide protons, and the relative contributions from dipolar interactions and water exchange processes, we have measured apparent T_1 relaxation time constants for different proteins and experimental conditions (pH and temperature). We distinguish three different experimental scenarios depicted in Figure 1.4a: (i) amide-proton selective spin inversion; (ii) water-flip back (wfb) spin inversion, and (iii) non-selective spin inversion. In the first case, only amide proton spins are inverted, while in the second case all protein protons are inverted leaving only the water protons at equilibrium, and finally in the third case the water protons are also inverted. The measured inversion-recovery curves of amide ^{1}H polarization have been fitted to a mono-exponential function, as explained in the last section, and the apparent T_1 values are plotted in Figure 1.4b–d as a function of the peptide sequence. Figure 1.4b shows the LRE effects obtained for the small globular protein ubiquitin ($\tau_c \approx 3$ ns). Except for a loop region (residues 9–13) and the last four C-terminal residues, the measured relaxation time constants are quite uniform with a non-selective $T_1^{\text{non-sel}} \cong 900$ ms that is reduced to $T_1^{\text{sel}} \cong 200$ in the selective case. The main mechanisms for this 4.5-times acceleration of longitudinal relaxation are the dipolar interactions with neighboring non-amide protons, while water exchange processes only marginally contribute to the LRE effect. The observed situation is completely different for amides located in the highly flexible protein regions, where a non-selective T_1 of up to $T_1^{\text{non-sel}} \cong 2.5$ s is observed, a value that approaches the T_1 of

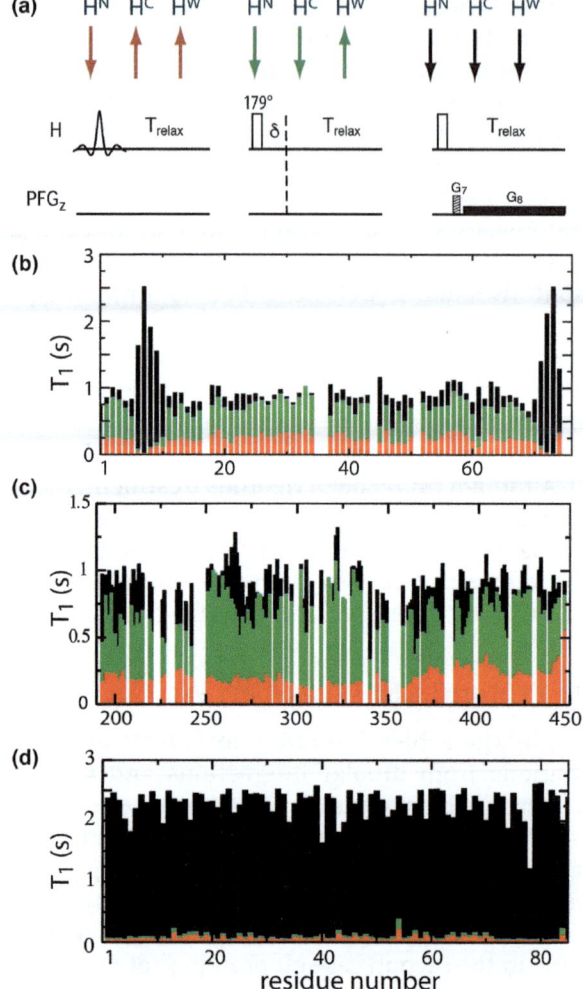

Figure 1.4 Residue-specific amide proton T_1 time constants measured for several proteins under different initial conditions. (a) Pulse schemes used for spin inversion of different sets of ^{1}H spins in inversion-recovery experiments: amide ^{1}H (H^N), aliphatic ^{1}H (H^C), and water ^{1}H (H^W). The displayed inversion-recovery block is followed by a 2D ^{1}H–^{15}N readout sequence and a recycle delay of 6 s. For amide proton-selective inversion (left panel, red bars) a REBURP pulse shape was applied with a bandwidth of 4.0 ppm centered at 9.0 ppm; for water-flip-back inversion (central panel, green bars) a 179° inversion pulse is followed by a short delay of 10 ms during which radiation damping brings the water back to equilibrium; for non-selective ^{1}H inversion (right panel, black bars) a broad-band inversion pulse is followed by a strong pulsed field gradient to spatially defocus residual water transverse magnetization. A low-power magnetic field gradient is applied during the entire relaxation delay T_{relax} to avoid radiation-damping effects. Apparent amide ^{1}H T_1 relaxation time constants measured for (b) the small globular protein ubiquitin (pH 7.5, 25 °C), and two intrinsically disordered proteins (IDPs),[26] (c) NS5A (pH 6.5, 5 °C) and (d) α-synuclein (pH 7.4, 15 °C).

water (3.4 s at 25 °C, see Figure 1.2a). Under these conditions, the selective and wfb inversion schemes perform the same, leading to relaxation times $T_1^{sel} = T_1^{wfb} = 40 - 70$ ms, clearly demonstrating that water exchange is the main mechanism for the observed LRE effect.

Figure 1.4 also shows the measured LRE effects for two so-called intrinsically disordered proteins (IDPs) that have been studied under different sample conditions: low pH (6.4) and low temperature (5 °C) for the first one—NS5A (Figure 1.4c)—and physiological pH (7.5) and higher temperature (15 °C) for the second one—α-synuclein (Figure 1.4d). IDPs have found widespread interest in recent years in structural biology in general, and in biomolecular NMR spectroscopy in particular.[12,13] These highly dynamic proteins or protein fragments, which are particularly abundant in eukaryotes and viruses, have been ignored since the early days of structural investigation of proteins. Although they have no stable structure, IDPs are involved in many cellular signaling and regulatory processes, where structural flexibility presents a functional advantage in terms of binding plasticity and promiscuity.[14] During recent years, NMR spectroscopy has become the technique of choice to characterize residual structure in IDPs and their interaction with binding partners. Owing to the low chemical shift dispersion in the NMR spectra of IDPs, and the often-encountered low solubility and limited lifetime of NMR samples, fast multidimensional data acquisition techniques are of primary importance for NMR studies of IDPs.

Interestingly, despite their high degree of internal flexibility, the LRE effects observed in IDPs are similar to those discussed above for a globular protein. Under conditions that do not favor solvent exchange (Figure 1.4c), but where the local tumbling correlation times are in the slow tumbling regime, the dipolar relaxation mechanism dominates (similar to the structured parts of ubiquitin), resulting in a significant reduction in the apparent average T_1 time constants from $T_1^{non\text{-}sel} \cong 1$ s to $T_1^{sel} \cong 200$ ms. This observation also indicates that a few protons at close proximity are sufficient for efficient spin diffusion, and thus LRE effects. At higher pH and temperature the solvent exchange rates increase and become the main source of amide proton relaxation. Consequently, the relaxation behavior of amide protons in α-synuclein (Figure 1.4d) is very similar to that observed for the flexible loop and C-terminus in ubiquitin with $T_1^{non\text{-}sel} \cong 2.3$ s and $T_1^{sel} = T_1^{wfb} \cong 60$ ms. Note that this impressive LRE translates into a sensitivity gain of up to a factor of 6 if optimal short recycle delays are chosen, and if the pulse sequence performs perfectly in terms of selective spin manipulation.

To conclude this section, significant LRE effects are observed for amide protons in proteins of different size and structure, and under a variety of experimental conditions, as a result of two complementary relaxation mechanisms, which are dipolar interactions and solvent exchange processes. Therefore LRE pulse schemes are expected to be of great value for all types of amide proton-based NMR experiments.

1.2.5   LRE for Protons Other Than Amides

The concepts introduced for LRE, and exemplified above for amide protons in proteins, can also be exploited for NMR experiments involving other proton spins in slowly tumbling molecules. The requirements to obtain significant LRE effects are the following: (i) the protons of interest can be excited (manipulated) by either shaped pulses or *via* a coupling-based flip-back scheme, and (ii) the selective excitation leaves neighboring protons as well as water protons, in the case of exchangeable sites, close to their thermodynamic equilibrium. These requirements are fulfilled for amide, aromatic, and aliphatic H^α and methyl protons in proteins, and for imino, base and sugar protons in nucleic acids (DNA, RNA). Especially for imino ^1H the relaxation behavior is very similar to the one discussed above for amide protons, with dipolar interactions as well as solvent exchange contributing to the observed LRE effects.[15] Imino protons can be easily manipulated by shaped pulses as they resonate in a well-separated frequency range (~10–15 ppm). Examples of LRE-optimized NMR experiments for different types of protons will be given later.

1.3   BEST: Increased Sensitivity in Reduced Experimental Time

In this section we will describe the basic features and experimental performance of longitudinal relaxation enhanced pulse schemes that use band-selective shaped ^1H pulses for selective manipulation of a subset of protons in the molecule. Such experiments have been termed BEST, an acronym for *Band-selective Excitation Short-Transient*.[16] As explained in more detail in the previous section, such BEST experiments yield enhanced steady-state ^1H polarization of the excited spins at short recycle delays, allowing for faster repetition of the pulse sequence, and thus reducing the overall experimental time requirements.

1.3.1   Properties of Band-selective Pulse Shapes

The performance of BEST-type experiments in terms of LRE efficiency and overall sensitivity critically depends on the choice and appropriate use of shaped pulses. Therefore, we will briefly discuss the basic properties of some of the most prominent pulse shapes. Major efforts were made in the early 1990s in developing band-selective pulses using numerical optimization methods. These band-selective pulse shapes are characterized by a top-hat response in frequency (chemical shift offset) space that is uniform (constant rotation angle) over the chosen bandwidth (excitation/inversion band), and close to zero (no effect) for spins resonating outside this spectral window. A narrow transition region exists in between where spin evolution is un-defined. Examples of such numerically optimized "top-hat" pulse shapes are the BURP pulse family,[17] the Gaussian pulse cascades,[18,19] the SNOB pulses,[20] and polychromatic (PC) pulses.[21] An additional feature of shaped pulses is that they have been optimized for a particular rotation angle, typically 90° or 180°, and some of them, so-called excitation (90°) or inversion (180°) pulse shapes, only perform such a rotation when starting from pure spin polarization (H_z), while their action on spin coherence is *a priori* undefined.

The shapes and excitation (or inversion) profiles of the band-selective pulses used in BEST-type experiments (PC9, EBURP-2, REBURP and Q3) are shown in Figure 1.5a. We have recently demonstrated by numerical simulations of the spin evolution during such pulse shapes that their effect on the spin density operator can be described reasonably well (first order approximation) by a sequence of free evolution delays and an additional time period accounting for the effective pulse rotation of $\beta = 90°$ or 180° over the chosen frequency band: $\tau_1 - R(\beta) - \tau_2$. These binary (delay/pulse rotation) replacement schemes,[22] as we like to call them, are also shown in Figure 1.5a, with

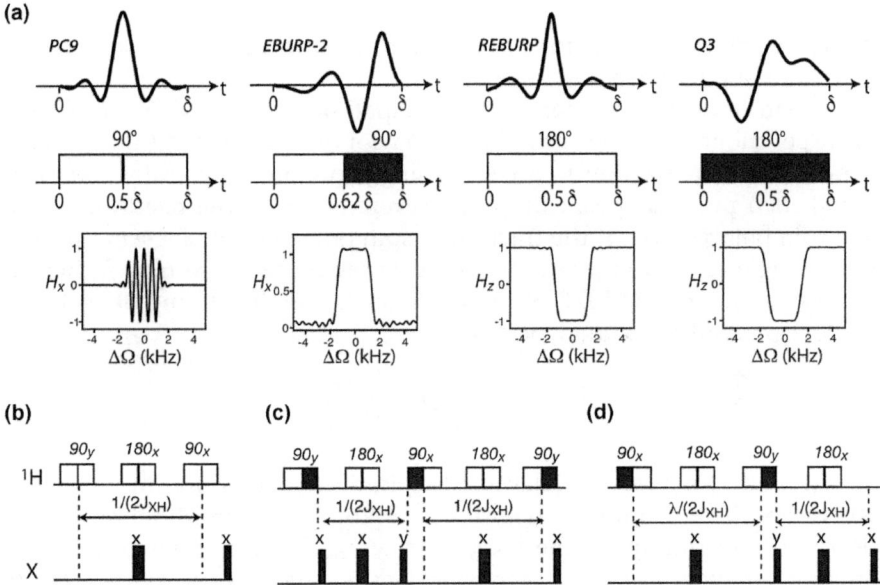

Figure 1.5 Properties of band-selective pulse shapes used in BEST and SOFAST experiments. (a) The amplitude-modulated time-domain profiles of PC9,[21] EBURP-2,[17] REBURP,[17] and Q3[19] are plotted together with the corresponding binary replacement schemes obtained from numerical simulations.[22] Chemical shift and heteronuclear coupling evolution during the pulse shape is represented by open squares, while the time during which the actual spin rotation is occurring is represented by a black square. In addition the corresponding frequency-response of the spins is shown for each pulse shape. The binary replacement schemes make it straightforward to account for spin evolution in the basic BEST-type pulse sequence elements: (b) INEPT, (c) sensitivity-enhanced reverse INEPT (SE REVINEPT), and (d) single-transition-to-single-transition polarization transfer (ST2-PT).

empty boxes representing the free evolution delays and filled boxes corresponding to the pulse rotation. Clearly, the resulting schemes are very different for pulse shapes optimized to perform the same type of action, a selective 90° or 180° rotation. While, for example, for the symmetric PC9 and REBURP pulses, chemical shift and coupling evolution is active during the entire pulse length, no spin evolution at all is observed for Q3. Similar results have recently been obtained by a factorization into individual Euler-angle rotation operators of the propagator describing the action of the shaped pulse on the spin system.[23] With these binary pulse schemes in hand, it becomes straightforward to replace standard hard pulses in complex pulse sequences by appropriate shaped pulses, and to properly adjust coherence transfer and chemical shift editing delays taking into account spin evolution during the shaped pulses as described by the binary schemes. In particular, this allows application of such pulse shapes even in cases where several coherence transfer pathways need to be realized at the same time. BEST-implementations for the most common building blocks (INEPT, sensitivity-enhanced reverse INEPT, and single-transition-to-single-transition transfer) used for ^{1}H–X heteronuclear coherence transfer are shown in Figure 1.5b–d.

A second important property of the shaped pulse in the context of BEST-type experiments is its off-resonance behavior, as even a slight perturbation of the spin magnetization in a fast-pulsing experiment, typically employing several such pulse shapes, can have a severe effect on the resulting steady-state spin polarization. If the fraction of spin polarization of a set of protons after a single repetition (scan) of the pulse sequence is given by f, then the steady-state polarization H_z^{ss} of these protons at the beginning of each scan can be expressed in an analytical form as:

$$H_z^{ss} = \frac{1 - \exp(T_{rec}/T_1)}{1 - f\exp(T_{rec}/T_1)} \tag{1.7}$$

with T_{rec} and T_1 the recycle delay and longitudinal relaxation time constants, respectively. The proton polarization relevant for longitudinal spin relaxation at the beginning of the recycle delay T_{rec} is then given by fH_z^{ss}. The computed ^{1}H polarization is plotted in Figure 1.6a as a function of T_{rec} for different T_1 values assuming a 5% perturbation of the equilibrium proton polarization after a single scan ($f = 0.95$). If we focus on a recycle delay of $T_{rec} = 200$ ms, a typical value for BEST-type experiments, we see that even such a slight perturbation results in a dramatic reduction in the steady-state polarization for protons characterized by long T_1 values of a few seconds. This is notably the case for the bulk water, while it is less of an issue for others, *e.g.* aliphatic protein protons. In order to check experimentally the effect of different pulse shapes on the water protons, we have measured the water polarization under steady-state conditions after applying a series of identical shaped pulses with a nominal excitation bandwidth of 4 ppm every 100 ms at varying offsets from the water frequency. Our experimental data

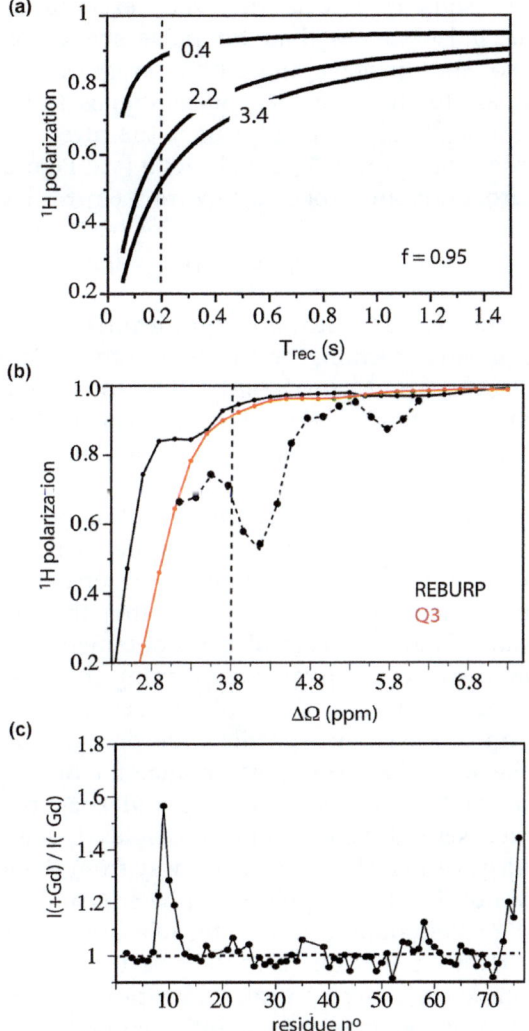

Figure 1.6 (a) Effect of a slight perturbation (5%) by a single scan on the steady-state ^{1}H polarization as a function of the recycle time T_{rec}, calculated from eqn (1.7) for T_1 time constants varying from 400 ms to 3.4 s. (b) Off-resonance performance of two commonly used refocusing (180°) pulse shapes, REBURP (black curves) and Q3 (red curve). The steady-state water ^{1}H polarization has been measured after applying the shaped pulse (nominal band width of 4 ppm) 32-times with an inter-pulse delay of 100 ms. The off-resonance profile is obtained by repeating the measurement for different shaped pulse offsets with respect to the water frequency. Straight lines correspond to measurements on a last generation cryoprobe, while the black dashed line shows the result obtained on an old cryoprobe. (c) Intensity ratios measured for individual amide ^{1}H sites in ubiquitin (pH 7.4, 20 °C) in ^{1}H–^{15}N BEST-TROSY spectra measured on samples with and without a paramagnetic water relaxation compound (0.5 mM gadodiamide).

indicate that off-resonance effects (non-zero excitation) are more pronounced for 180° pulse shapes than for pulse shapes optimized for 90°
rotations. Off-resonance profiles measured on a last generation triple-
resonance cryoprobe for the two 180° pulse shapes REBURP and Q3 are
shown in Figure 1.6b (straight lines). Under the given experimental conditions (25 °C), the water proton T_1 equals 3.4 s (see Figure 1.2a), and thus
even slight perturbations are expected (Figure 1.6a) to have a measurable
effect on the steady-state water polarization. For both pulse shapes the off-
resonance behavior is close to optimal ($W_z^{ss} > 0.9$) at frequency offsets
$\Delta\Omega > 3.8$ ppm. Consequently, for optimal sensitivity in situations where
solvent exchange processes contribute significantly to the LRE effect, as
discussed in the previous section, the frequency offset of the shaped pulses
should be set at least 3.8 ppm × BW (ppm)/4 away from the water resonance,
with BW being the excitation bandwidth chosen for the shaped pulses.

Finally, it is worth mentioning that the situation may be less optimal for
older (cryo)probes or other hardware imperfections. The dashed black curve
in Figure 1.6b displays the results obtained for REBURP on an older NMR
spectrometer equipped with a cryogenic probe. This time, significant
perturbations of the water ^{1}H polarization are observed, notably at certain
frequency offsets. A closer inspection reveals that the measured curve is
similar to the result of numerical simulations obtained for B_1 fields that are
about 15% below their nominal value, explaining that the observed effect
is owing to a slight detuning of the probe under fast-pulsing conditions.
Recording such an off-resonance profile thus provides an easy and convenient way to check the hardware performance for BEST-type fast-pulsing
experiments. If a problem of probe detuning under fast-pulsing conditions
has been identified several solutions are possible: (i) the frequency offset
of the shaped pulses can be shifted further away from the water resonance;
(ii) the power level of the shaped pulses can be adjusted to account for the
detuning, or (iii) a small amount of a water relaxation agent can be added
to the sample to reduce the water ^{1}H T_1 (Figure 1.2a). The effect of adding
0.5 mM gadodiamide on the cross peak intensities in a BEST ^{1}H–^{15}N spectrum of ubiquitin measured on such a slightly detuned NMR probe is shown
in Figure 1.6c. As expected from the curves in Figure 1.6a, for solvent-
accessible regions, an up to 60% sensitivity increase is observed for the Gd^{3+}-
containing sample, which is explained by the significantly larger steady-state
water polarization achieved under these experimental conditions. However,
for certain amide protons a slight signal decrease is observed, most likely
owing to a paramagnetic relaxation-induced line broadening. Note that no
significant signal enhancement could be observed for the Gd^{3+}-containing
sample if the same BEST experiments were performed on a well-tuned probe.

1.3.2   BEST-HSQC *versus* BEST-TROSY

The basic BEST pulse sequence blocks, introduced in Figure 1.5b–d can be
combined to set up a variety of pulse sequences that correlate amide

protons with other backbone and side-chain ^{13}C and ^{15}N nuclei in 2D, 3D, or even higher dimensional NMR experiments. We distinguish two types of BEST experiments: BEST-HSQC[16,24] and BEST-TROSY.[15,25,26] They differ in the way amide ^{15}N and ^{1}H coherence is evolving and frequency labeled. In a scalar-coupled heteronuclear two-spin system, such as ^{1}H–^{15}N, each spin has two allowed single-quantum (SQ) transitions, as explained in more detail in any standard NMR textbook. This results in peak doublets in the NMR spectrum under conditions of free spin evolution (Figure 1.7a). A first option to avoid scalar-coupling-induced line splitting consists of heteronuclear decoupling that averages the two single-transition frequencies, and results in a single resonance line. In this HSQC method, used since the early days of 2D spectroscopy,[27] both SQ transitions contribute to the detected NMR signal. A second, alternative approach selects only one out of the two transitions by so-called spin-state selection techniques.[28,29] This is realized in TROSY-type experiments,[30] introduced in the late 1990s, where the ^{1}H and ^{15}N transitions with the most favorable relaxation properties (narrow lines) are selected (Figure 1.7a). This line narrowing effect results in improved spectral resolution in TROSY compared to HSQC spectra, and may under certain experimental conditions also compensate for the reduced intrinsic sensitivity of TROSY experiments owing to the spin-state selection. Note that, although only one out of four peaks in the multiplet is selected in TROSY, the signal loss associated with this selection process is a factor of 2 (and not 4). TROSY-type experiments are preferably performed at high magnetic field strength B_0 where line narrowing resulting from CSA-dipolar cross-correlation[31–33] is most effective.

BEST implementations of basic HSQC and TROSY experiments are displayed in Figure 1.7b and 1.7c, respectively. They both start with an initial INEPT-type ^{1}H–^{15}N transfer. The relaxation delay Δ stands for additional chemical shift editing and coherence transfer steps. The two sequences differ mainly in the ^{1}H–^{15}N back transfer step, which is performed by a sensitivity-enhanced INEPT (SE-REVINEPT) sequence in HSQC, and a single-transition to single-transition coherence transfer sequence (ST2-PT) in TROSY. Another difference is the absence of composite ^{15}N decoupling during ^{1}H detection in BEST-TROSY, a feature that becomes of practical interest if long signal acquisition times are required where composite decoupling under fast-pulsing conditions may lead to significant probe heating.

An additional specific feature of TROSY, that becomes particularly interesting for BEST-TROSY experiments, is the simultaneous detection of coherence transfer pathways originating from both ^{1}H and ^{15}N polarization:

$$^1H \text{ pathway: } H_z \xrightarrow{\text{INEPT}} \pm 2H_zN_x = \pm(H^\alpha N_x - H^\beta N_x) \xrightarrow{\text{ST2-PT}} \pm N^\beta H_x \quad (1.8a)$$

$$^{15}N \text{ pathway: } N_z \xrightarrow{\text{INEPT}} N_x = (H^\alpha N_x + H^\beta N_x) \xrightarrow{\text{ST2-PT}} N^\beta H_x \quad (1.8b)$$

with the single-transition spin states selected by the ST2-PT sequence given in bold letters. The phase ϕ_0 of the last ^{1}H pulse in the INEPT sequence needs to be adjusted in order to add the contributions from the two pathways. Most interestingly, the TROSY sequence has a built-in compensation mechanism for relaxation-induced signal loss during the relaxation delay Δ (Figure 1.7c).[25] In fact, ^{1}H polarization that builds up during Δ is transferred by the ST2-PT sequence into enhanced off-equilibrium ^{15}N polarization, which will be available for the next scan as long as it survives relaxation during the recycle delay (T_{rec}). Note that an additional ^{15}N 180° pulse has been added to the TROSY sequence after the signal detection period in order to create ^{15}N polarization that is of the same sign as equilibrium polarization.

^{15}N polarization enhancement pathway:

$$\cdots \xrightarrow{\Delta} H_z \xrightarrow{\text{ST2-PT}} -N_z \xrightarrow{180°N} +N_z$$

(1.8c)

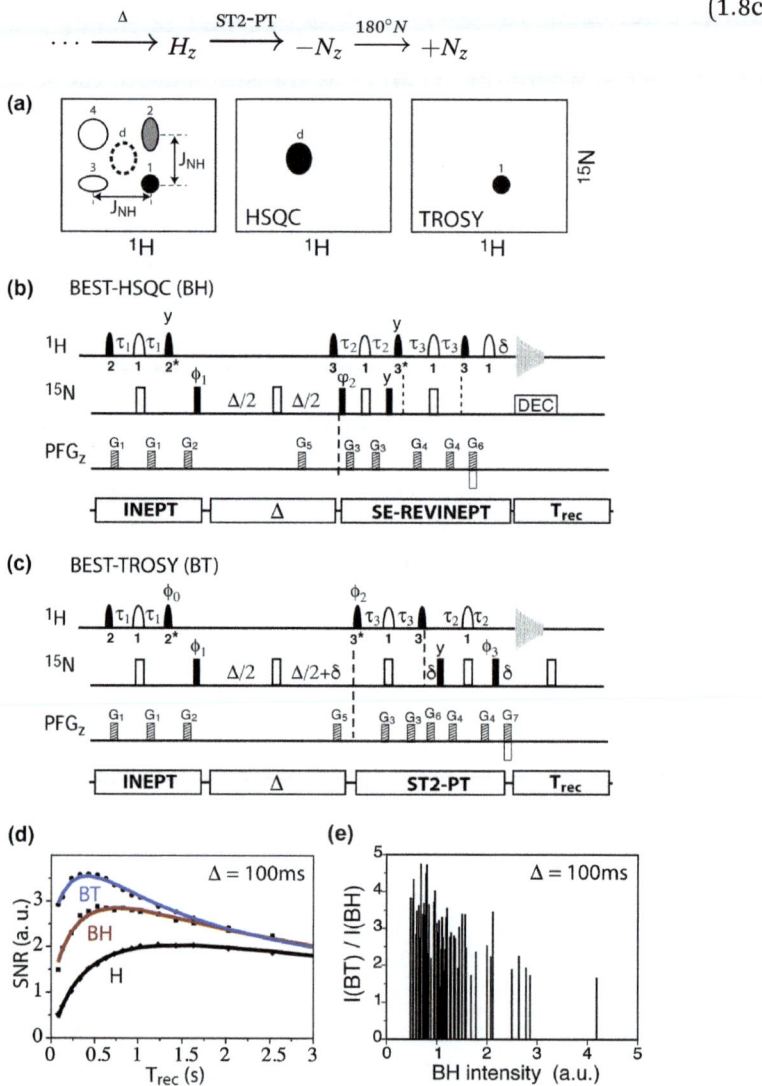

As a consequence of this third coherence transfer pathway, the steady-state ^{15}N polarization in TROSY depends on the longitudinal relaxation rate constants T_1^H and T_1^N, as well as the pulse sequence delays Δ and T_{rec}. The expected ^{15}N polarization enhancement can be calculated by the following expression:

$$\lambda_N = \frac{N_z^{ss}}{N_z^0} \propto 1 - \left(\left(1 - \frac{\gamma_H}{\gamma_N} \left(1 - \exp(-\Delta/T_1^H) \right) \right) \exp(-T_{rec}/T_1^N) \right) \quad (1.9)$$

^{15}N polarization enhancement becomes particularly pronounced for short T_1^H and long T_1^N, as well as short recycle delays T_{rec}. This is exactly the situation encountered in BEST-TROSY experiments at high magnetic field strengths. For a protein with an 8 ns correlation time at 800 MHz, typical relaxation time constants are $T_1^H = 50$–300 ms (Figure 1.4) and $T_1^N \approx 1$ s. For a typical recycle delay $T_{rec} = 200$ ms and assuming an overall relaxation delay $\Delta = 100$ ms, this results in enhancement factors varying form $\lambda_N = 2.5$ to $\lambda_N = 7.3$. The steady-state ^{15}N polarization enhancement is expected to further increase with the magnetic field strength and the size of the protein (rotational correlation time).[25]

Experimental sensitivity curves measured on a sample of ubiquitin (5 °C, 800 MHz, $\tau_c \cong 8$ ns) with the 1D pulse sequences of Figure 1.7b and 1.7c (with $\Delta = 100$ ms), as well as a conventional HSQC sequence are shown in Figure 1.7d. These curves, obtained by integrating the intensities measured in the 1D spectra, reflect intrinsic sensitivity differences of the pulse schemes without taking into account the additional line narrowing in TROSY experiments. It is interesting to note that the sensitivity optimum

---

Figure 1.7 (a) Different ways of reducing the multiplet pattern observed in a scalar coupled two-spin systems to a single cross peak, either using hetero-nuclear decoupling (HSQC) or single-transition spin state selection (TROSY). Pulse sequences for (b) BEST-HSQC and (c) BEST-TROSY experiments. The numbers under the ^{1}H pulse symbols indicate the shaped pulse type used: (1) REBURP, (2) PC9, and (3) EBURP-2. An asterisk indicates time and phase reversal of the corresponding pulse shape. They typically cover a bandwidth of 4 ppm (centered at 8.5 ppm). To account for spin evolution during the shaped pulses, the transfer delays are set to $\tau_1 = 1/(4J_{NH}) - 0.5\delta_1 - 0.5\delta_2$, $\tau_2 = 1/(4J_{NH}) - 0.5\delta_1$, and $\tau_3 = 1/(4J_{NH}) - 0.5\delta_1 - 0.6\delta_3$, with δ_1, δ_2, δ_3 the pulse lengths of PC9, EPURP-2, and REBURP, respectively. A generic relaxation delay Δ accounts for the ^{1}H, ^{15}N relaxation behavior in different experiments, where this delay is replaced by additional transfer and chemical shift editing periods. (d) Sensitivity curves measured for ubiquitin (800 MHz, 20 °C) using BEST-HSQC (blue); BEST-TROSY (red), or a conventional HSQC sequence with a relaxation delay $\Delta = 100$ ms. The intensity points have been obtained from integration of the recorded 1D amide ^{1}H spectra. (e) Signal intensity ratios for individual cross peaks measured in BEST-HSQC (BH) and BEST-TROSY (BT) 2D iHN(CA) spectra of a 138-residue protein, and plotted as a function of the BH peak intensity.[25] The observed correlation indicates a tendency toward a stronger enhancement effect (S/N gain) for weaker correlation peaks.

shifts to slightly shorter T_{rec} values for BEST-TROSY with respect to BEST-HSQC. The sensitivity advantage of BEST-TROSY (BT) with respect to BEST-HSQC (BH) becomes even further pronounced when comparing individual peak intensities in 2D ^{1}H–^{15}N correlation spectra, as illustrated in Figure 1.7e for a 16 kDa protein (20 °C, 800 MHz, $\tau_c \cong 10$ ns). The intensity ratios (BT/BH) vary from about 1.5 in the most flexible protein regions to more than 4. An interesting observation is the correlation between the peak intensity measured in the BH spectrum and the intensity gain from BT, indicating that weak peaks are enhanced more than already stronger peaks. As a result, the peak intensity distribution for proteins displaying heterogeneous conformational flexibility becomes more uniform in BEST-TROSY spectra compared to the BEST (or conventional) HSQC counterparts.

In summary of this section, the appropriate use of band-selective pulse shapes on the ^{1}H channel, the basis of BEST experiments, allows us to set up a variety of NMR experiments for the study of proteins and nucleic acids. A non-exhaustive list of BEST-type experiments, proposed to date for the study of proteins or nucleic acids is provided in Table 1.1. These pulse schemes provide an advantage in terms of experimental sensitivity and data acquisition speed. Under fast pulsing conditions, the sensitivity improvement achieved by BEST pulse schemes compared to conventional hard-pulse-based analogues may reach up to one order of magnitude in the most favorable cases. As a rule of thumb, at moderate field strength (<700 MHz) and tumbling correlation times $\tau_c < 10$ ns, BEST-HSQC implementations are typically advantageous in terms of experimental sensitivity. At higher magnetic field strengths and/or for larger molecules, BEST-TROSY outperforms BEST-HSQC both in terms of spectral resolution and experimental sensitivity, with the additional advantage that no ^{15}N decoupling is applied during signal detection allowing for long acquisition times without the need to care about duty cycle requirements of the NMR hardware.

1.3.3   BEST-optimized ^{13}C-detected Experiments

^{13}C direct detection is a routine tool for NMR studies of small molecules, mainly because of the higher spectral resolution achieved in the ^{13}C spectrum with respect to the corresponding ^{1}H spectrum. Recently, a series of ^{13}CO-detected experiments has been proposed and shown to provide useful additional NMR tools for protein studies,[34] especially in the context of molecules with a high degree of intrinsic disorder (IDPs) or proteins containing paramagnetic centers where ^{13}CO detection offers some advantages: (i) the ^{13}CO line width is not affected by solvent exchange; (ii) signals from proline residues can be observed; (iii) ^{13}CO detection is less affected by paramagnetic line broadening. The major inconvenience of ^{13}C-detection is the lower intrinsic sensitivity due to the ∼four-fold smaller gyromagnetic ratio of ^{13}C with respect to ^{1}H (see eqn (1.1)). This situation has been

Table 1.1 Non-exhaustive list of LRE-optimized ¹H-detected experiments that have been proposed in the literature for NMR investigations of proteins or nucleic acids.

LRE technique	Application	Purpose	Refs.
SOFAST	Proteins	2D $^1H–^{15}N$ or $^1H–^{13}C$	Schanda 2005[38,39], Amero 2009[42]
ST-SOFAST	Proteins	2D $^1H–^{15}N$ or $^1H–^{13}C$	Kern 2008[48]
J-SOFAST	Proteins	2D $^1H–^{15}N$ or $^1H–^{13}C$	Kupce 2007[49], Mueller 2008[50]
ultraSOFAST	Proteins	2D $^1H–^{15}N$ or $^1H–^{13}C$	Gal 2007[54], Kern 2008[48]
SOFAST	RNA (iminos)	2D $^1H–^{15}N$	Farjon 2009[15]
SOFAST	RNA (bases)	2D $^1H–^{13}C$	Sathyamoorthy 2014[58]
BEST-HSQC	Proteins	2D $^1H–^{15}N$	Pervushin 2002[10]
BEST-HSQC	Proteins	3D Backbone assignment expts: HNCO, HNCA, HNCACB, HN(CO)CA, HN(CO)CACB, iHNCA, iHNCACB, HNN	Schanda 2006[16], Lescop 2007[24], Kumar 2010[59]
BEST-HSQC	Proteins	Backbone J coupling and RDC measurements: $N–H^N$, $C^\alpha–H^\alpha$, $CO–H^N$, $N–CO$...	Rasia 2011[60]
BEST-HSQC	Proteins	Measurement of RDCs between amide ^1H	Schanda 2007[61]
BEST-TROSY	Proteins	2D $^1H–^{15}N$	Pervushin 2002[10], Favier 2011[25]
BEST-TROSY	RNA (iminos)	2D $^1H–^{15}N$	Farjon 2009[15]
BEST-TROSY	Proteins	Backbone assignment: 3D HNCO, HNCA, HNCACB, HN(CO)CA, HN(CO)CACB, HN(CO)CA, HN(CO)CACB: iHNCA, iHNCACB, (H)N(COCA)NH, (HN)CO(CA)NH	Pervushin 2002[10], Favier 2011[25], Solyom 2013[26]
BEST-TROSY	Proteins	Bidirectional HNC experiments with enhanced sequential correlation pathway: HNCA+, HNCO+, HNCACB+	Gil-Caballero 2014[62]
BEST-TROSY	IDPs	Proline-selective $^1H–^{15}N$ correlation experiments	Solyom 2013[26]
BEST-TROSY	RNA (iminos)	Trans-hydrogen-bond correlation experiment: HNN-COSY	Farjon 2009[15]
HET-SOFAST/BEST	Proteins	Quantification of LRE effect from water exchange or $^1H–^1H$ cross-relaxation	Schanda 2006[57], Rennella 2014[56]
Flip-back	Proteins	2D $^1H–^{15}N$	Deschamp 2006[63]
Flip-back	Proteins	Backbone assignment: 3D HNCO, HNCA	Diercks 2005[64]
Flip-back	Proteins	Aromatic side chain assignment	Eletsky 2005[65]

significantly improved over recent years by the development of ^{13}C-optimized cryogenic triple-resonance probes and pulse sequence improvements. Among those tools, the use of ^{1}H as a starting polarization source for the coherence transfer pathway has been proposed in order to increase experimental sensitivity.[35,36] Such ^{1}H-start and ^{13}C-detect experiments are amenable to BEST-type LRE optimization, as recently demonstrated for the 2D H^{N-BEST} CON experiment,[37] which starts from amide ^{1}H polarization, but only detects ^{15}N and ^{13}CO chemical shifts in a 2D correlation spectrum. This CO–N spectrum is of particular interest for the study of IDPs at close to physiological temperature and pH where solvent-exchange line broadening of amide ^{1}H may prevent their detection in a ^{1}H-detected experiment. Other ^{13}C-detected experiments starting from aliphatic ^{1}H polarization have been optimized in terms of LRE by using a coupling-based flip back scheme, as depicted in Figure 1.2c.

1.4 SOFAST-HMQC: Fast and Sensitive 2D NMR

So far, we have described LRE optimization in heteronuclear single-quantum correlation experiments, HSQC and TROSY. In this section, we will focus on heteronuclear multiple-quantum correlation (HMQC) experiments, and in particular on 2D ^{1}H–^{15}N and ^{1}H–^{13}C HMQC, where BEST-optimization can be combined with Ernst-angle excitation to further enhance the steady-state ^{1}H polarization for very short recycle delays ($T_{rec} \ll T_1^{H}$). This type of experiments is called band-Selective Optimized Flip Angle Short Transient (SOFAST) HMQC.[38,39]

1.4.1 Ernst-angle Excitation

Let us consider a simple one-pulse experiment with a nominal pulse flip-angle β. As a consequence of such a simple pulse scheme, part of the spin polarization ($\cos \beta$) is preserved, and can thus be directly used for the subsequent transient without the need for any relaxation period, while a fraction $\sin \beta$ gives rise to the detected NMR signal. In practice, the steady-state polarization will depend on the recycle delay T_{rec} and the effective longitudinal relaxation time T_1, as described by eqn (1.7) with the factor $f = \cos \beta$. The resulting SNR is then given by:

$$\text{SNR}(\beta, T_{rec}, T_1) \propto \frac{H_z^{ss} \sin \beta}{\sqrt{T_{rec} + t_{seq}}} = \frac{(1 - \exp(T_{rec}/T_1)) \sin \beta}{(1 - \cos \beta \exp(T_{rec}/T_1))\sqrt{T_{rec} + t_{seq}}}$$

$$(1.10)$$

Sensitivity curves SNR(T_{rec}) according to eqn (1.10) for $T_1 = 200$ ms and different flip angles β are plotted in Figure 1.8a. Choosing smaller flip angles ($\beta < 90°$) shifts the sensitivity maximum to shorter recycle delays, with the highest overall sensitivity obtained for $\beta \approx 60°$.

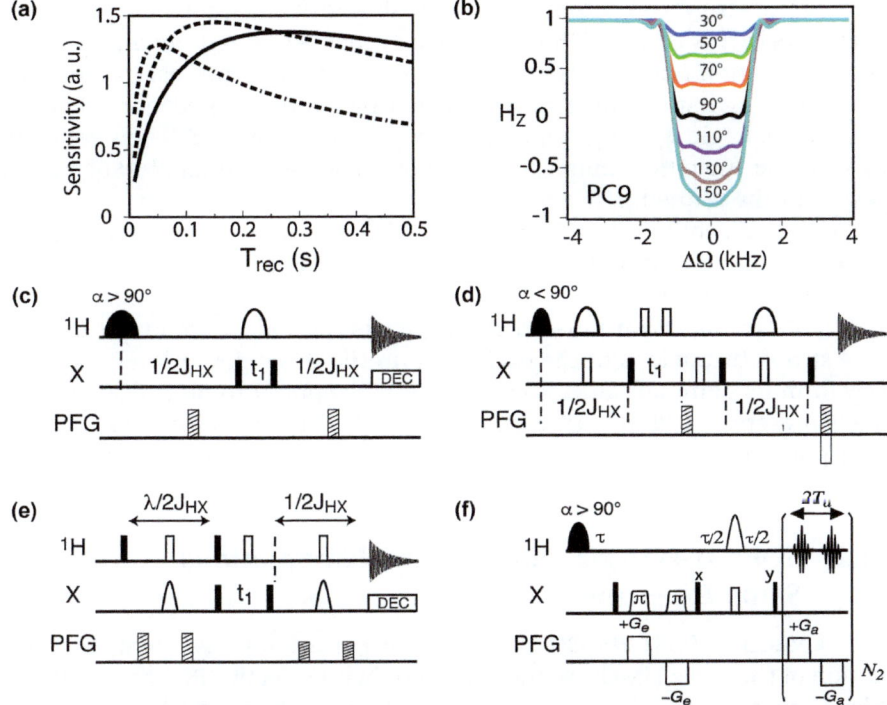

Figure 1.8 SOFAST-HMQC experiments. (a) Sensitivity curves computed according to eqn (1.10) for flip angles of $\beta = 90°$ (straight line), $\beta = 60°$ (dashed line), and $\beta = 30°$ (dotted-dashed line). A relaxation time $T_1 = 200$ ms and a pulse sequence duration of $t_{seq} = 20$ ms have been assumed for the calculation. (b) Excitation profile of PC9 pulse of 3 ms duration measured for different flip angles in the range 20°–150°. Different implementations of SOFAST-HMQC: (c) standard SOFAST,[38,39] (d) single-transition (ST) SOFAST,[48] (e) J-SOFAST,[49,50] and (f) single-scan ultra-SOFAST.[48,54]

The flip angle providing the highest sensitivity for a given recovery time T_{rec} is called the Ernst angle.[1] Closer inspection of eqn (1.10) shows that highest sensitivity is achieved if the following relation is satisfied:

$$\cos \beta^{opt} \cong \exp(-T_{rec}/T_1) \tag{1.11}$$

with the assumption that $t_{seq} \ll T_{rec}$. In a single-pulse experiment, Ernst-angle excitation is realized by adjusting the nutation angle of the excitation pulse. It becomes, however, a non-trivial task when dealing with multi-pulse sequences employing a series of 90° and 180° pulses. Although even complex pulse sequences may still be designed in a way that some of the ^{1}H spin polarization is restored at the end, their practical utility is very limited. Because the magnetization component that is finally restored follows a complex trajectory during the pulse sequence, longitudinal and transverse spin relaxation effects, as well as pulse imperfections, reduce its magnitude, and thus severely compromise the overall sensitivity of the experiment. Exceptions to this rule are HMQC-type experiments that require only a single proton excitation pulse followed by one (or several) 180° pulses. In this case, Ernst angle excitation can still be realized by taking into account that the effective flip angle is now given by $\beta^{\mathrm{opt}} = \alpha^{\mathrm{opt}} + n(180°)$ with α^{opt} being the flip angle of the excitation pulse and n the number of additional 180° pulses present in the sequence.[40]

In order to combine BEST-type LRE with Ernst angle excitation, a pulse shape is required that performs band selective excitation for a wide range of flip angles (power levels). This is not the case for the majority of pulse shapes (BURP, SNOB, Gaussian pulse cascades, *etc.*) reported in the literature that show strong distortions of the excitation profile if the power level is moved away form its nominal value. An exception is the polychromatic PC9 pulse,[21] which preserves a "top-hat" excitation profile for a range of power levels corresponding to on-resonance flip angles between 0° and 150° (Figure 1.8b).

1.4.2 SOFAST-HMQC: Different Implementations of the Same Experiment

The standard SOFAST-HMQC experiment (Figure 1.8c)[39] is a simple modification of the basic HMQC sequence where band-selective PC9 and REBURP pulse shapes replace the 90° and 180° hard pulses, respectively, to enhance the ^{1}H steady-state polarization and the overall experimental sensitivity under fast-pulsing conditions. This sequence also achieves good water suppression by a built-in "Watergate" sequence (PFG-180 sel-PFG),[41] as long as the water resonates outside the chosen excitation band, and homonuclear ^{1}H decoupling during the ^{1}H–X transfer delays and the t_1 chemical shift labeling period. In general, SOFAST-HMQC requires some phase cycling ($n_{PC} \geq 2$) in order to remove artefacts *e.g.* residual solvent signals from the spectra. For the highest sensitivity of the experiment, the flip angle should be adjusted to $\alpha \approx 120°$ with a recycle delay $T_{\mathrm{rec}} \approx T_1$. Alternatively, if data acquisition speed is the major objective ($T_{\mathrm{rec}} = T_{\mathrm{det}}$), a larger flip angle ($\alpha \approx 150°$) typically further increases experimental sensitivity. Under favourable experimental conditions (high magnetic field strength, cryogenic probe, protein concentration of a few hundred μM), 2D SOFAST-HMQC fingerprint spectra can be recorded in less than 5 s for molecular systems ranging from small globular proteins (Figure 1.9a), large molecular assemblies (Figure 1.9c), to nucleic acids (Figure 1.9d). This has opened new perspectives for site-resolved real-time NMR investigations of molecular kinetics, such as protein folding, oligomerization and assembly.[42–45] It also allows the study of proteins and cell metabolites in living cells under sample conditions characterized by short lifetimes.[46,47]

An alternative SOFAST-HMQC pulse sequence is shown in Figure 1.8d. The single-transition (ST) version[48] selects one of the ^{1}H doublet components (TROSY line) in the direct detection dimension, while heteronuclear

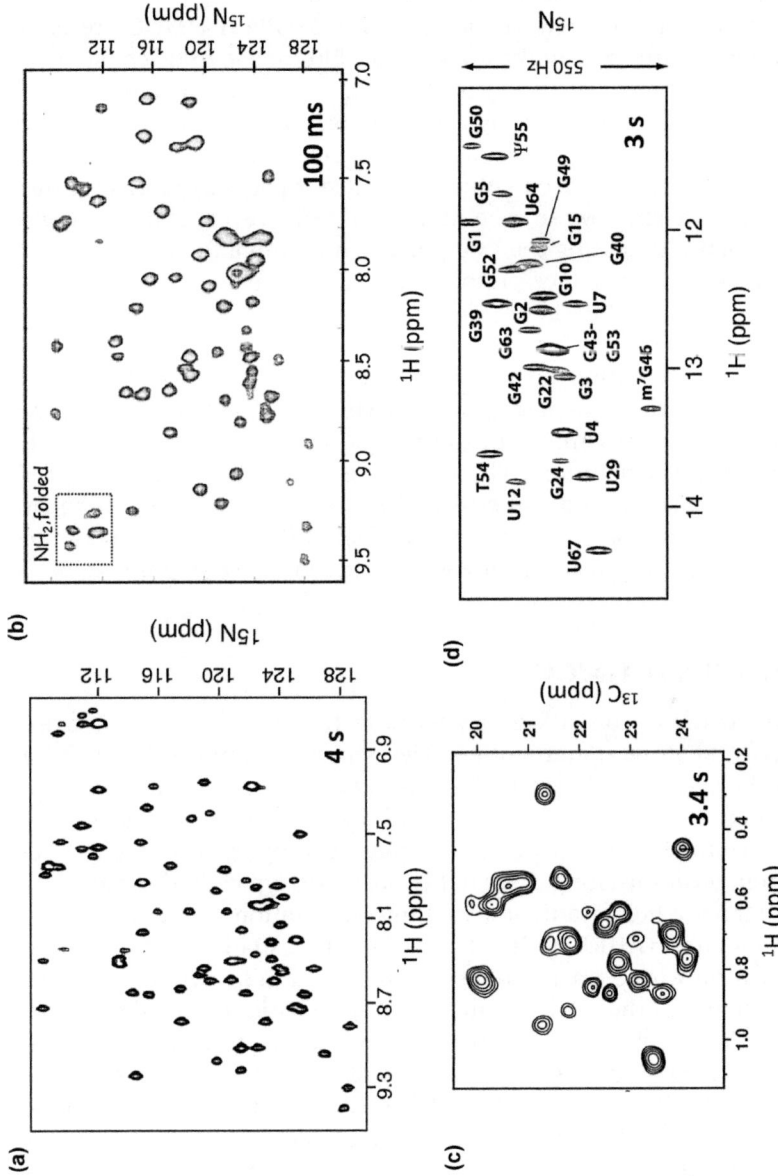

Figure 1.9 2D SOFAST-HMQC spectra of different biomolecular systems. (a) amide ^1H–^{15}N correlation spectrum of the small protein ubiquitin (8.5 kDa, 0.2 mM, 25 °C, 18.8 T); (b) ultraSOFAST-HMQC spectrum of a 2 mM ubiquitin sample recorded in a single scan; (c) methyl ^1H–^{13}C correlation spectrum of a large molecular assembly, the TET2 protease (468 kDa, 80 μM, 0.9 mM, 37 °C, 18.8 T); (d) imino ^1H–^{15}N correlation spectrum of tRNA^{Val} (26 kDa, 25 °C, 18.8 T). Figure reproduced with permission from E. Rennella and B. Brutscher, *ChemPhysChem* 2013, **14**, 3059 (ref. 43).

decoupling is still applied in the indirect (t_1) dimension. The basic idea behind this scheme is that both quadrature components present after the t_1 free evolution delay are converted into detectable in-phase and anti-phase ^{1}H coherence. As a consequence of the single-transition spin state selection, the sensitivity of ST-SOFAST is a factor $\sqrt{2}$ lower as compared to the standard SOFAST version. Despite this reduced sensitivity, ST-SOFAST-HMQC presents some interesting features that make it a valuable alternative to be considered for practical applications: (i) reduced RF load owing to the absence of composite X-spin decoupling during signal detection; (ii) suppression of strong signals from the solvent or sample impurities in a single scan by the use of pulsed-field-gradient-based coherence-transfer-pathway selection; (iii) as the pulse sequence employs an even number of 180° ^{1}H pulses, Ernst-angle excitation is achieved by setting the PC9 pulse to a nominal flip angle $\alpha <90°$, for which a cleaner off-resonance performance is observed.

Finally, the J-SOFAST-HMQC[49,50] (Figure 1.8e) uses a coupling-based flip-back scheme, instead of band-selective pulses, for ^{1}H excitation. In order to achieve Ernst-angle excitation, the transfer delay of the first INEPT block is adjusted to $\lambda/2J_{HX}$ with $\lambda = \cos \alpha$. J-SOFAST-HMQC presents an attractive alternative for ^{1}H–^{13}C experiments of samples at natural-abundance or with a low level of isotope enrichment. This sequence becomes also of interest if the selection of observed and unperturbed protons is preferably done by exploiting the characteristic ^{13}C or ^{15}N chemical shift ranges by means of band-selective inversion pulses, as already discussed above (Figure 1.2c).

1.4.3 UltraSOFAST-HMQC

The speed of 2D data acquisition can be even further enhanced by combining SOFAST-HMQC with gradient-encoded single-scan multidimensional NMR.[51–53] In the so-called ultraSOFAST-HMQC experiment[48,54] (Figure 1.8f), the incremented time delay t_1 is replaced by a spatial chemical shift encoding, where nuclear spins in the sample are progressively excited according to their position along a spatial coordinate, typically the z-axis, by the use of a magnetic field gradient acting in combination with a frequency-swept radiofrequency (CHIRP) pulse. This results in a spatial winding of the spin magnetization with a position-dependent phase $C\Omega_1 z$, where Ω_1 is the resonance frequency of the nuclear spin, and C is a spatio-temporal constant depending on the sample length and some user-defined acquisition parameters. A spatial frequency decoding or unwinding of the resulting helix is achieved during the signal detection period by the use of a second magnetic field gradient. As a result of this acquisition gradient G_a, a spin echo is created whose position in time depends on the resonance frequency Ω_1 as well as on the strength of G_a. This reading process can be repeated numerous (N_2) times by oscillating the sign of the readout gradients, yielding a set of indirect time-domain spectra as a function of the t_2 time evolution. Fourier transformation along the t_2 dimension only results in the desired 2D

NMR spectrum. For a more thorough discussion of single-scan NMR, we refer the reader to Chapter 2 by Gal.

The pulse sequence in Figure 1.8f is based on ST-SOFAST-HMQC,[48] thus avoiding the need for heteronuclear decoupling during the readout process. This allows the use of weaker acquisition gradients, resulting in reduced filter bandwidth, and therefore increases the overall sensitivity of the experiment. Interestingly, in contrast to the conventional ST-SOFAST-HMQC experiment, in the case of spatial frequency encoding the spin-state selection process does not lead to any additional sensitivity loss, making this sequence the optimal choice under all circumstances. An amide ^{1}H–^{15}N spectrum of the small protein ubiquitin, recorded in a single scan, is shown in Figure 1.9b.

As a major application, ultraSOFAST-HMQC allows the recording of ^{1}H–^{15}N (and ^{1}H–^{13}C) correlation spectra with repetition rates of up to a few s^{-1}, thus enabling real-time studies of molecular kinetics occurring on time scales down to a few seconds.[54,55]

1.5 Conclusions

Fast-pulsing techniques provide a convenient way to reduce the acquisition time of multidimensional NMR spectra without compromising, often even significantly improving, experimental sensitivity, and without the need for non-conventional data processing tools. Therefore, there is no good reason for not using fast-pulsing techniques whenever possible, and in situations where LRE effects are expected to be sizeable. The only drawback may be the loss of some ^{1}H signals with unusual resonance frequencies, owing for example to an important ring current shift contribution. The use of typical recycle delays of 200 to 300 ms in BEST experiments reduces the data acquisition time required for a high-resolution 3D data set to a few hours, which often corresponds to or is less than the time needed to obtain sufficient SNR in the final spectrum (sensitivity limited regime). The situation is of course different for higher dimensional (>3D) experiments that are of particular interest for the study of larger proteins, as well as IDPs. Here, the time requirements, even for BEST-type experiments, are dictated by sampling requirements, especially if high spectral resolution in all dimensions is desired. In order to shift these high dimensional experiments again from the sampling-limited to a sensitivity-limited regime, we need to combine the fast-pulsing techniques described here with one of the non-uniform sparse data sampling methods that will be described in subsequent chapters.

LRE effects can also be quantified by measuring effective amide ^{1}H T_1 relaxation time constants under various experimental conditions as shown in Figure 1.4, or by comparing peak intensities in a reference SOFAST/BEST spectrum with those measured after perturbation of either the bulk water or aliphatic protons, as implemented in the HET-SOFAST/BEST experiments.[56,57] This experiment provides valuable information on solvent exchange rates, and the efficiency of dipolar-driven spin diffusion that can be related to the local compactness at the site of a given amide proton.

Finally, fast-pulsing multidimensional SOFAST- and BEST-type experiments offer new opportunities for site-resolved real-time NMR studies of protein folding and other kinetic processes that occur on a seconds to minutes time scale, and that are difficult or even impossible to investigate by other high-resolution methods.[43]

References

1. R. Ernst, G. Bodenhausen and G. Wokaun, *Principles of Nuclear Magnetic Resonance in One and Two Dimensions*, Oxford University Press, Oxford, 1987.
2. J. Jeener, In Ampère International Summer School II Basko Polje, 1971.
3. J. Cavanagh, W. J. Fairbrother, A. G. Palmer, N. J. Skelton, *Protein NMR Spectroscopy*, Academic Press, San Diego, 1996.
4. H. G. Hocking, K. Zangger and T. Madl, *ChemPhysChem*, 2013, **14**, 3082.
5. S. Cai, C. Seu, Z. Kovacs, A. D. Sherry and Y. Chen, *J. Am. Chem. Soc.*, 2006, **128**, 13474.
6. F.-X. Theillet, A. Binolfi, S. Liokatis, S. Verzini and P. Selenko, *J. Biomol. NMR*, 2011, **51**, 487.
7. N. Sibille, G. Bellot, J. Wang and H. Déméné, *J. Magn. Reson.*, 2012, **224**, 32.
8. A. Eletsky, O. Moreira, H. Kovacs and K. Pervushin, *J. Biomol. NMR*, 2003, **26**, 167.
9. S. Hiller, G. Wider, T. Etezady-Esfarjani, R. Horst and K. Wüthrich, *J. Biomol. NMR*, 2005, **32**, 61.
10. K. Pervushin, B. Vögeli and A. Eletsky, *J. Am. Chem. Soc.*, 2002, **124**, 12898.
11. P. Schanda, *Prog. Nucl. Magn. Reson. Spectrosc.*, 2009, **55**, 238.
12. H. J. Dyson and P. E. Wright, *Nat. Rev. Mol. Cell Biol.*, 2005, **6**, 197.
13. V. N. Uversky, C. J. Oldfield and A. K. Dunker, *Annu. Rev. Biophys.*, 2008, **37**, 215.
14. P. Tompa, *Trends Biochem. Sci.*, 2002, **27**, 527.
15. J. Farjon, J. Boisbouvier, P. Schanda, A. Pardi, J. P. Simorre and B. Brutscher, *J. Am. Chem. Soc.*, 2009, **131**, 8571.
16. P. Schanda, H. Van Melckebeke and B. Brutscher, *J. Am. Chem. Soc.*, 2006, **128**, 9042.
17. H. Geen and R. Freeman, *J. Magn. Reson.*, 1991, **93**, 93.
18. L. Emsley and G. Bodenhausen, *Chem. Phys. Lett.*, 1990, **165**, 469.
19. L. Emsley and G. Bodenhausen, *J. Magn. Reson.*, 1992, **97**, 135.
20. E. Kupce, J. Boyd and I. D. Campbell, *J. Magn. Reson. B*, 1995, **106**, 300.
21. E. Kupce and R. Freeman, *J. Magn. Reson. A*, 1994, **108**, 268.
22. E. Lescop, T. Kern and B. Brutscher, *J. Magn. Reson.*, 2010, **203**, 190.
23. Y. Li, M. Rance and A. G. Palmer, *J. Magn. Reson.*, 2014, **248**, 105.
24. E. Lescop, P. Schanda and B. Brutscher, *J. Magn. Reson.*, 2007, **187**, 163.

25. A. Favier and B. Brutscher, *J. Biomol. NMR*, 2011, **49**, 9.
26. Z. Solyom, M. Schwarten, L. Geist, R. Konrat, D. Willbold and B. Brutscher, *J. Biomol. NMR*, 2013, **55**, 311.
27. G. Bodenhausen and D. J. Ruben, *Chem. Phys. Lett.*, 1980, **69**, 185.
28. A. Meissner, J. Ø. Duus and O. W. Sørensen, *J. Magn. Reson.*, 1997, **97**, 92.
29. M. D. Sørensen, A. Meissner and O. W. Sørensen, *J. Biomol. NMR*, 1997, **10**, 181.
30. K. Pervushin, R. Riek, G. Wider and K. Wüthrich, *Proc. Natl. Acad. Sci. U.S.A.*, 1997, **94**, 12366.
31. B. Brutscher, *Concepts Magn. Reson.*, 2000, **12**, 207.
32. L. S. Yao, A. Grishaev, G. Cornilescu and A. Bax, *J. Am. Chem. Soc.*, 2010, **132**, 10866.
33. L. S. Yao, A. Grishaev, G. Cornilescu and A. Bax, *J. Am. Chem. Soc.*, 2010, **132**, 4295.
34. W. Bermel, I. Bertini, I. Felli, M. Piccioli and R. Pierattelli, *Prog. Nucl. Magn. Reson. Spectrosc.*, 2006, **48**, 25.
35. W. Bermel, I. Bertini, V. Csizmok, I. C. Felli, R. Pierattelli and P. Tompa, *J Magn Reson*, 2009, **198**, 275.
36. W. Bermel, I. Bertini, I. C. Felli, R. Pierattelli, V. Uni, V. L. Sacconi and S. F. Florence, *J. Am. Chem. Soc.*, 2009, **131**, 15339.
37. S. Gil, T. Hošek, Z. Solyom, R. Kümmerle, B. Brutscher, R. Pierattelli and I. C. Felli, *Angew. Chem., Int. Ed.*, 2013, 11808.
38. P. Schanda and B. Brutscher, *J. Am. Chem. Soc.*, 2005, **127**, 8014.
39. P. Schanda, E. Kupce and B. Brutscher, *J. Biomol. NMR*, 2005, **33**, 199.
40. A. Ross, M. Salzmann and H. Senn, *J. Biomol. NMR*, 1997, **10**, 389.
41. M. Piotto, V. Saudek and V. Sklenar, *J. Biomol. NMR*, 1992, **2**, 661.
42. C. Amero, P. Schanda, M. A. Dura, I. Ayala, D. Marion, B. Franzetti, B. Brutscher and J. Boisbouvier, *J. Am. Chem. Soc.*, 2009, **131**, 3448.
43. E. Rennella and B. Brutscher, *ChemPhysChem*, 2013, **14**, 3059.
44. E. Rennella, T. Cutuil, P. Schanda, I. Ayala, F. Gabel, V. Forge, A. Corazza, G. Esposito and B. Brutscher, *J. Mol. Biol.*, 2013, **425**, 2722.
45. P. Schanda, V. Forge and B. Brutscher, *Proc. Natl. Acad. Sci. U. S. A.*, 2007, **104**, 11257.
46. A. Motta, D. Paris and D. Melck, *Anal. Chem.*, 2010, **82**, 2405.
47. Y. Ito and P. Selenko, *Curr. Opin. Struct. Biol.*, 2010, **20**, 640.
48. T. Kern, P. Schanda and B. Brutscher, *J. Magn. Reson.*, 2008, **190**, 333.
49. E. Kupce and R. Freeman, *Magn. Reson. Chem.*, 2007, **45**, 2.
50. L. Mueller, *J. Biomol. NMR*, 2008, **42**, 129.
51. M. Mishkovsky and L. Frydman, *Annu. Rev. Phys. Chem.*, 2009, **60**, 429.
52. Y. Shrot and L. Frydman, *J. Chem. Phys.*, 2008, **128**.
53. A. Tal and L. Frydman, *Prog. Nucl. Magn. Reson. Spectrosc.*, 2010, **57**, 241.
54. M. Gal, P. Schanda, B. Brutscher and L. Frydman, *J. Am. Chem. Soc.*, 2007, **129**, 1372.
55. M.-K. Lee, M. Gal, L. Frydman and G. Varani, *Proc. Natl. Acad. Sci. U. S. A.*, 2010, **107**, 9192.
56. E. Rennella, Z. Solyom and B. Brutscher, *J. Biomol. NMR*, 2014, **60**, 99.

57. P. Schanda, V. Forge and B. Brutscher, *Magn. Reson. Chem.*, 2006, **44**, S177.
58. B. Sathyamoorthy, J. Lee, I. Kimsey, L. R. Ganser and H. Al-Hashimi, *J. Biomol. NMR*, 2014.
59. D. Kumar, S. Paul and R. V. Hosur, *J. Magn. Reson.*, 2010, **204**, 111.
60. R. M. Rasia, E. Lescop, J. F. Palatnik, J. Boisbouvier and B. Brutscher, *J. Biomol. NMR*, 2011, **51**, 369.
61. P. Schanda, E. Lescop, M. Falge, R. Sounier, J. Boisbouvier and B. Brutscher, *J. Biomol. NMR*, 2007, **38**, 47.
62. S. Gil-Caballero, A. Favier and B. Brutscher, *J. Biomol. NMR*, 2014, **60**, 1.
63. M. Deschamps and I. D. Campbell, *J. Magn. Reson.*, 2006, **178**, 206.
64. T. Diercks, M. Daniels and R. Kaptein, *J. Biomol. NMR*, 2005, **33**, 243.
65. A. Eletsky, H. S. Atreya, G. Liu and T. Szyperski, *J. Am. Chem. Soc.*, 2005, **127**, 14578.

CHAPTER 2

Principles of Ultrafast NMR Spectroscopy

MAAYAN GAL[a,b]

[a] Migal, Galilee Institute of Research Ltd., 11016 Kiryat Shmona, Israel;
[b] Faculty of Sciences and Technology, Tel-Hai Academic College, Upper Galilee, Israel
Email: MaayanG@migal.org.il

2.1   Introduction

Owing to its power and usefulness for analyzing the structure, function, and dynamics of molecules, NMR is a leading analytical tool in a wide range of scientific disciplines. This array of applications have grown following the introduction of the Fourier Transform (FT) NMR experiments[1] and received an enormous boost after the development of nD NMR spectroscopy. Yet, two major issues have prevented multidimensional NMR from breaching new frontiers. These are the relatively low sensitivity of NMR spectroscopy and the long experimental times associated with the acquisition of >2D NMR spectra. Among the various techniques designed to cut loose from the latter limitation and accelerate the acquisition of nD NMR spectra, ultrafast (UF) NMR spectroscopy is the only method that can—at least in principle—collect an nD NMR spectral grid within a single transient. The principles that stand at the heart of single-scan UF NMR spectroscopy are described in this chapter.

New Developments in NMR No. 11
Fast NMR Data Acquisition: Beyond the Fourier Transform
Edited by Mehdi Mobli and Jeffrey C. Hoch
© The Royal Society of Chemistry 2017
Published by the Royal Society of Chemistry, www.rsc.org

2.1.1   One- and Two-dimensional FT NMR

A breakthrough came in the 1970s when the basic scheme of the 1D NMR experiment changed to a pulse-excited Fourier Transform, which enabled time domain data acquisition. The most basic single pulse NMR experiment then resulted with the acquisition of Free Induction Decay (FID) being induced from the ensemble of excited spins. This well-known FID signal of the basic FT single-pulse 1D NMR experiment can be described by:[1,2]

$$S(t) = \sum_{\omega} I(\omega) e^{i\omega t} e^{\left(\frac{-t}{T_2}\right)} \tag{2.1}$$

Eqn (2.1) describes a sum of precessing spin magnetizations with a spectral distribution $I(\omega)$ that decay back to a null transverse equilibrium magnetization with a uniform relaxation time T_2. Considering that the soul goal of the latter experiment is to find $I(\Omega)$ from the single FID, FT can be applied to yield:

$$I(\omega) = \sum_{\text{Digitized } t} S(t) e^{-i\omega t} \tag{2.2}$$

Not long after, two-dimensional (2D) NMR experiment was proposed by Jeener,[3] an experiment whose goal is to measure and assign connectivities between different nuclear sites by spreading the frequency domain data onto a 2D map $I(\omega_1, \omega_2)$. This breakthrough in the history of NMR is the basis of what enables us today to study large and complex biomolecular systems. As a natural extension of eqn (2.1), the FID of a 2D NMR experiment contains the evolution of two time-domain variables t_1 and t_2 and can be cast into the following:

$$S(t_1, t_2) = \sum_{\omega_2} \left[\sum_{\omega_1} I(\omega_1, \omega_2) e^{i\omega_1 t_1} e^{-\frac{t_1}{T_2}} \right] e^{i\omega_2 t_2} e^{-\frac{t_2}{T_2}} \tag{2.3}$$

Although there is a similarity between eqn (2.1) and (2.3), the latter holds a great difference. Unlike 1D NMR experiments in which the time-domain data can be physically acquired within a single scan, 2D (and higher-dimensional) NMR experiments require a way for indirectly detecting the evolution of the spins along orthogonal time domains t_1 and measure the resonance frequencies of the indirectly detect dimension ω_1. For doing so, the so-called t_1 parameter is systematically modulated by discrete incrementation of a certain delay throughout a series of individual scans, such that $t_1 = n \cdot \Delta t_1$ where n is the number of increments/experiments and Δt_1 is the incremented delay. The directly detected time variable t_2 is observed during the final acquisition period in each of the scans and is directly digitized as in a 1D NMR experiment. Applying this experimental protocol that is described by eqn (2.3) serves as the basis of the 2D spectral grid we are looking for and the 2D connectivity is revealed by applying FT on the now two-dimensional time-domain:

$$I(\omega_1, \omega_2) = \sum_{t_2} \left[\sum_{t_1} S(t_1, t_2) e^{-i\omega_1 t_1} \right] e^{-i\omega_2 t_2} \tag{2.4}$$

Although eqn (2.4) resembles its counterpart in eqn (2.2), there is a large difference between the two. Whereas 1D NMR is an experiment that can be completed within single transient, 2D data acquisitions will—regardless of sensitivity considerations—demand the collection of at least several number of scans to sample the t_1 time domain adequately. The need to sample at least several data points and taking into account that each increment may need to be separated by a relatively long recycle delay and be repeated several times for signal averaging and/or phase cycling manipulations is what sets the conclusion that one major disadvantage of multidimensional FT NMR is the relatively long acquisition times required for obtaining these spectra.

2.2 Principles of UF NMR Spectroscopy

UF NMR spectroscopy breaks the aforementioned need to execute an array of experiments in order to collect 2D (or higher dimensional) NMR spectral data and can deliver it within a single transient. This ability derives from the application of an ancillary inhomogeneous broadening—in general coming from the application of a magnetic field gradients—to encode the spins' evolution along an ancillary—usually a spatial—domain, rather than in a temporal one as in eqn (2.4). By devising a way of performing this encoding in a single pulse sequence, the full $I(\omega_1, \omega_2)$ correlation can be collected in a single scan.

2.2.1 Magnetic Field Gradients

Magnetic field gradients play an imperative role in Magnetic Resonance Imaging, a special branch of NMR. In MRI, gradients are used to establish spatial localization of the spins of interest.[4–6] UF NMR spectroscopy adapted some MRI principles for obtaining spatial information by means of magnetic field gradients and, hence, it is helpful to review the effect gradients have on the spin ensemble in order to later understand the principles of UF NMR spectroscopy.

Magnetic field gradient coils are widespread in current NMR hardware. In most cases gradients are designed to generate a field that changes linearly along a certain axis, as:

$$G_r = \frac{\partial B_r}{\partial r}, \quad r = x, y, z \tag{2.5}$$

In practice, applying such magnetic field gradient on a sample is equivalent to imparting different precession frequencies to the spins, according to their position along the r-axis. Figure 2.1 shows an example of this effect for a sample containing a single chemical site (*e.g.*, H_2O) with an arbitrary chemical shift offset Ω, in the absence (left) and presence (right) of a magnetic field gradient acting along the main z-axis for spins that are spread over a range of positions $-L/2 \leq z \leq +L/2$. When applying an external

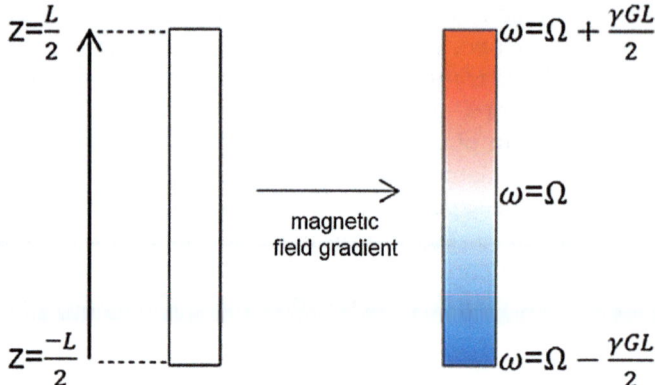

Figure 2.1 The effect of a magnetic field gradient on the spins' precession frequencies. Under the influence of only the external field (left) $B_z = B_0$ the spins frequency offset is constant and has no spatial dependency along the sample. Following the activation of an external magnetic field gradient (right) the spins frequency becomes spatially dependent according to their position z, imparting on them a precession frequency according to eqn (2.6).

magnetic field gradient, G, its effect adds up to the chemical shift of the spins according to:[5,7]

$$\omega = \Omega + \gamma G z \qquad (2.6)$$

where Ω is the chemical shift (*e.g.*, in Hz), γ is the gyromagnetic ratio (*e.g.*, in Hz Gauss^{-1}), G is the magnetic field gradient (*e.g.*, in Gauss cm^{-1}) and z is the spin's distance (*e.g.*, in cm) centered as zero in the middle of the sample.

It is imperative to understand the effect of such magnetic field gradient on the spin ensemble. The equations below describe a single chemical site with $\Omega = 0$, whose excitation is followed by the action of a magnetic field gradient along the z-axis, similar to what is depicted in Figure 2.1. After an excitation pulse, all the spins in the sample are flipped to the transverse plane, for instance, along the spin-space x-axis of the rotating frame. An evolution period τ in the presence of the magnetic field gradient will then impart a spatial-dependent phase on the spins based on the chemical shift defined in eqn (2.6). Considering that $\Omega = 0$, this evolution will result in a magnetization pattern of the form:

$$M = M_x \cos(\omega\tau) + M_y \sin(\omega\tau)$$

$$= M_x \cos(\gamma G z \tau) + M_y \sin(\gamma G z \tau) = M_x \cos(kz) + M_y \sin(kz) \qquad (2.7)$$

Here the wavenumber k, which is defined as the gradient's integral over time, is defined by the equation:

$$k = \int_0^\tau G(t)\, \mathrm{d}t \qquad (2.8)$$

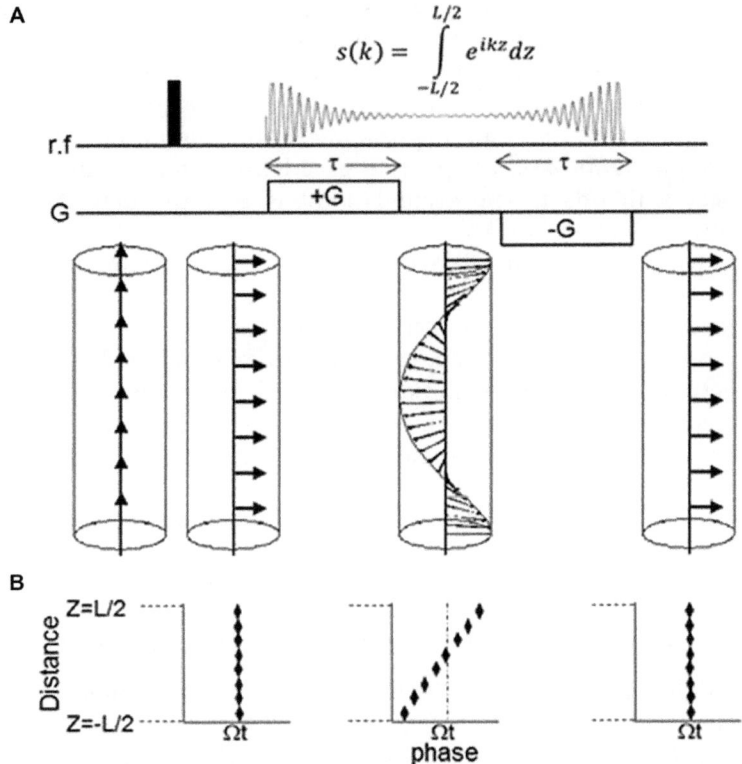

Figure 2.2 Illustration of the gradient-echo effect. Following an excitation pulse, the spin magnetizations are assumed uniformly aligned along the x-axis of their Bloch spheres regardless of their z (vertical) coordinates. This alignment is dephased by the action of a magnetic field gradient G over a time τ, resulting in a null signal at the end of the first gradient period. A second, opposite gradient can refocus the spins leading to a recovery of observable signal. This gradient-driven winding-unwinding process is known as gradient-echo. (A) 1D single-pulse gradient-echo sequence. (B) Spin-packets' spatial phase dependencies. Notice that in an idealized scenario, all the phenomena here described will be identical regardless of the sites' chemical shift offsets.

Adapted with permission from M. Gal and L. Frydman, Multidimensional NMR spectroscopy in a single scan, *Magn. Reson. Chem.*, 2015, **53**, 971. Copyright 2015 John Wiley & Sons Ltd.

This wavenumber k, whose units are given in reciprocal space (*e.g.*, cm^{-1}), describes how many turns of magnetization the gradient pulse has imparted per unit length in the sample, after having acted for an interval τ. Figure 2.2 shows a graphical representation of the gradient's effect described in eqn (2.7) and (2.8). Following the excitation pulse, the magnetic field gradient imparts on the spins a spatially dependent, helical magnetization pattern, along the z-axis gradient. This helical spins pattern at the end of the first gradient pulse results in a macroscopic null net signal. However, as suggested by eqn (2.7) and (2.8), this user-defined helical pattern could also be reversed simply by reversing the gradient's sign for the same duration. This will lead to the recovery of the non-zero macroscopic signal. There are a few points worth mentioning with regard to the just-described gradient-echo. The first is that the echo reappearance at the end of the two opposite-consecutive gradients would happen at exactly the same instant regardless of chemical shift offset. The second point is that the definition of the wavenumber k enables its value to be made arbitrarily large simply by increasing G, the gradient strength, and thereby to be made arbitrarily short in time. These speed and efficiency considerations are important in the acquisition of single scan UF 2D NMR spectra.

These aforementioned gradient-echo principles stand at the heart of UF NMR spectroscopy. During excitation, the gradients enable one to address the spins in a temporally incremented manner according to their position in the sample and thus impart a spatially dependent phase; during the acquisition the gradients will lead to chemically selective gradient echo trains, whose timing will no longer be uniform but depend on the chemical shift offset of the site in question.

2.2.2 Generic Scheme of UF 2D NMR Spectroscopy

Eqn (2.9) defines the generic scheme of single-scan two-dimensional NMR experiment.

$$\underset{\substack{(\text{indirect domain}, t_1 \, \alpha z)}}{\text{Preparation} - \text{Spatial Encoding}} - \text{Mixing} - \underset{\substack{(\text{direct domain}, N_2 \text{ times})}}{\text{Spatial Decoding}} \qquad (2.9)$$

In single-scan 2D UF NMR spectroscopy, the conventional time-incremented parameter t_1 is replaced by spatiotemporal excitation that encodes the interaction we are trying to measure (*e.g.*, chemical shift, J-coupling, *etc.*) along the gradient's physical axis. This gradient-driven excitation leads to the encoding of these interactions as magnetization helices in a similar way to what is described in Figure 2.2. However, a key part of UF NMR encoding is that, in contrast to the influence of a gradient echo where the helical winding/k values are common for all chemical sites in the sample (Figure 2.2), a different winding at the end of the spatial encoding will be imparted to each of the chemical sites. This one-to-one correlation between the winding/k values and the interaction we are after is the core principle of UF NMR spatial excitation. In a similar way to the conventional temporal encoding, this stored "indirect-domain" encoding in the shape of site-specific helical winding is passed on to the detected nuclei by the application of a user-defined mixing sequence and spatially decoded during acquisition. Spatial decoding can be carried out, once again, with the aid of a magnetic field gradient that will refocus the spins' helix. However, owing to the fact that the different chemical sites were imparted with different helical pitches, the instant at which each site will create its own gradient-driven echo will be different. Considering that each of the helical windings represents a specific indirect-domain interaction, the moment of each echo appearance will enable us to determine the interaction value we are looking for. Moreover, since the action of this "unwinding" gradient can be implemented very rapidly and efficiently, it can be used to wind and unwind the spins' magnetization vectors numerous times (N_2 in eqn (2.9)) over the course of the real physical-time parameter t_2. Each time these windings/unwindings occur, an indirect-domain spectrum will be read; however, given the additional modulation occurring as a function of t_2, the frequencies corresponding to the direct-domain evolution can also be read out by a suitable FT. Hence, a full 2D correlation can be gathered in a single scan. These spatiotemporal encoding and decoding principles are discussed in further detail in the following paragraphs.

2.2.3 Spatial Encoding

As mentioned, the goal of the spatiotemporal encoding in UF 2D NMR is to induce distinct helical windings proportional to each of the chemical shift offsets in the sample. It follows from the argument above that the conventional indirect-domain temporal evolution that can be abbreviated as:

$$M_x \rightarrow M_x\cos(\omega_1 t_1) + M_y\sin(\omega_1 t_1) \tag{2.10}$$

where t_1 is the temporal incremented parameter, which needs to be replaced with a spatial representation holding the same information but encoded by the spin position along a physical axis z axis:

$$M_x \rightarrow M_x\cos(C\omega_1 z) + M_y\sin(C\omega_1 z) \tag{2.11}$$

Eqn (2.11) represents a spatial modulation whereby the indirect time-domain variable t_1 in eqn (2.10) has been replaced by a spatial coordinate, where $C = t_1/z$ is defined as spatiotemporal parameter under our control and the z's extending over the complete sample length L. The pattern in eqn (2.11) represents a helix of spin magnetizations, which, unlike the analogous gradient-driven winding that was described in eqn (2.7), has a winding pitch dictated by the unknown chemical shift ω_1. The result of the spatiotemporal encoding on two distinct chemical sites is illustrated in Figure 2.3. Distinguishing among the chemical site frequencies *via* the application of an unwinding gradient then becomes possible owing to the different helical windings provided by the encoding process.

The question thus arises of what kind of manipulations can induce such a unique chemical shift-dependent pattern. There are in fact several ways to achieve such a special spatial pattern that can be divided into discrete and continuous excitation of the spins along the gradient axis. In the discrete scheme, the sample is divided into frequency-discriminated sub-slices under the action of a magnetic field gradient. The spins in each sub-slice are progressively excited during the application of a magnetic field gradient by a user-defined band-selective RF pulse that is given a different offset in each

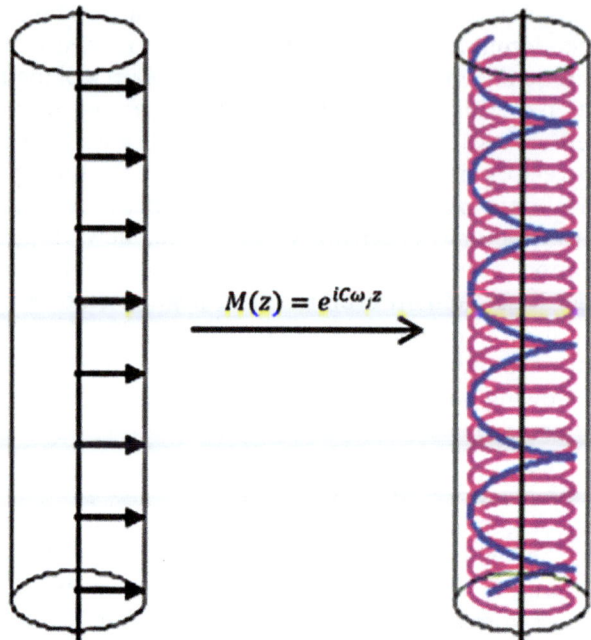

$$M(z) = e^{iC\omega_j z}$$

Figure 2.3 Graphical representation of two site-specific magnetization helices. The scheme illustrates the spatial pattern of two chemical sites with Ω_j ($j = 1, 2$) at the end of the spatial-encoding. The phase's spatial dependency of each spin is illustrated as a helix along the gradient axis. The different winding enables us to measure the interaction we are looking for.

excitation block. A negative gradient acting immediately after each positive one cancels out gradient-related spin-evolution. In this manner, each chemical site in the sample will end up having a different helix of magnetization.[8,9]

A few disadvantages are associated with the discrete excitation scheme, including the high number of fast-oscillating gradients. Hence, a more practical approach is the continuous excitation scheme, demanding only a few gradient oscillations over the course of the encoding process.[10] Key in these continuous encoding schemes is the application of linearly swept (or chirped) rf pulses,[11–13] acting while in the presence of a constant magnetic field gradient spreading out the spin offsets along the gradient's axis. Understanding a chirp-driven spin pattern is enlightening when trying to set up a UF NMR experiment. An illustrative example of the effect of a chirp pulse on the spins in the presence of magnetic field gradients and a worked example of how to set up excitation parameters can be found elsewhere.[13,14]

2.2.4 Decoding the Indirect Domain Information

The sole goal of spatial excitation within the single-scan UF NMR experiment is to deliver a pattern of site-specific magnetization helices. Conventional mixing could then transfer the different coherences to direct-detected nuclei and be decoded. Thus, the aim of spatial decoding at the end of eqn (2.9) is to translate each of the magnetization helices into an observable spin-echo whose timing is unique and provides the value of the interaction (*e.g.*, the chemical shift) we are looking for. The gradient echo principle depicted in Figure 2.2 suggests a simple way to rapidly read out these intervening indirect-domain frequencies: applying a magnetic field gradient during acquisition will refocus all magnetization helices, resulting in a chemical-shift-based echo. In addition, given the fact that each site is associated with a specific helix pitch, the helices will unwind at their own site-specific time, leading to a train of chemically specific echoes reflecting, in a one-to-one fashion, the $I(\omega_1)$ NMR spectrum being sought. It is enlightening to rearrange these arguments into equations in order to describe the signal during this site-specific echo. If, given a sample with a discrete $I(\Omega_i)$ spectral distribution, one applies a spatial encoding process of the kind previously described, the resulting spatially related spins' signal will be proportional to:

$$S_+(z) \propto [M_x(z) + M_y(z)] = \sum \frac{I(\Omega_i)}{L} e^{-iC\Omega_i z} e^{-Cz/T_2} \qquad (2.12)$$

The subsequent application of an acquisition gradient G_a will read out the intervening frequencies directly from the observable macroscopic time-domain signal, according to:

$$S(k) \approx \int_L dz S_+(z) e^{ikz} \approx \frac{1}{L} \sum I(\Omega_i) \int_L dz [e^{-iC\Omega_i z} e^{-Cz/T_2}] e^{ikz} \qquad (2.13)$$

where once again we have defined the wavenumber $k = \gamma_a \int_0^t G_a(t') dt'$. Note that apart from a $t_1 \leftrightarrow z$ redefinition, this equation represents a Fourier analysis that is very similar to that of conventional 2D temporal NMR acquisition. However, unlike the digital nature of the FT commonly applied on the time-domain signal in a usual 2D NMR experiment, a continuous integration is here carried out by G_a. In other words an "analogue" FT is carried out in real time by this gradient during acquisition. Indeed this gradient refocuses the spin helices and creates echoes at the site-specific moments when all a site's spin-packets are aligned throughout a sample. Aside from acting as a kind of continuous FT rather than a discrete FT carried out numerically on a computer, the overall result of this gradient-based readout will be the same as what we are used to in FT NMR: a series of peaks centered at "frequencies" $k_i = C\omega_i$, whose line shapes will be dictated by the usual T_2 effects, and whose overall integrated intensities will be proportional to $I(\Omega_i)$. This in turn calls for the need to redefine the k wavenumber domain as the equivalent of an NMR frequency axis: $\omega_1 \leftrightarrow -k/C$. Such rescaling enables a simple translation of the wavenumber values into frequency units. The effect of the generic spatial encoding–decoding process on two chemical sites with different Ω chemical shifts is shown in Figure 2.4.

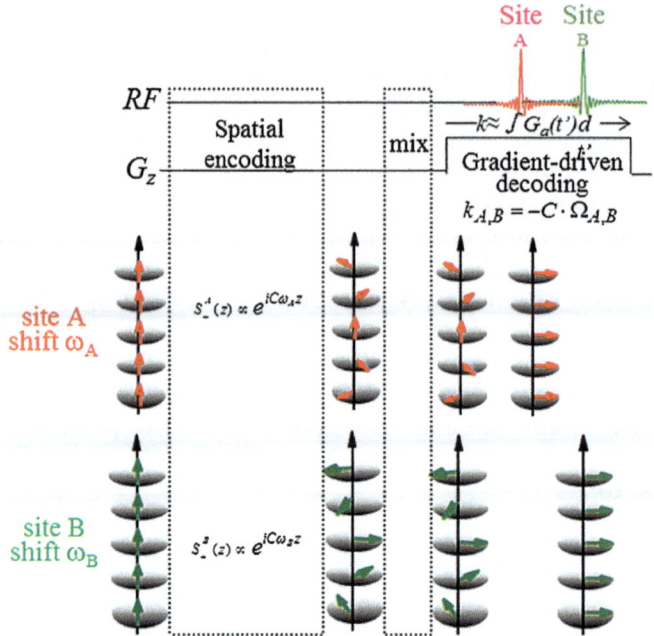

Figure 2.4 The spatio-temporal encoding and decoding of two distinctive sites' magnetization patterns. At the end of the spatial encoding each chemical site will have its own associated linear helical-winded pattern of magnetization, reflecting the site's chemical shift. Following a mixing process, a read-out of this chemical shift is achieved by applying an additional magnetic field gradient. This action creates site-specific spin echoes, reflecting the pre-mixing, indirect-domain spectrum.
Adapted with permission from M. Gal and L. Frydman, Multidimensional NMR spectroscopy in a single scan, *Magn. Reson. Chem.*, 2015, **53**, 971. Copyright 2015 John Wiley & Sons Ltd.

2.2.5 The Direct-domain Acquisition

The read-out of the spatiotemporally encoded indirect domain in Figure 2.4 is illustrated for two distinct chemical sites under the action of a magnetic field gradient. The question that now arises is how such a read-out could also reveal the direct domain spectra, completing the acquisition of 2D NMR spectra within a single scan. Indeed, the same principles of gradient-echo discussed in Section 2.2.1 and illustrated in Figure 2.2, such as the ability of concluding them very quickly and reversing them very efficiently, can now be applied in order to wind and unwind each of the site-specific spin-echoes numerous times. Along these winding-unwinding cycles, each of the site-specific echoes will be modulated by the direct dimension frequency and be bounded by the relaxation time T_2 of the detected spin. Moreover, this read-out cycle provides an opportunity to unravel the fate of a particular Ω_1 indirect-domain frequency twice: once during the $+G_a$ gradient application, and another during the $-G_a$ action. Considering that the gradient

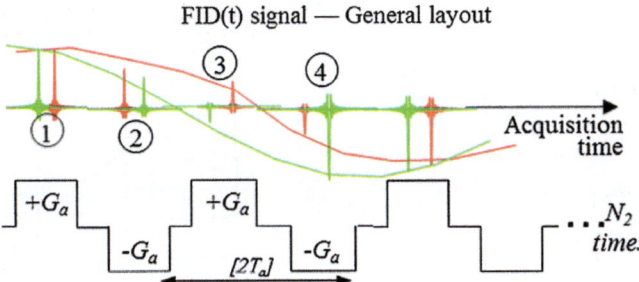

FID(t) signal — General layout

Figure 2.5 Extending the single gradient-driven refocusing process illustrated in Figure 2.4, to a multi echo process capable of yielding the full $S(k/\Omega_1, t_2)$ interferogram within a single scan. Two different sites are represented by the red and green color corresponding to their ω_2 frequency. In the actual experiment, the FID(t) signal arising during the course of this gradient echo train is read out continuously.

Adapted with permission from M. Gal and L. Frydman, Multidimensional NMR spectroscopy in a single scan, *Magn. Reson. Chem.*, 2015, 53, 971. Copyright 2015 John Wiley & Sons Ltd.

duration (T_a) in each of these cycles can be relatively short, with $T_a \leq 500$ μs, and that the typical T_2 decays are much longer, multiple $S(k/\omega_1)$ indirect-domain spectra can be monitored over an acquisition time $0 \leq t_2 \leq t_2^{max}$. Specifically, N_2 gradient oscillations of $\pm G_a$ with a half-period of T_a will provide approximately $2N_2$ direct-domain points for each k/ω_1 continuously sampled frequency element, with a dwell time T_a. Figure 2.5 depicts the collection of such a FID(t) signal while subjecting the spins to a square-wave oscillating gradient. Notice that each echo is modulated according to its corresponding ω_2 chemical shift: these color-coded echoes show faster modulations for the green than for the red species, representing a higher off-resonance frequency for the former spins. From here the full $I(\omega_1, \omega_2)$ 2D NMR spectrum being sought can then be retrieved by 1D FT *vs.* t_2 (FT$_{t_2}$) processing of the peaks appearing in each of the $S(k/\omega_1)$ spectra, while a single-scan acquisition suffices for the full characterization.

2.3 Processing UF 2D NMR Experiments

2.3.1 Basic Procedure

Unlike conventional FT NMR where the 2D spectral grid is collected by running separate experiments, in UF NMR experiments a single raw data FID contains the full 2D information. Hence, in order to process the collected FID, the raw data needs to be rearranged as exemplified in Figure 2.6 onto a 2D grid in the $(k/\Omega_1, t_2)$-space. The k/Ω_1 axis represents the data collected under the continuous application magnetic field gradient. The zig-zag pattern along this dimension is the result of the N_2 cycles of positive and negative gradient amplitudes. The t_2 axis represents the modulation of the spin-echo owing to the different ω_2's frequencies during acquisition for

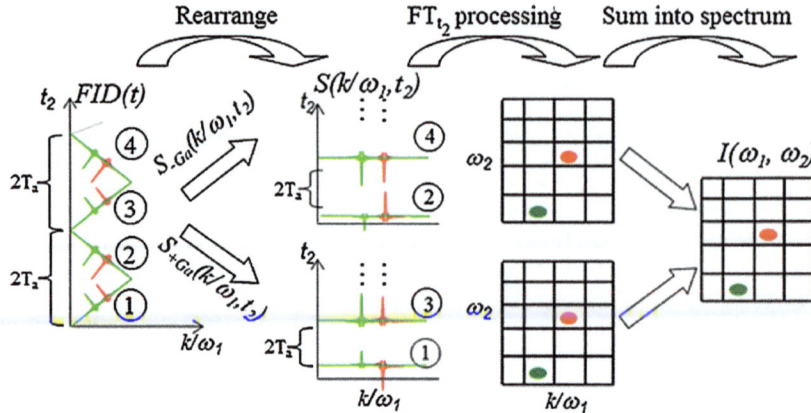

Figure 2.6   Processing the single-scan 1D FID(t) interferogram arising in UF NMR into a 2D $I(\omega_1, \omega_2)$ spectrum. Processing involves rearranging the digitized points into two data sets, 1D FT processing along t_2, and summation of the two resulting 2D spectra for SNR improvement. Weighting, phasing, *etc.*, are also processing possibilities.
Adapted with permission from M. Gal and L. Frydman, Multidimensional NMR spectroscopy in a single scan, *Magn. Reson. Chem.*, 2015, **53**, 971. Copyright 2015 John Wiley & Sons Ltd.

$t_2 = 2T_a$. A closer look at the data following the 2D rearrangement onto the k/t_2 grid shows that the echoes—the k-points—are not uniformly distributed along the t_2 domain. Although there are known processing schemes that can be applied in order to extract the ω_2 frequencies[15] this non-linear t_2 distribution is not suitable for the commonly used FFT algorithm. However, data sets containing uniformly distributed points along t_2 can be prepared by separating the raw-data FID into two: an $S_{+G_a}(k/\omega_1, t_2)$ and $S_{-G_a}(k/\omega_1, t_2)$, containing data points that were collected under the action of the $+G_a$ and $-G_a$, respectively. Indeed, applying the latter operation will result in two distinct data sets with uniformly distributed points with a dwell time of $\Delta t_2 = 2T_a$ along the direct dimension t_2. Fourier transformation of each of the new grids along t_2 will yield a pair of 2D spectra of $I_{\pm G_a}(\omega_1, \omega_2)$. Considering that both of these 2D spectra contain identical spectral information, it is in principle possible to combine them in order to obtain a final 2D spectrum with an additional $\sqrt{2}$ enhancement in the overall signal-to-noise ratio (SNR).

Figure 2.6 depicts the aforementioned processing scheme. It shows initially how the complex single raw FID can be viewed as being collected as a function of two independent variables, a wavenumber k and a real time parameter, t_2. The wavenumber can be written as

$$k = \gamma_a \int_0^t G_a(t') \mathrm{d}t' = \gamma_a G_a \left\{ t - 2T_a \mathrm{int}\left[\frac{t+T_a}{2T_a} \right] \right\} (-1)^{\frac{t+T_a}{2T_a}} \qquad (2.14)$$

representing a gradient that oscillates back and forth, and by doing so reveals the full indirect dimension spectrum $I(\omega_1)$ every interval T_a. The direct-domain evolution time t_2 advances monotonically for all the peaks in this $I(\omega_1)$ spectrum with every $+G_a/-G_a$ oscillation. Consequently, the collected data as a function of k and t_2 can be written as:

$$\text{FID}(t) = S(k, t_2) =$$

$$\int_L dz \left\{ \sum_{\text{acting } \Omega_2} \left[\sum_{\text{acting } \Omega_1} I(\Omega_1, \Omega_2) \exp(iC\Omega_1 z) \exp\left(-\frac{Cz}{T_2}\right) \right] \exp(i\Omega_2 t_2) \exp\left(-\frac{t_2}{T_2}\right) \right\}$$

$$(2.15)$$

2.3.2   SNR Considerations in UF 2D NMR

An important factor in the overall sensitivity of a given NMR experiment will be the band pass filter applied in order to allow a defined range of (ω_1, ω_2) frequencies to be collected. In conventional NMR experiments the filter bandwidth (fb) is determined according to direct spectral width during acquisition:

$$\text{SW}_2 = (\Delta t_2)^{-1} = 2\text{fb} \qquad (2.16)$$

where Δt is the physical dwell time. This value is usually on the order of a few kilohertz. By contrast, in UF 2D NMR spectroscopy, the activation of magnetic field gradient during acquisition spreads the frequency ranges that need to be collected and sets the required filter bandwidth according to:

$$\text{fb} = \gamma G_a L/2 \pm \Omega \qquad (2.17)$$

where L represents the effective coil length, γ the gyromagnetic ratio of the detected nucleus and the chemical shift range being targeted. Depending on the desired SW and thus on the value of T_a and G_a, this value can be on the order of 10s–100s of kilohertz. This represents an inherent sensitivity loss of UF in comparison to conventional 2D NMR experiments, which can be traced to the need in ultrafast NMR to sample multiple domains simultaneously. Even neglecting the relatively small Ω–driven range, the ensuing penalty in signal-to-noise ratio (SNR) will be given by:

$$\text{SNR}_{(\text{UF/conventional})} = \sqrt{\frac{\text{fb}_{\text{UF}}}{\text{fb}_{\text{conventional}}}} = \sqrt{\gamma G_a L \Delta t_2} = \sqrt{\# N_1 \text{ elements}} \quad (2.18)$$

where $\# N_1$ is the number of frequencies that can be resolved in the indirect domain. Eqn (2.17) and (2.18) reflect the notion that the SNR arising from a UF NMR experiment will be related to the acquisition gradient strength. The main purpose of executing interleaved or interlaced FT acquisitions is to enable the latter value to be lower. The sensitivity gain in interlaced FT should exceed that of a co-processed (positive/negative) UF data set by a factor of $\sqrt{2}$, arising from the reduction of G_a by a factor of 2. These considerations are somewhat different in interleaved FT UF NMR. Unlike interlaced FT, which in principle is still implemented in single-scan data, interleaved FT requires the collection of at least two scans. Considering a user-defined experimental time, the improvement in sensitivity should in theory be proportional to the number of interleaved scans, in the usual $n^{1/2}$ dependence. However, setting up interleaved experiments may lead to additional gains owing to the more ideal behavior of the gradients involved.

2.4 Discussion

Multidimensional NMR gives researchers of multidisciplinary fields the ability to study small and very large macromolecules in their native state at atomic resolution. However, this ability does not come for free but with the price of a long data collection time that keeps getting longer in order to meet the desired resolution in modern high field instruments. Advanced NMR pulse sequences that are oriented towards shortening the acquisition times without compromising resolution have been developed over the years. These can roughly be categorized into relaxation-enhanced methods that shorten the time between two adjacent scans (*e.g.*, SOFAST described in Chapter 1)[16–18] and a variety of non-uniformly sampled experiments that can deliver the high-dimensional correlations being sought (the subject of the remainder of this volume).[19–24] Unlike the latter, single-scan UF NMR spectroscopy is capable of delivering nD NMR spectra within a single scan in a linear fashion.[25] The latter characteristic is imperative for the execution of experimental schemes having high dynamic range, such as NOESY-based NMR experiments.

The UF NMR approach still faces limitations that in practice prevent it from delivering actual single-scan spectra. Foremost among the factors hindering UF NMR-based applications has been the issue of SNR. Several UF-based acquisition schemes were developed in cases where sensitivity dictates the collection of more than one scan. Interleaved UF NMR spectroscopy can be applied, thus lowering the gradient amplitude during acquisition and hence the filter band width.[13] This was found to be especially powerful when combined with schemes reducing the effective T_1 and thus the optimal recovery delay between two adjacent scans.[26] Other ways UF NMR spectroscopy can be exploited is by embedding it with other NUS techniques. This could especially benefit high dimensionality experiments, for instance, the incorporation of UF acquisition into an nD sequence such that two of the domains will be sampled linearly in a single transient, which could in principle shorten the experimental duration to that of an $(n-2)$D NMR experiment. In addition, in schemes that utilize higher spin-polarization, such as DNP, single-scan UF NMR experiments were already proven to be useful in exploiting the relatively short life-time of the hyperpolarized sample in order to collect 2D NMR data sets.[27–29]

All of these possible alternatives are thus important for research that requires faster NMR data collection and better resolution.

Acknowledgements

M. G. thanks Prof. Lucio Frydman—the inventor of the UF NMR method—for an exciting and wonderful period during his PhD thesis. He also thanks all the colleagues and collaborators that he has had the chance to work with over the years.

References

1. R. R. Ernst and W. A. Anderson, *Rev. Sci. Instrum.*, 1966, **37**, 93.
2. R. R. Ernst, G. Bodenhausen and A. Wokaun, *Principles of Nuclear Magnetic Resonance in One and Two Dimensions*, Oxford, 1987.
3. J. Jeener, in *Ampere International Summer School II*, Basko Polje, 1971.
4. P. Mansfield, *J. Phys. C: Solid State Phys.*, 1977, **10**, 55.
5. P. T. Callaghan, *Principles of Nuclear Magnetic Resonance Microscopy*, Oxford University Press, New York, 1991.
6. F. Schmitt, M. K. Stehling and R. Turner, *Echo-Planar Imaging Theory, Technique and Application*, Springer, Berlin, 1998.
7. F. Schmidt, M. K. Stehling, R. Turner and P. A. Bandettini, Springer Verlag, Berlin, 1998.
8. Y. Shrot and L. Frydman, *J. Chem. Phys.*, 2009, **131**, 224516.
9. L. Frydman, A. Lupulescu and T. Scherf, *Proc. Natl. Acad. Sci.*, 2002, **99**, 15858.
10. Y. Shrot, B. Shapira and L. Frydman, *J. Magn. Reson.*, 2004, **171**, 162.
11. M. Garwood and L. DelaBarre, *J. Magn. Reson.*, 2001, **153**, 155.
12. N. S. Andersen and W. Kockenberger, *Magn. Reson. Chem.*, 2005, **43**, 795.
13. M. Gal and L. Frydman, *Magn. Reson. Chem.*, 2015, **53**, 971.
14. M. Gal, L. Frydman, in *eMagRes*, John Wiley & Sons, Ltd, 2007.
15. M. Mishkovsky and L. Fydman, *J. Magn. Reson.*, 2004, **173**, 344.
16. K. Pervushin, B. Vogeli and A. Eletsky, *J. Am. Chem. Soc.*, 2002, **124**, 12898.
17. P. Schanda, E. Kupce and B. Brutscher, *J. Biomol. NMR*, 2005, **33**, 199.
18. S. Cai, C. Seu, Z. Kovacs, A. D. Sherry and Y. Chen, *J. Am. Chem. Soc.*, 2006, **128**, 13474.
19. J. C. Hoch, M. W. Maciejewski, M. Mobli, A. D. Schuyler and A. S. Stern, *Acc. Chem. Res.*, 2014, **47**, 708.
20. M. Mobli, M. W. Maciejewski, A. D. Schuyler, A. S. Stern and J. C. Hoch, *Phys. Chem. Chem. Phys.*, 2012, **14**, 10835.
21. S. G. Hyberts, H. Arthanari, S. A. Robson and G. Wagner, *J. Magn. Reson.*, 2014, **241**, 60.
22. S. G. Hyberts, D. P. Frueh, H. Arthanari and G. Wagner, *J. Biomol. NMR*, 2009, **45**, 283.

23. T. Ueda, C. Yoshiura, M. Matsumoto, Y. Kofuku, J. Okude, K. Kondo, Y. Shiraishi, K. Takeuchi and I. Shimada, *J. Biomol. NMR*, 2015, **62**, 31.
24. K. Kazimierczuk and V. Y. Orekhov, *Angew. Chem., Int. Ed.*, 2011, **50**, 5556.
25. M. Mishkovsky, M. Gal and L. Frydman, *J. Biomol. NMR*, 2007, **39**, 291.
26. M. Gal, T. Kern, P. Schanda, L. Frydman and B. Brutscher, *J. Biomol. NMR*, 2009, **43**, 1.
27. L. Frydman and D. Blazina, *Nat. Phys.*, 2007, **3**, 415.
28. P. Giraudeau, Y. Shrot and L. Frydman, *J. Am. Chem. Soc.*, 2009, **131**, 13902.
29. R. Panek, J. Granwehr, J. Leggett and W. Kockenberger, *Phys. Chem. Chem. Phys.*, 2010, **12**, 5771.

CHAPTER 3

Linear Prediction Extrapolation

JEFFREY C. HOCH,*[a] ALAN S. STERN[b] AND
MARK W. MACIEJEWSKI[a]

[a] UConn Health, Department of Molecular Biology and Biophysics, 263
Farmington Ave., Farmington, CT 06030-3305, USA; [b] Rowland Institute at
Harvard, 100 Edwin H. Land Blvd., Cambridge, MA 02139, USA
*Email: hoch@uchc.edu

3.1 Introduction

Linear prediction (LP) extrapolation has been a fixture of NMR data
processing almost from the very beginnings of multidimensional NMR. The
basic concept behind LP extrapolation is very simple to describe: Given a free
induction decay (FID) **d** with M samples $d_0, \ldots, d_{M-1}$ collected at uniform
intervals, LP models each sample as a linear combination of the preceding
samples:

$$d_k = \sum_{j=1}^{m} a_j d_{k-j}, \quad k = m, \ldots M - 1, \tag{3.1}$$

where a_j are the *LP coefficients*, sometimes called the *LP prediction filter*, and
the number of coefficients m is the order of the filter. Eqn (3.1) permits
extrapolation of the FID beyond M points. Taking $k = M$ gives a value for d_M;
this value can then be used to determine the value for d_{M+1}, and so on for as
long as desired. Naturally errors in such an extrapolation will become larger

New Developments in NMR No. 11
Fast NMR Data Acquisition: Beyond the Fourier Transform
Edited by Mehdi Mobli and Jeffrey C. Hoch
© The Royal Society of Chemistry 2017
Published by the Royal Society of Chemistry, www.rsc.org

as more points are predicted. Nevertheless, LP extrapolation is still quite useful for extrapolating the indirect dimensions of multidimensional experiments, where collecting long data records is frequently impractical. Assuming the LP prediction filter is stable (discussed below), the time series it generates is much more realistic than padding the time series with zeroes. The sinc-wiggle artifacts that are encountered with zero-filling are greatly reduced without the need to apply drastic line-broadening apodization.

This chapter is adapted from Chapter 4 of "NMR Data Processing".[1] Much more detail about the mathematics of LP, as well as connections to related methods, can be found there.

3.2 History of LP Extrapolation in NMR

How LP extrapolation became so ubiquitous in NMR spectroscopy is somewhat shrouded in history as there is no single, seminal paper to point to, and some of the key players have left the scene. Computer codes appeared in common NMR data processing software packages before papers describing LP extrapolation were published. Adding to the confusion, a method equivalent to LP extrapolation was reported using different terminology, obscuring the connection to LP.

The advent of LP extrapolation in NMR roughly coincided with the development of off-line computer programs for NMR data processing in the mid-1980s. Prior to that time, NMR data processing was almost exclusively performed using "on-line" computers connected to NMR spectrometers running the experiments. It's not clear whether the off-line implementations of LP extrapolation, in the programs FTNMR (from Hare Research, founder and principle developer Dennis Hare[†]) and NMR1 and NMR2 (from New Methods Research, founder George Levy and principle developer Marc-Andre Delsuc) predated implementations in the on-line software used to control commercial Fourier Transform NMR spectrometers (from Bruker, Varian, and Nicolet), which also occurred in the mid-1980s. The first LP codes in FTNMR were adapted from code developed in the laboratory of Andrew Byrd (R. A. Byrd, personal communication). LP codes in the Bruker software were implemented in the late 1980s by a graduate student from Dieter Ziessow's laboratory (C. Anklin, personal communication).

The earliest reference to "linear prediction extrapolation" in the NMR literature appears to be by Zeng *et al.*[2] from the lab of James Norris at Argonne National Lab. However, in a paper entitled "Phase-sensitive spectral analysis by maximum entropy extrapolation", Ni and Scheraga[3] described what appears be the first published application of LP extrapolation to NMR. They termed their approach "maximum entropy extrapolation" because they

---

[†]Hare, who was an important figure in the development of off-line NMR data processing, ultimately sold his NMR software company Hare Research and opened a brew pub in Washington state. He passed away in 2007; a newsletter (http://www.lopezislandmedical.org/docs/HM%20Spg14web.pdf) report indicates he used his private plane to ferry cancer patients from Lopez Island for treatment.

employed an algorithm developed by J. P. Burg[4] for computing LP filter coefficients as part of his maximum entropy method.

3.2.1 Broader History of LP

It should come as no surprise that the concepts underlying LP extrapolation have a history that stretches to 1795.[5] In addition, the appearance of LP *models* in the NMR literature predates the use of LP *extrapolation*. That is because the class of times series that obey the LP (eqn 3.1) coincides with the class of sums of exponentially damped (or exponentially growing) sinusoids and therefore includes ideal, noise-free FIDs. The LP model forms the basis of a host of parametric methods of spectrum analysis, methods that yield tables of frequencies, amplitudes and phases describing the sinusoids that make up the experimental FID. The methods bear such acronyms as LPSVD, LPQRD, and LPTLS, where the letters following "LP" refer to the method used to fit the parameters. Much of the development of algorithms for computing LP filter coefficients was carried out by investigators interested in parametric LP (described below), not LP extrapolation.

3.3 Determining the LP Coefficients

Expanding eqn (3.1) yields:

$$d_0 a_m + d_1 a_{m-1} + \cdots + d_{m-1} a_1 = d_m$$
$$d_1 a_m + d_2 a_{m-1} + \cdots + d_m a_1 = d_{m+1}$$
$$\vdots$$
$$d_{L-1} a_m + d_L a_{m-1} + \cdots + d_{M-2} a_1 = d_{M-1}$$
(3.2)

where $L = M - m$. In matrix form this is $\mathbf{D}\mathbf{a} = \mathbf{d}'$, with

$$\mathbf{D} = \begin{pmatrix} d_0 & d_1 & \cdots & d_{m-1} \\ d_1 & d_2 & \cdots & d_m \\ & & \vdots & \\ d_{L-1} & d_L & \cdots & d_{M-2} \end{pmatrix}, \mathbf{a} = \begin{pmatrix} a_m \\ a_{m-1} \\ \vdots \\ a_1 \end{pmatrix}, \text{ and } \mathbf{d}' = \begin{pmatrix} d_m \\ d_{m+1} \\ \vdots \\ d_{M-1} \end{pmatrix}. \quad (3.3)$$

Since $L > m$, this is an over-determined set of linear equations with a least-squares solution given by

$$\mathbf{a} = (\mathbf{D}^\dagger \mathbf{D})^{-1} \mathbf{D}^\dagger \mathbf{d}' \quad (3.4)$$

where $\mathbf{D}^\dagger$ is the complex conjugate of the transpose of $\mathbf{D}$. Several techniques have been proposed for inverting the matrix $\mathbf{D}^\dagger \mathbf{D}$, including Householder QR decomposition (QRD), Cholesky decomposition, and singular value decomposition (SVD). All three are standard methods, and their perform-ance has been compared.[6] SVD is better suited for detecting and rejecting noise components than the other methods.

3.4   Parametric LP and the Stability Requirement

Any complex-valued signal that is a sum of m exponentially damped (or growing) sinusoids will satisfy an LP equation with complex coefficients and prediction order equal to m. To see why, consider a signal consisting of a single decaying sinusoid:

$$d_k = (Ae^{i\phi})e^{-\pi Lk\Delta t}e^{2\pi ik\Delta tf} \tag{3.5}$$

where A is the amplitude, ϕ is the phase, Δt is the sampling interval, L is the linewidth, and f is the frequency. Defining $\alpha = Ae^{i\phi}$ and $\beta = e^{-\pi L\Delta t + 2\pi i\Delta tf}$ this can be rewritten as:

$$d_k = \alpha\beta^k. \tag{3.6}$$

Using this form for **d**, eqn (3.1) can be rewritten as:

$$\alpha\beta^k = \sum_{j=1}^{m} a_j\alpha\beta^{k-j} \tag{3.7}$$

Division by $\alpha\beta^{k-m}$ yields:

$$\beta^m = \sum_{j=1}^{m} a_j\beta^{m-j} \tag{3.8}$$

This will hold if and only if β is root of the characteristic polynomial:

$$P(z) = z^m - a_1 z^{m-1} - \cdots - a_{m-1}z - a_m \tag{3.9}$$

Efficient algorithms exist for rooting the characteristic polynomial based on Newton's method of successive approximation employing a Taylor series expansion.[6]

For any set of m decaying sinusoids with line widths L_j and frequencies f_j, and corresponding quantities β_j, there is a polynomial of order m whose roots are given by the m values of β_j:

$$P(z) = (z - \beta_1)(z - \beta_2)\cdots(z - \beta_m) \tag{3.10}$$

The coefficients of this polynomial form an LP filter of order m that is sat-isfied by each of the sinusoids (and also by their sum, since the LP equation is linear). Conversely, from the definition of β, the values of L_j and f_j are given by:

$$L_j = \frac{-1}{\pi\Delta t}\ln|\beta_j| \text{ and } f_j = \frac{1}{2\pi\Delta t}\text{phase}(\beta_j) \tag{3.11}$$

where $\text{phase}(\beta_j)$ is $\arctan(\text{real}(\beta_j)/\text{imag}(\beta_j))$, in radians. The amplitude and phase of the sinusoids can be determined from α similarly. The roots of the characteristic polynomial along with eqn (3.11) form the basis of parametric LP analysis. Instead of using the LP equation to extrapolate the FID followed by Fourier transformation, the result of parametric LP analysis is a table giving the frequencies, amplitudes, and phases of the sinusoids comprising the signal. Parametric LP approaches have been reviewed by de Beer and van Ormondt[7] and Gesmar and Led.[8]

Nothing in the preceding analysis precludes the values of any L_j from being negative, which corresponds to a growing rather than damped exponential. An LP filter that accepts exponentially growing components is *unstable*, and using it to extrapolate a signal would produce a result that grows without bound. Since experimental NMR signals normally decay (or neither grow nor decay, for constant-time dimensions), there is generally no reason to use unstable LP filters. Nonetheless, the presence of noise can frequently result in components with negative line widths. Simple regularization procedures have been devised to convert an unstable LP filter into a stable filter suitable for extrapolation. For positive line widths L_j, corresponding to damped sinusoids, the roots of the characteristic polynomial lie within the unit circle in the complex plane, $|\beta_j| < 1$. Stationary (neither damped nor growing) sinusoids have $|\beta_j| = 1$, and exponentially growing sinusoids have $|\beta_j| > 1$. The two most common methods for eliminating exponentially growing components from an LP filter involve reflecting the offending root about the unit circle (replacing β_j with $1/\beta_j^*$, where β_j^* is the complex conjugate of β_j) or displacing the root to the unit circle (replacing β_j with $\beta_j/|\beta_j|$). Both these approach preserve the frequency of the corresponding component. The revised roots are then recombined to give the stabilized LP filter.

The Burg method for computing the LP coefficients seeks to simultaneously minimize the forward and backward prediction error, which means that it implicitly assumes that the signals are stationary, *i.e.* neither growing nor decaying. The results for decaying NMR signals are uniformly poor, and we recommend against using the Burg algorithm. NMRPipe[9] and RNMRTK[10] use SVD for computing LP coefficients.

3.5 Mirror-image LP for Signals of Known Phase

When the number of data samples is small, as frequently occurs in the indirect dimensions of multidimensional NMR data, accurate LP extrapolation can be difficult because the order of the prediction filter is limited by the number of samples. It is possible to effectively double the number of samples by exploiting the fact that the phase of the signals in indirect dimensions is frequently known *a priori*, and the fact that signals with different decay rates can be extrapolated for short time periods using a single average decay value.[11] The decay of signals can be approximately canceled by weighting the data by a growing exponential with a rate equal to the negative of the average decay rate for all signal components, and a phase shift can be applied to adjust the signal's known phase to 0. Then the altered data $\tilde{\mathbf{d}}$

data can be reflected to negative time values using $\tilde{d}_{-k} = \tilde{d}_k{}^*$; that is, the negative time values are given by the complex conjugate of the positive time values. This effectively doubles the length of the data, allowing the LP prediction order to be twice as large. The result can be a significant improvement in the stability of LP extrapolation for short data records, although there are more robust methods for treating short data records.[12]

3.6 Application

Figure 3.1 illustrates the power of LP extrapolation for minimizing truncation artifacts and enhancing resolution without amplifying noise. Panels E and F illustrate how LP extrapolation enables the use of more gentle apodization, resulting in higher resolution while minimizing truncation artifacts (apparent in Panels B and D). Note, however, the distorted peak heights resulting from LP extrapolation in Panel F. Figure 3.2 compares the use of LP extrapolation with (Panel C) and without (Panel B) mirror-image extension of the data in t_1 (^{15}N) prior to LP extrapolation, showing the dramatic improvements that accrue for short (32 point) data records. The smaller

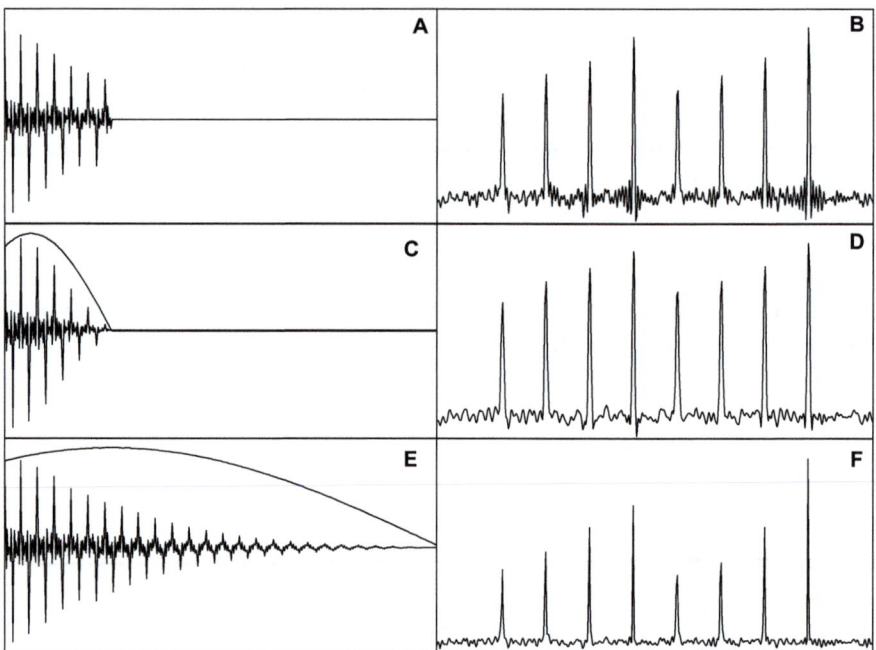

Figure 3.1   Synthetic data (A, C, E) and corresponding DFT spectra (B, D, F). (A) synthetic data consisting of 128 samples, padded with zeros to 512. (B) Data from (A) apodized using a 60°-shifted sine-bell that decays to zero after 128 samples. (C) 128 samples from (A) using LP extrapolation to 512 points, followed by apodization with a 60°-shifted sine-bell that decays to zero after 512 samples.

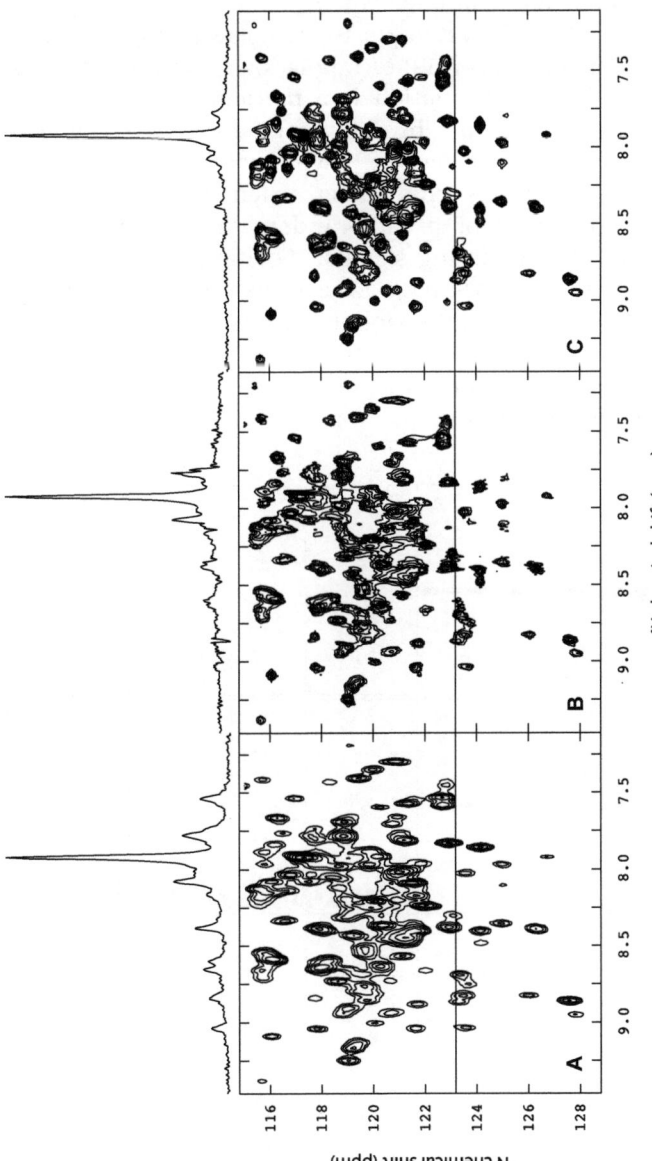

Figure 3.2 500 MHz ^{1}H–^{15}N NSQC spectra for human Prolactin (23 kDa). In all spectra the ^{1}H (f_2) dimension is processed using 70°-shifted sine-bell apodization, zero-filling from 128 to 256 points and DFT. 1D cross-sections parallel to the ^{1}H axis shown at the top are for a ^{15}N frequency The ^{15}N (f_1) dimension is processed using 32 complex samples with (A) 70°-shifted sine-bell apodization that decays to zero and DFT (B) LP extrapolation from 32 to 32 samples and DFT (B) LP extrapolation from 32 to 128 samples, 70°-shifted sine-bell apodization that decays to zero at 128 samples and DFT, and (C) mirror-image LP extrapolation from 32 to 128 samples, 70°-shifted sine-bell apodization that decays to zero at 128 samples and DFT.

number of peaks in the 1D cross-sections parallel to the ^{1}H axis shown above panel C are an indication of the dramatically improved resolution in f_1 (^{15}N). However the assumption that a signal contains only exponentially-damped sinusoids (which is implicit in LP) can also be a weakness, when the assumption does not hold. For example, experimental signals invariably include noise, and the LP model attempts to fit noise using exponentially decaying sinusoids. For this reason LP extrapolation is prone to generating false positive signals.[12] LP extrapolation is a particular sort of nonlinear amplifier: it amplifies components in data that have frequencies corresponding to the roots of the characteristic polynomial formed by the LP coefficients. Figure 3.3 illustrates this effect when LP extrapolation is performed on data consisting of pseudo-random "noise", using LP coefficients fitted to an FID containing signals. Violating the assumption of known phase when using mirror-image LP can also lead to problems, as shown in Figure 3.4. In this case, the violation leads to slight shifts in the *frequencies* of some signal components, rather than false positive signals, a particularly troublesome problem for applications of NMR that require measuring small frequency shifts.

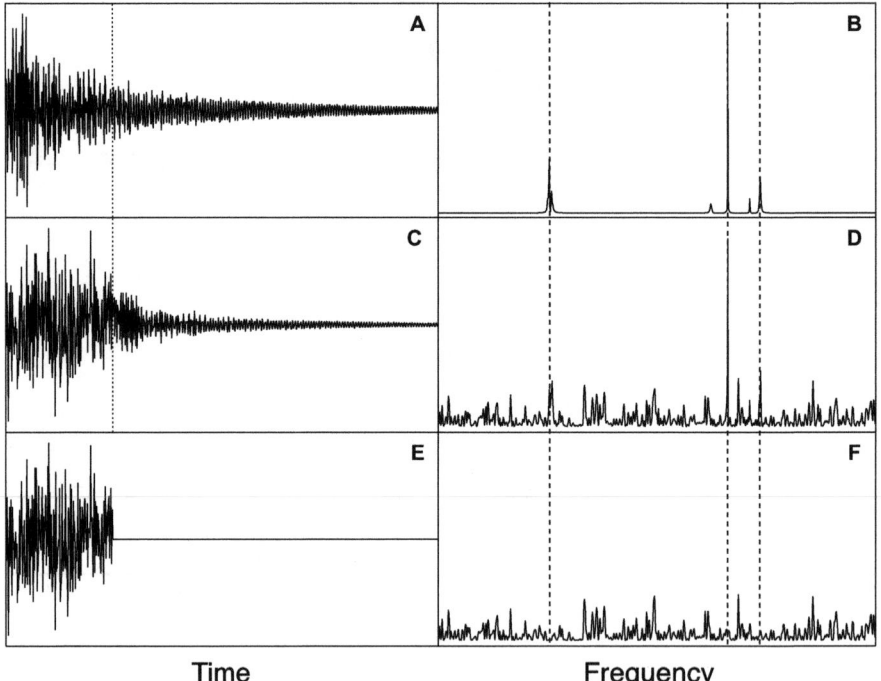

Time   Frequency

Figure 3.3   (A) LP extrapolation of an FID for a small molecule from 64 to 256 samples. (B) DFT of (A). (C) LP coefficients computed for the FID in panel A are used to extrapolate 64 samples of pseudo-random noise. (D) DFT of (C). (E) The 64 samples of pseudo-random noise zero-filled to 256. (F) DFT of (E).

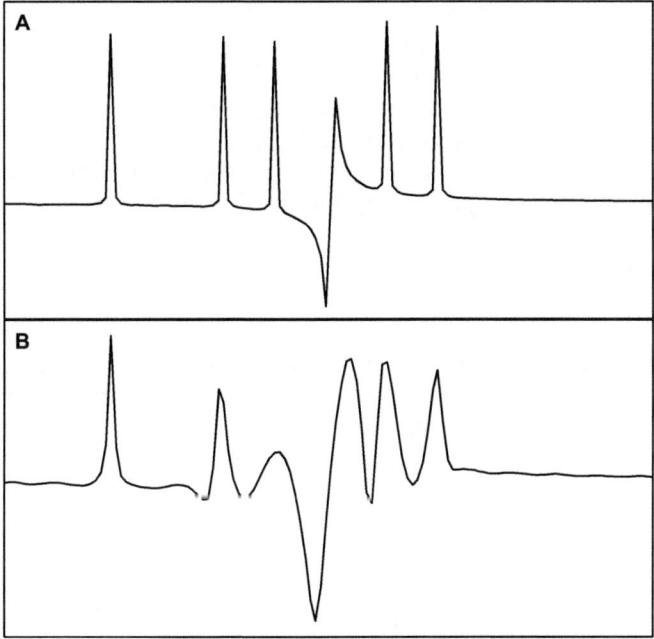

Figure 3.4 DFT spectra for synthetic data consisting of in-phase components and a single component that is 90° out-of-phase. (A) DFT of a synthetic FID, extrapolated using LP from 64 to 256 samples. (B) Mirror-image LP extrapolation as in (A), assuming in-phase signals.

3.7 Best Practices

LP extrapolation can significantly enhance NMR spectra provided that the data does not greatly violate the underlying assumptions (for example the signals decay exponentially and the signal components are larger than the noise level). The inevitable presence of noise means that the prediction filter order should generally be greater than the number of expected signal components. For the same reason LP extrapolation should not be extended to very long times from short data records, so that the extrapolation error does not become significant. A mitigating factor is that apodization frequently deemphasizes the extrapolated part of the data relative to the empirical part. Olejniczak and Eaton described fairly conservative practices for employing LP extrapolation,[13] but they can be generalized and expanded based on the results of a more recent critical comparison of different LP processing schemes:[12]

1. Utilize LP extrapolation for indirect dimensions that are sample-limited but not signal-limited, *i.e.* the signal has not decayed to below the noise level at the end of the sampled interval. A corollary is to limit the number of samples used to fit the LP coefficients to the part of the data record containing a substantial excess of signal over noise.

2. Limit the number of extrapolated points; no larger than three times the number of sampled points (quadrupling the length of the data record) is a good rule-of-thumb unless special care is taken to ensure fidelity (for example testing the accuracy in recovering the frequency of a synthetic signal added to the empirical data).

3. Use a prediction filter order that is comfortably larger than the expected number of signal components; this effectively rules out the use of LP extrapolation for the non-sparse direct dimension of multi-dimensional experiments (except when only a few points are to be extrapolated, for example when using backward LP extrapolation to replace corrupted samples at the beginning of an FID). A corollary is that when processing data with two or more indirect dimensions, it is best to use LP extrapolation on interferograms that have only a single time dimension, thus maximizing sparsity. This is accomplished by applying a Fourier transformation (without LP extrapolation) in one indirect dimension, then applying LP extrapolation in a different indirect dimension, Fourier transforming that dimension, and then inverting the Fourier transform applied earlier before using LP extrapolation in the first dimension.

4. When the number of samples is less than 32 and the phase of the signals is known, utilize mirror-image LP. For data sizes larger than 32 samples, the advantages of mirror-image LP are typically insignificant.

Acknowledgements

We thank Clemens Anklin, Andrew Byrd, Frank Delaglio, Marc-Andre Delsuc, and George Levy for discussions about the early history of LP extrapolation in NMR. J. H. gratefully acknowledges support from the US National Institutes of Health (grants GM104517 and GM111135).

References

1. J. C. Hoch and A. S. Stern, *NMR Data Processing*, Wiley-Liss, New York, 1996.
2. Y. Zeng, J. Tang, C. A. Bush and J. A. Norris, *J. Magn. Reson.*, 1989, **83**, 473–483.
3. F. Ni and H. A. Scheraga, *J. Magn. Reson.*, 1986, **70**, 506–511.
4. J. P. Burg, Ph.D. Thesis, Stanford University, 1975.
5. R. Prony, *J. Ec. Polytech.*, 1795, **1**, 24–76.
6. W. H. Press, B. P. Flannery, S. A. Teukolsky and W. T. Vetterling, *Numerical Recipes in Fortran*, Cambridge University Press, Cambridge, 1992.
7. R. De Beer and D. van Ormondt, *NMR: Basic Princ. Prog.*, 1992, **26**, 201–248.
8. J. J. Led and H. Gesmar, *Chem. Rev.*, 1991, **91**, 1413–1426.

9. F. Delaglio, S. Grzesiek, G. W. Vuister, G. Zhu, J. Pfeifer and A. Bax, *J. Biomol. NMR*, 1995, **6**, 277–293.

10. A. S. Stern and J. C. Hoch, http//:rnmrtk.uchc.edu.

11. G. Zhu and A. Bax, *J. Magn. Reson.*, 1990, **90**, 405–410.

12. A. S. Stern, K. B. Li and J. C. Hoch, *J Am Chem Soc*, 2002, **124**, 1982–1993.

13. E. T. Olejniczak and H. L. Eaton, *J. Magn. Reson.*, 1990, **87**, 628–632.

CHAPTER 4

The Filter Diagonalization Method

A. J. SHAKA* AND VLADIMIR A. MANDELSHTAM*

Department of Chemistry, University of California, Irvine, California 92697-2025, USA
*Email: ajshaka@uci.edu; mandelsh@uci.edu

4.1   Introduction

In this section we will briefly highlight the historical developments of the Filter Diagonalization Method (FDM) and its applications to experimental and calculated data sets, and go through the main ideas behind the algorithm.

FDM is an algorithm for estimating the spectrum from time-domain data sampled on an equidistant grid. With this conventional sampling schedule, both a conventional Fourier Transform (FT) spectrum and an FDM "spectrum" can be calculated and compared, and like the FT, FDM can be applied to 2D, 3D, 4D, or higher-dimensional NMR data. For data sets with very high signal-to-noise ratios (SNR), and time-domain signals that have essentially fully decayed into the noise floor, there is little difference between the 1D FT and 1D FDM spectra for typical NMR data.[1] Significant differences arise in the following cases: (i) the spectrum does *not* resemble a typical high-resolution NMR spectrum of narrow nearly Lorentzian peaks on a floor of relatively featureless noise, *e.g.*, in the case of imaging; (ii) the SNR in the spectrum is low; or (iii) the time-domain data has not been sampled for a long-enough duration that the signal has decayed. As we shall see, FDM is most appropriate to try to obtain *good*

New Developments in NMR No. 11
Fast NMR Data Acquisition: Beyond the Fourier Transform
Edited by Mehdi Mobli and Jeffrey C. Hoch
Published by the Royal Society of Chemistry, www.rsc.org

NMR spectra more quickly, by short-circuiting the time-consuming repetitive sampling of multidimensional data. FDM as described herein will not make it possible to get faster results that are accurate when the SNR is mediocre, as can be verified by mock data sets with varying levels of random noise. However, when the SNR is already decent but the truncation of the data compromises the resolution of the FT spectrum, FDM can have a dramatic effect. FDM is definitely not appropriate as a method to try to dig out signals from low SNR data sets, or to obtain accurate line shapes of very non-Lorentzian features, such as static powder patterns in solid-state NMR.

Filter diagonalization first appeared in a paper by Wall and Neuhauser[2] as a way to obtain the eigenenergies of a quantum system from time autocorrelation functions emerged from a quantum dynamics simulation. Two aspects in such a context are essential: (a) aside from numerical roundoff, the computed data has practically infinite SNR; and (b) computing a long time signal that is adequate for the required energy resolution would be computationally expensive. A deep and profitable connection between the number of basis functions used to determine energy eigenstates and the analysis of a time autocorrelation function by FDM was made, and it emerged, rather surprisingly, that some energy eigenstates (namely those involved in the dynamics manifested in the time autocorrelation function) could be extracted quite accurately even if the total number of energy eigenstates were extremely large. In other words, the Hamiltonian matrix could be extremely large, and yet a limited number of energies in the region of interest could be obtained with high accuracy using relatively short propagation times. Of the totality of matrix elements in the matrix representation of the Hamiltonian, it was possible to filter out only those relevant to the problem, and then diagonalize a much smaller matrix to obtain the energy eigenvalues of interest. Thus, the method had two steps: to *filter* and then to *diagonalize*. Follow-on work by Mandelshtam and Taylor[3] reformulated FDM to a more conventional form of a signal-processing method for solving the *harmonic inversion problem* (HIP) and showed how the algorithm could be made highly numerically efficient. As NMR spectra in liquids often show line shapes that are close to Lorentzian (the HIP assumption), the application to experimental data was only natural.[1]

FDM has been known as a numerical method for almost 20 years, so a natural question is "Why hasn't the method become mainstream yet?" The short answer is "It has", at least in theoretical fields of both physics and chemistry where a number of articles on filter diagonalization are published every year and several 1D FDM codes are made available to the community. With respect to NMR itself, there are several reasons why FDM is not mainstream: (i) most routine users of 1D and 2D NMR work up the data on the instrument itself, and use whatever software is provided with the NMR spectrometer—at this time FDM is not supplied on either Agilent or Bruker spectrometers; (ii) higher-dimensional FDM code has not progressed from experimental code to robust implementation, and it is in 3D and 4D spectra that FDM has the most theoretical advantage, with the result that the most impressive gains remain out of reach; (iii) groups focusing on a series of structure determinations often want to obtain data sets, and employ processing methods, in a repetitive way, to obtain (hopefully) data, spectra, and structures that are consistent with those obtained previously by others in the group, locking in whatever processing technology was employed earlier; (iv) there is no generally accepted repository of NMR data sets to compare different strategies, and hence no easy way to establish best practice for given types of NMR data. Different authors may therefore inadvertently introduce bias by publishing examples that are especially suited to the processing scheme they are proposing. The same data set is unavailable to others, making comparisons difficult or impossible.

Before presenting a mathematically rigorous formulation of FDM, it is worthwhile to give a very simple example, namely a single exponentially damped sinusoid with frequency ν_0, corresponding to a Lorentzian peak at position ν_0 of the NMR spectrum. In this case the discretely instantaneously-sampled free-induction decay (FID), $c_n \equiv C(\tau_n)$, where τ is the dwell time, can be written:

$$c_n = A_0 e^{i\phi_0} e^{2\pi i n \tau \nu_0} e^{-n\tau\gamma_0}$$

or in a more compact form:

$$c_n = du^n$$

which emphasizes that, for the special case of a single damped sinusoid, there are only two complex numbers, d and u, to be determined or, equivalently, four real numbers: A, ϕ, ν_0 and γ_0. One way to obtain these parameters is to perform a non-linear least squares fit of the data. While such an approach has its merit, even with much more complicated data,[4] there are two problems that are encountered in a nonlinear optimization. First, the number of local minima of a general non-linear multi-parametric function that is being optimized grows exponentially with the number of fitting parameters, quickly making the problem of global optimization intractable. A second, more subtle problem is that even when a "very good" fit of a noisy data by, say, M Lorentzians is found, it may well be physically meaningless when there are many distinct but equivalently good fits, so that the computed parameters are extremely sensitive to variations in the data or variations in the optimization protocol. In particular, such a fit is usually very sensitive to the choice of M with no existing recipe for choosing it. In general, one does not know *a priori* how many peaks should be present, and noise makes the number of "peaks" essentially as large as the data set allows. With non-linear least squares, correct close guesses for the "noise peaks" is impractical, making the entire exercise frustrating.

Suppose, however, that there is very little noise and reason to believe there is only one resonance in the NMR spectrum. How many time points would we need to obtain the peak? With FT processing, we would still need to sample the FID for a sufficiently long time, *e.g.*, $N\tau \approx 5/\gamma_0$, to obtain a narrow peak without additional broadening from truncation, and to have small sinc-function wiggles in the baseline. On the other hand, to solve for the two complex-valued unknowns, u and d, we need only two complex-valued data points, which, for example, could be c_0 and c_1:

$$u = c_1/c_0; \quad d = c_0$$

That is, in this case a non-linear optimization is avoided and the solution, obtained by solving a *linear system* is unique! As u and d contain all the information for a single Lorentzian peak, then two noise-free data points can, in principle, do the job. However, any corruption of either point will be essentially catastrophic. Making the problem more complicated, suppose we had a noise-free anti-phase doublet. Then $c_0 = 0$ while c_1 would be finite. In this case the simple two-point fit would "fail" as we would divide by zero. However, we cannot expect to obtain two peaks from two data points, if we had four points, we would recover the two peaks exactly. Sadly, we do not know the number of peaks beforehand, and noise will always result in extra "peaks". Therefore, we never attempt to restrict the number of presumed signal peaks, nor is it a parameter of the method. The number of peaks for a given calculation is simply $M = N/2$, assuming N is even. Some of these may be signal, some may be noise, and some may result from non-ideal line shape. In the general case of M peaks, simple division of one point by another will not yield the desired results. As every point in the time domain contains a contribution from each of the M peaks, we must first somehow disentangle the contributions, and then find the parameters such that:

$$c_n = \sum_{k=1}^{M} d_k u_k^n \quad (n = 0, \ldots, N-1) \tag{4.1}$$

where for a kth peak the spectral parameters $u_k = \exp(2\pi i\tau\nu_k - \tau\gamma_k)$ and $d_k = A_k e^{i\phi_k}$, $(k = 1, \ldots, M)$ define the peak frequency ν_k, its width γ_k, integral A_k and phase ϕ_k. This identifies the *harmonic inversion problem* (HIP). It arises often in diverse branches of science and engineering. At first glance, HIP may seem unremarkable and somewhat arbitrary. However, it is by far the most popular among all possible parametric forms. The reason for the ubiquity of the HIP lies in its unique property of having a linear algebraic solution (see for example ref. 5) and so allowing the consideration of a much larger parameter space than is feasible in generic non-linear optimization problems, and without the problems of local minima that can hinder other forms. Numerous linear-algebra based techniques exist, starting with the oldest Prony method,[39] but reviewing them all would be beyond the scope of the present chapter. Sometimes it is handled by making a Singular Value Decomposition (SVD)[6,7] of a "data matrix" constructed from the time-domain data, or by a related Linear Prediction (LP) method.[8] Practically speaking, the time to perform the SVD scales as N^3 and rapidly becomes prohibitive for FIDs with thousands of points, even using modern computers. Albeit related to LP, the approach in FDM is slightly different, and vastly more efficient. The key observation is fairly obvious: points in the frequency spectrum near one peak and far away from another will be dominated by the parameters of the nearby peak. If the FID is shifted by one point in time and an FT conducted, there will be a *phase shift* of the spectrum that will depend on the position and width of the nearby peak much more than other peaks. This suggests that the strategy could be much more efficient in the frequency domain. Indeed, one can attempt to estimate the parameters in a number of frequency windows Δf_{win}, and still obtain quite accurate results. Of course, one still must conduct something like SVD to disentangle the residual overlap between the peaks but, by reducing one huge matrix problem in the time domain to a series of much smaller matrix problems in the frequency domain, there is nearly linear scaling as the size of the data set increases, rather than the cubic scaling of a brute-force approach.

FDM is free from many of the well-known problems of LP; it can handle any size data set, can directly calculate multidimensional spectra, and does not use any kind of extrapolation of the time-domain data. Rather, the parameters are extracted efficiently from the FID and used to calculate the spectrum directly, with enhanced digital resolution.

4.2   Theory

This section provides a formal and self-contained formulation of FDM for solution of the HIP and related spectral estimation problem in 1D as well as in nD cases.

4.2.1   Solving the Harmonic Inversion Problem: 1D FDM

A hidden subtlety in the formulation of the HIP [eqn (4.1)] arises if the total number of terms M is *fixed* and so is independent of N. This seemingly natural assumption, motivated by the underlying idea that the number of *signal* peaks is fixed once the sample for analysis is chosen, is generally flawed. The problem becomes ill-defined, making the solution extremely sensitive to the variations (noise) in the input data. Even starting with a very special case of an N-point FID consisting of exactly M damped sinusoids, where $M < N/2$, a general infinitesimal perturbation of the input signal c_n will promptly destroy this special property, and can immediately require $M = N/2$ peaks to obtain an exact fit. A well-defined numerically stable formulation of the HIP is thus to set $M = N/2$ (considering, without serious sacrifice, only even N), corresponding to having the number of unknowns consistent with the number of equations.

Assuming $N = 2M$, the fact that the HIP [eqn (4.1)] has a unique solution is quite remarkable. This is a general result, independent of the algorithm used to numerically solve the HIP. However, such a unique solution may or may not be physically meaningful. A good example of a meaningful solution would correspond to a line list with isolated narrow Lorentzian peaks with the other entries in the line list, presumably representing noise and having small amplitudes d_k, or possibly large widths γ_k. For an opposite extreme case, recall the well-known variant of the HIP, namely the multiexponential fit,

$$c_n = \sum_{k=1}^{M} d_k e^{-n\tau\lambda_k} \tag{4.2}$$

where the data points c_n are all real, decay with n, and do not oscillate. The underlying physics may dictate that such a process should be characterized by a sum of several purely decaying exponentials, *i.e.* having real parameters d_k and λ_k, so one may then naturally assume that a good fit of c_n in terms of a small number of d_k and λ_k solves the problem. Such a good fit can actually be easily produced by either pruning the line list of a numerical solution of the corresponding HIP and leaving only a small number of dominating terms, or even by solving a non-linear optimization problem with a fixed number of terms. Apparently, the fact that eqn (4.2) has a unique solution in terms of $M = N/2$ complex pairs $\{d_k, \lambda_k\}$ is physically irrelevant. Although such a solution exists in infinite arithmetic, in finite arithmetic it cannot be reproduced but, most importantly, both the exact and an approximate solution of eqn (4.2) are extremely unstable. There actually are an infinite number of equivalently "good", but qualitatively distinct, fits of c_n by a sum of decaying exponentials that can be obtained by various existing methods. Thus, even though any decaying sequence c_n is usually not difficult to fit by several exponentials quite accurately, as the parameters of the fitting protocol are varied (*e.g.*, by varying the signal length N or by allowing for more exponentials to be included in the fit), the extracted decay constants λ_k and amplitudes d_k change abruptly, bearing little relation to those determined using a slightly different fitting protocol.

One of the simplest algorithms to solve eqn (4.1) is the *Matrix-Pencil Method* (MPM); it is known in the literature under various names (see *e.g.* ref. 9, 10) and is closely related to many other linear algebraic methods. Consider a time signal c_n ($n = 0, \dots, N-1$) with an even number of points ($N = 2M$). Define the data matrices $\mathbf{U}^{(p)} \in \mathbb{C}^{M \times M}$ ($p = 0,1,2$) and $\mathbf{C} \in \mathbb{C}^M$:

$$\mathbf{U}_{mn}^{(p)} = c_{m+n+p}; \quad \mathbf{C}_n = c_n \tag{4.3}$$

The solution of eqn (4.1) is then given by the generalized eigenvalue problem:

$$(\mathbf{U}^{(1)} - u_k \mathbf{U}^{(0)})\mathbf{B}_k = 0 \tag{4.4}$$

where the eigenvectors B_k satisfy the orthonormality condition:

$$\mathbf{B}_k^{\mathrm{T}} \mathbf{U}^{(0)} \mathbf{B}_{k'} = \delta_{kk'} \tag{4.5}$$

which follows from the symmetry of the data matrices, $[\mathbf{U}^{(p)}]^{\mathrm{T}} = \mathbf{U}^{(p)}$. The amplitudes are obtained by:

$$d_k = (\mathbf{C}^{\mathrm{T}}\mathbf{B}_k)^2 \tag{4.6}$$

It is eqn (4.4) that identifies the MPM. To the best of our knowledge, the compact formula for the amplitudes [eqn (4.6)] was first given in ref. 3. Although with exact arithmetic eqn (4.3)–(4.6) give the exact solution to the HIP [eqn (4.1)], such an approach is not practical when N is large. A much more practical algorithm, the FDM, is derived by using a transformation to a Fourier basis:

$$\tilde{\mathbf{U}}^{(p)} = Z^{\mathrm{T}}\mathbf{U}^{(p)}Z \tag{4.7}$$

where $\mathbf{U}^{(p)} \in \mathbb{C}^{K_{\mathrm{win}} \times K_{\mathrm{win}}}$ and the elements of the transformation matrix $Z \in \mathbb{C}^{M \times K_{\mathrm{win}}}$ are given by:

$$Z_{nj} = e^{-in\varphi_j}; \quad (j = 1,\ldots,K_{\mathrm{win}}, \quad n = 0,\ldots,M-1) \tag{4.8}$$

The values φ_j are real and are distributed uniformly within a small frequency interval: $2\pi f^{\mathrm{min}} < \varphi_j < 2\pi f^{\mathrm{max}}$.

The unusual form of the basis transformation [eqn (4.7)] is chosen so that the transformed matrices $\mathbf{U}^{(p)}$ are also symmetric, $[\tilde{\mathbf{U}}^{(p)}]^{\mathrm{T}} = \tilde{\mathbf{U}}^{(p)}$. The resulting generalized eigenvalue problem reads:

$$(\tilde{\mathbf{U}}^{(1)} - u_k\tilde{\mathbf{U}}^{(0)})\tilde{B}_k = 0; \quad \tilde{B}_k^{\mathrm{T}}\tilde{\mathbf{U}}^{(0)}\tilde{B}_{k'} = \delta_{kk'} \tag{4.9}$$

with the new eigenvectors related to those in eqn (4.4) by $B_k \approx Z\tilde{B}_k$. The approximation holds assuming that the eigenvectors B_k can be expanded in the narrow-band Fourier basis for the eigenvalues u_k corresponding to frequencies within the window, $f^{\mathrm{min}} < \nu_k < f^{\mathrm{max}}$. For these eigenvalues, the amplitudes d_k can be computed using:

$$d_k = (\mathbf{C}^{\mathrm{T}}Z\tilde{B}_k)^2 \tag{4.10}$$

We found empirically that the results are usually not sensitive to the parameters of the Fourier basis as long as the points φ_j are chosen appropriately. In most cases an equidistant grid is effective, with spacing:

$$\Delta\varphi = \frac{2\pi}{M\tau\aleph} \quad \text{or} \quad K_{\mathrm{win}} = \frac{M\tau\aleph}{f^{\mathrm{max}} - f^{\mathrm{min}}} \tag{4.11}$$

and essentially any value of parameter $\aleph \geq 1$ (*e.g.*, one can use $\aleph = 1.1$). The frequency window must be large enough so that K_{win} is not too small, *e.g.*, $K_{\mathrm{win}} = 100$ is not too small, but is small enough to guarantee that the linear algebra calculation is not time-consuming. Although it would defeat the purpose of the Fourier basis, one can in principle set $[f^{\mathrm{min}}; f^{\mathrm{max}}] = [-1/2\tau; 1/2\tau]$ (*i.e.* the full Nyquist range) and $K_{\mathrm{win}} = M = N/2$ ($\aleph = 1$) corresponding to a complete Fourier basis. In this case, the transformation matrix Z is unitary and with exact arithmetic the resulting spectral parameters using either eqn (4.4) or (4.9) would be indistinguishable.

Owing to the special Hankel structure of $\mathbf{U}^{(p)}$, eqn (4.7) can be evaluated efficiently (*i.e.*, avoiding an expensive matrix–matrix multiplication) using:

$$[\tilde{\mathbf{U}}^{(p)}]_{jj'} = \hat{S} \sum_{\sigma=0,1} \frac{e^{i\sigma[\tau M(\varphi_j - \varphi_{j'}) + \pi]}}{1 - e^{i\tau(\varphi_j - \varphi_{j'})}} \sum_{n=\sigma M}^{(\sigma+1)(M-1)} e^{-in\tau\varphi_j} c_{n+p} \tag{4.12}$$

where $\hat{S}$ defines a symmetrization operator over the indices j and j':

$$\hat{S} g_{jj'} = g_{jj'} + g_{j'j} \tag{4.13}$$

For diagonal elements $(j = j')$ a numerically practical expression reads:

$$[\tilde{\mathbf{U}}^{(p)}]_{jj} = \sum_{n=0}^{2M-2} e^{-in\tau\varphi_j}(M - |M - n - 1|)c_{n+p} \tag{4.14}$$

Usually, even for a small window (small K_{win}) the spectral parameters computed by FDM are practically indistinguishable from those computed using a complete basis. The reason for this property of FDM is that the Fourier-transformed matrices $\tilde{\mathbf{U}}^{(p)}$ are diagonally dominant, with $\tilde{U}_{jj}^{(1)}/\tilde{U}_{jj}^{(0)}$ being already a good approximation for the eigenvalues u_j. Given this property of FDM, the possibly large spectral width $[-1/2\tau, 1/2\tau]$ can be covered with small windows. and the whole "line list" (d_k, u_k) can be constructed by collecting the results from all the small-window calculations. As first noted in ref. 2, in order to describe the overall spectrum one has to implement multiple overlapping windows and gradually deemphasize the results at the edges of each window, where the parameters are likely to be somewhat less accurate. In Figure 4.1 we show an example of spectral construction by combining the results from several overlapping windows with 50% overlap. For each single window $[f_{\text{min}}^{(r)}, f_{\text{max}}^{(r)}]$ labeled by index r all the frequencies $f_k^{(r)}$ and amplitudes $d_k^{(r)}$ are retained and used to construct the spectrum $S^{(r)}(f)$ only inside this window. The overall spectrum is then constructed by:

$$S(f) = \sum_r g^{(r)}(f) S^{(r)}(f) \tag{4.15}$$

where $g^{(r)}(f)$ is an appropriate weighting function which is non-zero only inside the r-th window $[f_{\text{min}}^{(r)}, f_{\text{max}}^{(r)}]$. From our numerical tests we found that any reasonable choice satisfying $\sum_r g^{(r)}(f) = 1$ works well. For example, one can implement:

$$g^{(r)}(f) = \frac{1}{2}\left[1 - \cos\left(2\pi \frac{f - f_{\text{min}}^{(r)}}{f_{\text{max}}^{(r)} - f_{\text{min}}^{(r)}}\right)\right] \tag{4.16}$$

also used in Figure 4.1.

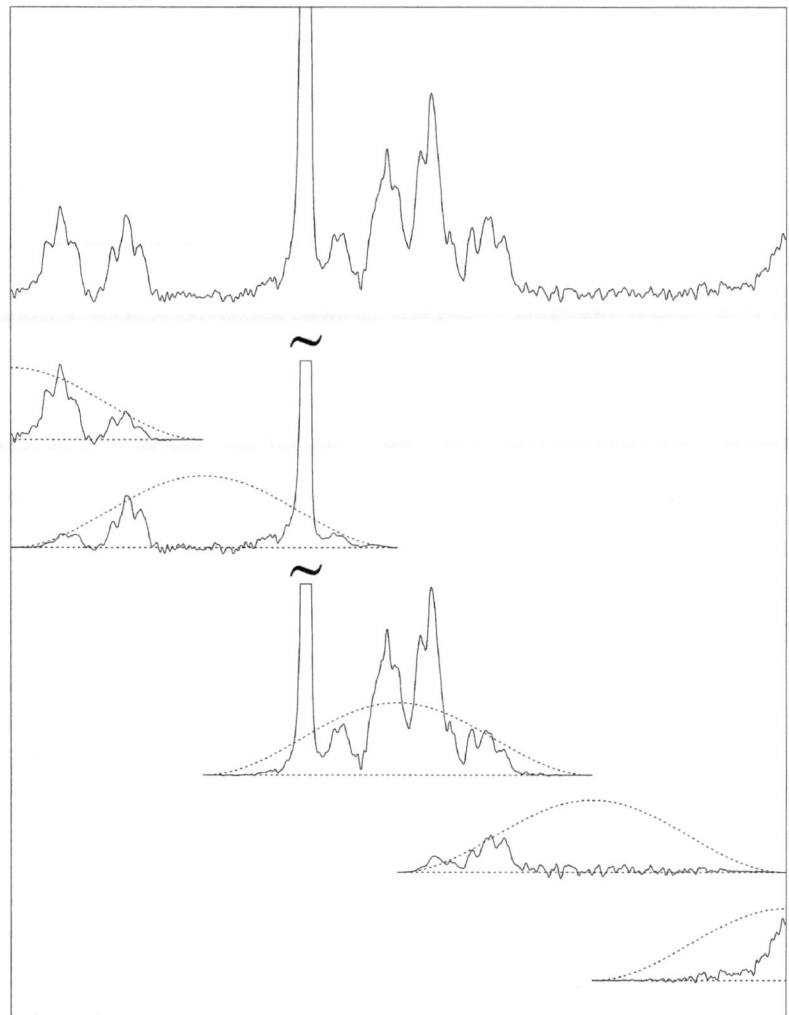

Figure 4.1 Decomposition of the spectrum (upper trace) into the overlapping windows using the cosine-weighting functions [eqn (4.16)] shown with dotted lines.

4.2.2 The Spectral Estimation Problem and Regularized Resolvent Transform

In this subsection we consider a problem related to the HIP, namely, given a finite time signal $c(n\tau)\equiv c_n$ $(n=0,\ldots,N-1)$, estimate its infinite time Discrete FT spectrum $S(f)$:

$$S(f) = -\frac{\tau c_0}{2} + \tau \sum_{n=0}^{\infty} e^{-2\pi in\tau f} c_n \equiv -\frac{\tau c_0}{2} + \tau \sum_{n=0}^{\infty} z^{-n} c_n \qquad (4.17)$$

where $z := e^{2\pi i\tau f}$.

If the signal c_n satisfies the harmonic form [eqn (4.1)] with $2M \leq N$, in exact arithmetic the spectrum $S(f)$ can be computed *exactly* by the following expression (*Resolvent Transform*):[11]

$$S(f) = -\frac{\tau c_0}{2} + \tau \mathbf{C}^{\mathrm{T}} \left(\mathbf{U}^{(0)} - \mathbf{U}^{(1)}/z \right)^{-1} \mathbf{C} \tag{4.18}$$

or using the Fourier basis:

$$S(f) = -\frac{\tau c_0}{2} + \tau \tilde{\mathbf{C}}^{\mathrm{T}} \left(\tilde{\mathbf{U}}^{(0)} - \tilde{\mathbf{U}}^{(1)}/z \right)^{-1} \tilde{\mathbf{C}} \tag{4.19}$$

Both eqn (4.18) and (4.19) generally require solution of an ill-conditioned linear system:

$$\mathbf{R}(f)\mathbf{X}(f) = \mathbf{C} \tag{4.20}$$

With:

$$\mathbf{R}(f) = \mathbf{U}^{(0)} - \mathbf{U}^{(1)}/z$$

And:

$$S(f) = \mathbf{C}^{\mathrm{T}} X(f) \tag{4.21}$$

The ill-conditioned linear system can be solved using the simplest regularization recipe[11] based on the Tikhonov regularization:

$$(\mathbf{R}^{\dagger}(f)\mathbf{R}(f) + q^2 \mathbf{I})\mathbf{X}(f) = \mathbf{R}^{\dagger}(f)\mathbf{C} \tag{4.22}$$

where q is the regularization parameter that controls the condition number of the initially ill-conditioned matrix $\mathbf{R}^{\dagger}\mathbf{R}$. The Regularized Resolvent Transform (RRT) then reads:

$$S^{(\mathrm{RRT})}(f) = \mathbf{C}^{\mathrm{T}}(\mathbf{R}^{\dagger}(f)\mathbf{R}(f) + q^2 \mathbf{I})^{-1}\mathbf{R}^{\dagger}(f)\mathbf{C} \tag{4.23}$$

and using Fourier basis:

$$S^{(\mathrm{RRT})}(f) = \tilde{\mathbf{C}}^{\mathrm{T}}(\tilde{\mathbf{R}}^{\dagger}(f)\tilde{\mathbf{R}}(f) + q^2 \mathbf{I})^{-1}\tilde{\mathbf{R}}^{\dagger}(f)\tilde{\mathbf{C}} \tag{4.24}$$

We note here that in the case of 1D spectral estimation the regularization does not pose any problems, while it becomes a critical issue for a multi-D spectral estimation. Apparently, in 1D, unless heavy regularization (large q) is applied to solve eqn (4.22), the RRT spectrum is practically indistinguishable from the FDM *ersatz* spectrum computed using the FDM parameters u_k and d_k:

$$S^{(\mathrm{FDM})}(f) = \sum_k d_k L_k(f) \tag{4.25}$$

where the complex Lorentzian line shape function (including aliasing) is correctly defined as:

$$L_k(f) := \frac{\tau}{1 - u_k/z} - \frac{\tau}{2} \tag{4.26}$$

4.2.3 Hybrid FDM

While the resolution of the FT spectrum is simply limited by the uncertainty principle $\delta f \sim (N\tau)^{-1}$, the resolution of the ersatz spectrum [eqn (4.25)] arising from the solution of the HIP is not just defined by the signal size N, but is a rather nontrivial function of N and several other factors, such as spectral line shapes, multiplet structure, and SNR. It is easy to imagine a situation in which, within a certain spectral region, a finite signal c_n is very poorly represented by the parametrization of eqn (4.1), either caused by non-Lorentzian line shapes or low SNR, or both, thereby making the corresponding HIP ill-posed and its solution not useful for spectral estimation within this region. However, assuming that the FT spectrum within this spectral region is still usable, or at least inadequate in a way that is familiar and easy to interpret, we would like to have a possibility to correct the ersatz spectral estimate such that it becomes at least as "good" as the FT spectrum. Our approach to this correction is to use a hybrid spectrum that unifies the strengths of FDM and FT.[2,18,19] In other words, if for some reason a genuine peak happens to be resolved in the FT spectrum then, by construction, the hybrid spectrum will recover this peak if it is missing in the ersatz spectrum (false negative). On the other hand, a spurious peak (false positive) appearing in the ersatz spectrum will be suppressed, to at least some degree in the hybrid spectrum. The idea of a hybrid spectrum was originally proposed by Wall and Neuhauser.[2] Here we present a different FDM-hybrid expression that we find particularly both efficient and well-behaved:

$$S^{(\text{hybrid})}(f) = -\frac{\tau c_0}{2} + \tau \sum_{n=0}^{\infty} z^{-n} \sum_k d_k u_k^n + \tau \sum_{n=0}^{N-1} g_n z^{-n} \left[c_n - \sum_k d_k u_k^n \right] \tag{4.27}$$

$$= S^{(\text{FDM})}(f) + S_N^{(\text{FT})}(f) - S_N^{(\text{FDM})}(f)$$

where g_n is an apodization function. The finite FT of the parametrically-represented time signal $c_n = \sum_k d_k u_k^n$ is given by the formula:

$$S_N^{(\text{FDM})}(f) = \tau \sum_k d_k \sum_{n=0}^{N-1} (u_k/z)^n g_n \tag{4.28}$$

In order to obtain an analytic formula for this expression, consider the cosine apodization function:

$$g_n = \cos(n\alpha) = \frac{1}{2}(\delta^n + \delta^{-n}) \tag{4.29}$$

with $\alpha = \pi/2N$ and $\delta = e^{i\alpha}$, so that $\cos(N\alpha) = 0$ and $\delta^N = i$. Substituting this into eqn (4.28) we obtain:

$$
\begin{aligned}
S_N^{(FDM)}(f) &= \frac{\tau}{2} \sum_k d_k \sum_{n=0}^{N-1} \left[(u_k/z\delta)^n + (u_k\delta/z)^n \right] \\
&= \frac{\tau}{2} \sum_k d_k \left[\frac{1 - i(u_k/z)^N}{1 - (u_k\delta/z)} + \frac{1 + i(u_k/z)^N}{1 - (u_k/z\delta)} \right] \\
&= \tau \sum_k d_k \frac{1 + (u_k/z)^{N+1}\sin\alpha - (u_k/z)\cos\alpha}{1 + (u_k/z)^2 - 2(u_k/z)\cos\alpha}
\end{aligned}
\tag{4.30}
$$

which is an expression that is easy to evaluate numerically. Other apodization functions that are combinations of polynomials and trigonometric functions can also be cast as closed expressions.

Given the parameter list $\{u_k, d_k\}$ produced by FDM, regardless of whether it is meaningful or not, eqn (4.27)–(4.30) provide working expressions for computing the hybrid FDM-FT spectral estimate.

4.2.4 Multi-D Spectral Estimation and Harmonic Inversion Problems

Consider a general complex valued D-dimensional time signal $c_{\vec{n}} \equiv c_{n_1 \cdots n_D} \equiv C(n_1\tau_1, \ldots, n_D\tau_D)$, defined on an equidistant rectangular time grid of size $N_{total} = N_1 \times \cdots \times N_D$.

One possible and natural generalization of 1D HIP is:

$$
c_{n_1 \cdots n_D} = \sum_{k=1}^M d_k u_{1k}^{n_1} \cdots u_{Dk}^{n_D}
\tag{4.31}
$$

where the unknown parameters are the complex amplitudes d_k and frequencies $\omega_{lk} = 2\pi\nu_{lk} - i\gamma_{lk}$ with $u_{lk} = e^{i\tau_l\omega_{lk}}$, $(l = 1, \ldots, D)$.

Once computed, these parameters can be used to construct various representations of the multi-D spectrum, such as:

$$
S(f_1, \ldots, f_D) = \sum_k d_k \prod_{l=1}^D L_{lk}(f_l)
\tag{4.32}
$$

where, as in eqn (4.33), we defined:

$$
L_{lk}(f) := \frac{\tau_l}{1 - u_{lk}/z_l} - \frac{\tau_l}{2}
\tag{4.33}
$$

with $z_l = e^{2\pi i \tau_l f_l}$, $(l = 1, \ldots, D)$, which corresponds to the multi-D FT of the time signal:

$$S(f_1, \ldots, f_D) = \sum_{\substack{n_l = 0 \\ l = 1, \ldots, D}}^{\infty} \left\{ \prod_{l=1}^{D} (i\tau_l) \left(1 - \frac{\delta_{n_l 0}}{2}\right) z_l^{-n_l} \right\} c_{n_1 \cdots n_D} \qquad (4.34)$$

The peaks in such a spectrum have an *a priori* poor appearance owing to their slowly decaying tails. A more useful is an absorption-mode spectral representation:

$$A(f_1, \ldots, f_D) = \sum_k d_k \prod_{l=1}^{D} \mathrm{Re}\, L_{lk}(f_l) \qquad (4.35)$$

which is advantageous to eqn (4.32) because of the faster-decaying tails and, even more importantly, because the peak integrals in such a spectrum are proportional to their amplitudes d_k.

Unfortunately, the multi-D HIP is only more cumbersome than the 1D HIP, and thus FDM can hardly provide a sensible multi-D line list, unless the data is represented well by the form of eqn (4.31) with high SNR. We will refer to such, perhaps unrealistic, cases as a *well-defined HIP*. Each entry in the line list then represents a peak. As opposed to this ideal situation for a *poorly defined HIP*, each spectral feature is often a result of interference of a number of Lorentzians and as such individual entries in the line list may not be meaningful. However, even for a poorly defined HIP, various spectral representations, including absorption-mode spectra, can still be constructed within the FDM framework, but by carefully avoiding the direct solution of eqn (4.31).

4.2.5   Spectral Estimation by Multi-D FDM

Similarly to the 1D case, the multi-D HIP [eqn (4.31)] can be cast into a linear algebraic problem. More precisely, consider a multi-D signal $c_{\vec{n}}$ that satisfies the form of eqn (4.31) exactly with:

$$M \leq M_{\text{total}} := 2^{-D} N_{\text{total}} \qquad (4.36)$$

that is, assuming infinite SNR. Define the data matrices $\mathbf{U}^{(p)} \in \mathbb{C}^{M_{\text{total}} \times M_{\text{total}}}$ $(p = 0, \ldots, D)$ and $\mathbf{C} \in \mathbb{C}^{M_{\text{total}}}$:

$$\mathbf{U}^{(p)}_{\vec{n},\vec{m}} = c_{\vec{n}+\vec{m}+\vec{p}}; \quad \mathbf{C}_{\vec{n}} = c_{\vec{n}} \qquad (4.37)$$

With:

$$\vec{0} := (0, \ldots, 0); \ \vec{1} := (1, 0, \ldots, 0); \ \vec{2} := (0, 1, 0, \ldots, 0); \ \vec{3} := (0, 0, 1, \ldots, 0) \ldots$$

It then follows that eqn (4.31) can be mapped exactly to a set of D simultaneous generalized eigenvalue problems:

$$\mathbf{U}^{(p)} \mathbf{B}_k = u_{pk} \mathbf{U}^{(0)} \mathbf{B}_k; \quad \mathbf{B}_k^{\mathrm{T}} \mathbf{U}^{(0)} \tilde{\mathbf{B}}_k = 1 \qquad (4.38)$$

with the amplitudes given by:

$$d_k = (\mathbf{B}_k^{\mathrm{T}}\mathbf{C})^2 \qquad (4.39)$$

The "exact mapping" means that there exists a set of simultaneous eigenvectors $\{\mathbf{B}_k, k = 1, \ldots, M\}$ that solve the generalized eigenvalue problems [eqn (4.38)] exactly.

The above equations are a generalization of the 1D Matrix-Pencil Method (MPM) described above, but while the 1D MPM is in principle a working method, even for the cases of relatively large 1D data sets and finite SNR, the multi-D MPM can hardly be practical. It cannot be practical at least because, for a general multi-D NMR data set, the data matrices would be enormous. This problem will be circumvented by using FDM, *i.e.*, by performing a local spectral analysis. However, there is another and more fundamental problem that makes the solution of the multi-D HIP highly nontrivial. The mapping between eqn (4.38) and (4.31) is exact only for the case of infinite SNR (*i.e.*, under the condition of eqn (4.36). For a general case of finite SNR, the multi-D HIP can be satisfied exactly only if the number of unknowns matches the number of data points:

$$M(D+1) = N_{\mathrm{total}} \qquad (4.40)$$

In the case of a finite SNR the two conditions [eqn (4.36) and (4.40)] are contradictory, unless $D = 1$. That is, a multi-D HIP is an ill-posed problem, while the 1D HIP is a well-posed problem. In other words, for $D > 1$ and a finite SNR, a set of exact simultaneous eigenvectors $\{\mathbf{B}_k, k = 1, \ldots, M\}$ does not exist, and therefore we must either try to solve these equations in a least squares fashion,[12] which is a very non-trivial task by itself, or assume that the eigenvectors are different:[13,14]

$$\mathbf{U}^{(p)}\mathbf{B}_{pk} = u_{pk}\mathbf{U}^{(0)}\mathbf{B}_{pk}; \quad \mathbf{B}_{pk}^{\mathrm{T}}\mathbf{U}^{(0)}\tilde{\mathbf{B}}_{pk} = 1 \qquad (4.41)$$

In the latter case, eqn (4.39) particularly can no longer be used, which implies that we do not have a complete, fully coupled solution of the multi-D HIP [eqn (4.31)]. Moreover, the eigenvalues u_{pk} ($p = 1, \ldots, D$) that describe a particular peak can be coupled only if they all have the same eigenvector $\mathbf{B}_{pk} = \mathbf{B}_k$. Yet, in spite of this difficulty, as we showed in ref. 13, 14, the solutions of eqn (4.41) can still be used in a meaningful way for constructing various spectral representations of the data. In what follows we first describe how a local spectral analysis using windowing and Fourier basis can be performed in order to reduce the originally huge eigenvalue problems to that of a manageable size, and then present working expressions for multi-D spectral estimation.

To this end, in the spirit of Figure 4.1, the D-dimensional frequency domain is decomposed into rectangular overlapping windows, which are small enough for the emerging matrices to be of manageable size. In each window defined by $[f_1^{\mathrm{min}}; f_1^{\mathrm{max}}] \times \cdots \times [f_D^{\mathrm{min}}; f_D^{\mathrm{max}}]$ an evenly spaced direct-product Fourier grid $\vec{\varphi}_j \equiv (\varphi_{1j}, \ldots, \varphi_{Dj})$ ($j = 1, \ldots, K_{\mathrm{win}}$) of total size K_{win} is considered. The grid spacing in each dimension is set as in eqn (4.11), defined by

$M_l = N_l/2$. A transformation of the data matrices $\mathbf{U}^{(p)}$ $(p = 0, \ldots, D)$ analogous to that in 1D FDM [eqn (4.7)] is implemented, resulting in a set of data matrices $\tilde{\mathbf{U}}^{(p)} \in \mathbb{C}^{K_{\text{win}} \times K_{\text{win}}}$:

$$\tilde{\mathbf{U}}_{jj'}^{(p)} = \sum_{\substack{\sigma_r = 0,1 \\ r = 1, \ldots, D}} \left\{ \prod_{r=1}^{D} \hat{S}_r \frac{e^{i\tau_r \sigma_r [M_{rj'}(\varphi_{rj} - \varphi_{rj'}) + \pi]}}{1 - e^{i\tau_r(\varphi_{rj} - \varphi_{rj'})}} \right\} \sum_{\substack{n_r = \sigma_r M_{rj'} \\ r = 1, \ldots, D}}^{\sigma_r(M_{rj'}-1)+M_{rj}-1} e^{-i\vec{n}\vec{\phi}_j} c_{\vec{n}+\vec{p}}, \quad (4.42)$$

where $\hat{S}_r$ defines the symmetrization operator over the subscripts rj and rj', as in eqn (4.13). The matrix elements corresponding to $\varphi_{rj} = \varphi_{rj'}$ are computed according to eqn (4.14). For example, for $\varphi_{1j} = \varphi_{1j'}$ we have:

$$\tilde{\mathbf{U}}_{jj'}^{(p)} = \sum_{\substack{\sigma_r = 0,1 \\ r = 2, \ldots, D}} \left\{ \prod_{r=2}^{D} \hat{S}_r \frac{e^{i\tau_r \sigma_r [M_{rj'}(\varphi_{rj} - \varphi_{rj'}) + \pi]}}{1 - e^{i\tau_r(\varphi_{rj} - \varphi_{rj'})}} \right\} \sum_{n_1 = 0}^{2M_{1j}-2} (M_{1j} - |M_{1j} - n_1 - 1|)$$

$$\times \sum_{\substack{n_r = \sigma_r M_{rj'} \\ r = 2, \ldots, D}}^{\sigma_r(M_{rj'}-1)+M_{rj}-1} e^{-i\vec{n}\vec{\phi}_j} c_{\vec{n}+\vec{p}} \quad (4.43)$$

with similar expressions to treat other singularities. Finally, for $\vec{\phi}_j = \vec{\phi}_{j'}$, *i.e.*, the diagonal elements of the U-matrices, we have:

$$\tilde{\mathbf{U}}_{jj}^{(p)} = \sum_{\substack{n_r = 0 \\ r = 1, \ldots, D}}^{2M_{rj}-2} \left\{ \prod_{r=1}^{D} (M_{rj} - |M_{rj} - n_r - 1|) \right\} e^{-i\vec{n}\vec{\phi}_j} c_{\vec{n}+\vec{p}} \quad (4.44)$$

Also define a vector $\tilde{\mathbf{C}} \in \mathbb{C}^{K_{\text{win}}}$ with elements:

$$\tilde{\mathbf{C}}_j = \sum_{\substack{n_r = 0 \\ r = 1, \ldots, D}}^{M_{rj}-1} e^{-i\vec{n}\vec{\phi}_j} c_{\vec{n}} \quad (4.45)$$

Similarly to the 1D FDM, when evaluated in a Fourier basis, the data matrices $\tilde{\mathbf{U}}^{(p)}$ acquire a structure that allows one to solve the corresponding generalized eigenvalue problems in a block-diagonal fashion:

$$\tilde{\mathbf{U}}^{(p)} \tilde{\mathbf{B}}_{pk} = u_{lk} \tilde{\mathbf{U}}^{(0)} \tilde{\mathbf{B}}_{pk}; \quad \tilde{\mathbf{B}}_{pk}^{\mathrm{T}} \tilde{\mathbf{U}}^{(0)} \tilde{\mathbf{B}}_{pk} = 1, \quad (l = 1, \ldots, D) \quad (4.46)$$

For the infinite SNR case with eqn (4.36) satisfied there exists a simultaneous eigenbasis $\mathbf{B}_{pk} = \mathbf{B}_k$, and the multi-D HIP can be solved exactly with amplitudes given by:

$$d_k = (\tilde{\mathbf{B}}_k^{\mathrm{T}} \tilde{\mathbf{C}})^2 \quad (4.47)$$

Although the numerical bottleneck of the D-dimensional FDM is usually associated with the solution of the generalized eigenvalue problems [eqn (4.46)], an intelligent programming of the expressions to compute the U-matrices is desirable. For example, the use of globally equidistant grids allows one to evaluate all the D-dimensional Fourier sums using fast FT algorithms, making the overall U-matrix construction for all the windows scale quasi-linearly with the data size. An additional saving is possible if one takes advantage of the fact that some of the Fourier sums are related to each other *via* simple recursion relations (see ref. 14, 15). Also note that, even unintelligently programmed, eqn (4.42)–(4.44) will scale as $\sim N_{\text{total}} \times K_{\text{win}}$ for a single window, which is still acceptable.

For the case of a finite SNR the eigenvectors cannot be rearranged to turn the overlap matrices $\tilde{\mathbf{B}}_{pk}^{\text{T}} \tilde{\mathbf{U}}_0 \tilde{\mathbf{B}}_{p'k'}$ into unit matrices. Although for this general case the solution of the multi-D HIP is at least non-trivial, this solution can be avoided, while one can still construct various useful spectra that may have resolution superior to that of an FT spectrum. For example, the infinite FT spectrum [eqn (4.34)] can be estimated using:

$$S(f_1,\ldots,f_D) = \sum_{k_1,\ldots,k_D} d_{k_1,\ldots,k_D} \prod_{l=1}^{D} L_{lk_l}(f_l) \tag{4.48}$$

where we defined:

$$d_{k_1,\ldots,k_D} := \tilde{\mathbf{C}}^{\text{T}} \tilde{\mathbf{B}}_{1k_1} \tilde{\mathbf{B}}_{1k_1}^{\text{T}} \tilde{\mathbf{U}}_0 \tilde{\mathbf{B}}_{2k_2} \cdots \tilde{\mathbf{B}}_{(D-1)k_{D-1}}^{\text{T}} \tilde{\mathbf{U}}_0 \tilde{\mathbf{B}}_{Dk_D} \tilde{\mathbf{B}}_{Dk_D}^{\text{T}} \tilde{\mathbf{C}} \tag{4.49}$$

Note that for the case of infinite SNR [eqn (4.36)] this formula is always exact, even for an arbitrary arrangement of the eigenvectors $\mathbf{B}_{lk}$, *i.e.* the explicit construction of the fully coupled line list is still not required. At first glance for the absorption-mode spectrum [eqn (4.35)] the situation is similar, for example one can use the expression:

$$A(f_1,\ldots,f_D) = \sum_{k_1,\ldots,k_D} d_{k_1,\ldots,k_D} \prod_{l=1}^{D} \operatorname{Re} L_{lk_l}(f_l) \tag{4.50}$$

which requires the same input as that in eqn (4.48), *i.e.* also resulting from a single purely phase-modulated multi-D data. Each term in the sum of eqn (4.50) has an absorption line shape, albeit it is complex-valued. However, taking the absolute value of A would produce a real-valued spectrum with absorption line shapes.

For the case of well-defined HIP (infinite SNR), eqn (4.50) is also exact. Despite its obvious advantage over the complex Lorentzian line shape [eqn (4.48)], the absorption-mode line shape is not yet optimum for distinguishing peaks on a contour plot. The star-shaped contours, with still prominent tails along the frequency axes, mean that overlapping tails may sum up to look like a "peak" and make interpretation more difficult. Usually, Lorentzian-to-Gaussian transformation is performed to give clean elliptical contours that make the contour plot easier to discern by eye. In FDM the Gaussian absorption-mode spectral representation $A_G(f_1,...,f_D)$ may be constructed by simply replacing the Lorentzian line shapes with Gaussians

$$L_{lk_l}(f_l) \rightarrow G_{lk_l}(f_l) := \frac{2\sqrt{\pi \log(2)}}{\gamma_{lk_l}} \exp\left[-4\pi^2 \log(2) \left(\frac{f_l - \nu_{lk_l}}{\gamma_{lk_l}}\right)^2\right] \tag{4.51}$$

Unfortunately, both eqn (4.50) and (4.51) may suffer from instabilities, unlike the other, much better behaved (but less useful) spectral representation [eqn (4.48)]. The reason for the instability is the presence of the non-analytic functions of u_{lk} (Re and Gaussian). Of course, one can produce absorption line shapes using the power spectrum $|S(f_1, ..., f_D)|^2$, but for the price of severely distorted amplitudes. Another much better spectral representation that also has absorption line shapes and is stable is the pseudo-absorption spectrum:

$$S^{(2)}(f_1, ..., f_D) = \sum_{k_1, ..., k_D} d_{k_1, ..., k_D} \prod_{l=1}^{D} L_{lk}^2(f_l) \tag{4.52}$$

In this representation the amplitudes are also distorted, but to a much lesser extent, namely, they are multiplied by the inverse peak width $(1/\gamma_{lk})$ in each dimension, and therefore the narrow peaks are overemphasized relative to the broad ones.

Clearly, even with the lack of a fully-coupled line list, FDM provides great flexibility in the choice of the spectral representation.

4.2.6 Regularization of the Multi-D FDM

So far we have ignored the difficulties associated with the solutions of the generalized eigenvalue problems [eqn (4.46)] arising in the multi-D FDM. For example, consider the problem:

$$\mathbf{UB} = u\mathbf{SB} \tag{4.53}$$

with possibly ill-conditioned data matrices $\mathbf{U}$ and $\mathbf{S}$. The way these matrices are constructed supports the assumption that these matrices have a common (or nearly common) null space. Empirically, we discovered that problems that emerge from solving 1D HIP are not severe and can be handled well by the QZ algorithm alone, which is designed to take care of this particular type of singularities. Unfortunately, for a multi-D HIP, unless it is very well defined (very high SNR), a straightforward numerical solution of eqn (4.53) by the QZ algorithm would usually give very unstable results. Thus, a more sophisticated regularization is required. Note that the problem of regularization also arises naturally in RRT, where one has to solve a possibly highly ill-conditioned linear system [eqn (4.20)]. The Tikhonov regularization [eqn (4.22)] is one plausible solution to the problem, which has proven to be sufficiently robust and efficient for both 1D as well as for multi-D spectral estimation. However, to the best of our knowledge, there is no mathematically rigorous regularization method for an ill-conditioned generalized eigenvalue problem [eqn (4.53)].

Assuming that matrices U and S have a common singular subspace, in principle, a meaningful solution may be obtained by first considering SVD of S:

$$S = V\Sigma W^{\dagger} \qquad (4.54)$$

where $VV^{\dagger} = I$, $WW^{\dagger} = I$ and $\Sigma = \text{diag}\{\sigma_k\}$ with real $\sigma_k > 0$. In the most commonly implemented regularization scheme the unwanted "noise subspace", believed to be associated with small singular values σ_k, is removed by using a truncated SVD, in which one defines the effective range space of S by keeping its largest P singular values, and then reevaluates U and S in this subspace. However, such a regularization is usually quite sensitive to the choice of P, and without knowing the "correct" result in advance the "wrong" choice of P may easily produce either false-positives or false-negatives, or both. We have discovered empirically[16] a more robust regularization scheme, in which using an apparent analogy with the Tikhonov regularization the ill-conditioned eigenvalue problem [eqn (4.53)] is replaced by:

$$S^{\dagger}UB = u(S^{\dagger}S + q^2I)B \qquad (4.55)$$

where q is the regularization parameter. Here the solutions of the regularized equation depend smoothly on q. The effect of the regularization is usually not easy to anticipate without actually trying a few different values of q, which must be large enough to eliminate spurious peaks (false-positives), and, on the other hand, not too large to avoid severe distortion (or even elimination) of the genuine peaks. Since the exact spectra are unknown, there is no easy way to find an optimal value of q (if any). An empirical indication that the results are converged and are optimized with respect to the choice of q is their stability with respect to changes of q within a relatively large range.

Most of our previously reported multi-D FDM results have been obtained using the above regularization. However, one can still achieve a greater flexibility with regularization by starting with SVD of S. The idea then would be to modify the singular values σ_k in a way that would reduce their contribution to the final result, if they are too small (*i.e.* "bad"), while minimizing the distraction of the "good" singular values. The degree of regularization is better to be controlled by a single adjustable parameter, q, with the final result being a smooth function of q. The latter condition is desirable as in this case examining the q-dependence may help in choosing an optimal value of q. For example, one can consider a regularization in which **S** is replaced by $\mathbf{S}_q = \mathbf{V}\boldsymbol{\Sigma}_q\mathbf{W}^\dagger$, where:

$$\boldsymbol{\Sigma}_q = \text{diag}\left\{\frac{\sigma_k^2 + q^2}{\sigma_k}\right\} \tag{4.56}$$

and then solve eqn (4.53). Unfortunately, this regularization gives results very similar to those of eqn (4.55).

To this end, we note that the most recent application of FDM to the Diffusion Ordered Spectroscopy (DOSY)[40] identified a noticeably better regularization recipe based on simply replacing the small singular values $\sigma_k < q$ with q, and leaving all the other singular values unchanged. Here q plays the same role of a regularization parameter as before. This regularization turns out to be superior not only to that of eqn (4.55) but also to all the other regularization recipes that were tested for the DOSY processing.

4.3 Examples

4.3.1 1D NMR

Usually in 1D NMR there is no difficulty in acquiring as much data as desired as the sampling rate can be fast and the memory to store the FID is essentially limitless now. As such, cases where 1D data sets are truncated are not common any longer. Nevertheless, the parametric representation of the spectrum can facilitate some useful data cleanup, an important issue when hundreds or thousands of spectra are to be automatically processed, as in a metabolomics study.[17] Figure 4.2 shows how peaks with wide line width can be almost completely removed from biological fluids, allowing narrower resonances from small-molecule metabolites to stand out.

When the physical spectrum consists of narrow isolated lines and the SNR is sufficiently high, the FDM ersatz spectrum is always superior to the FT spectrum, and the difference is especially prominent when the time domain data is truncated. This situation is well documented in the FDM literature. On the other hand, when SNR is low and/or the physical spectrum consists of overlapping and possibly non-Lorentzian lines that are not resolved in the FT spectrum of (a hypothetical) infinitely long time signal, parametrization of such a data as a sum of Lorentzians is hardly advantageous for either processing or interpretation. Nevertheless, in this latter case the FDM ersatz spectrum of a sufficiently long time signal simply coincides with the FT spectrum within the error bars defined by noise. However, when the same data is truncated, the two spectra may be different: both are unresolved, but the appearance of the non-converged FDM ersatz spectrum may be less appealing, especially because, unlike the unresolved FT spectrum, it fails in a way that is not easy to either predict or interpret. As an example, consider the 500 MHz NMR spectra of the fluorinated ribose derivative **1** shown in Figure 4.3 with various numbers of complex points in the FID, a 1.4 ppm

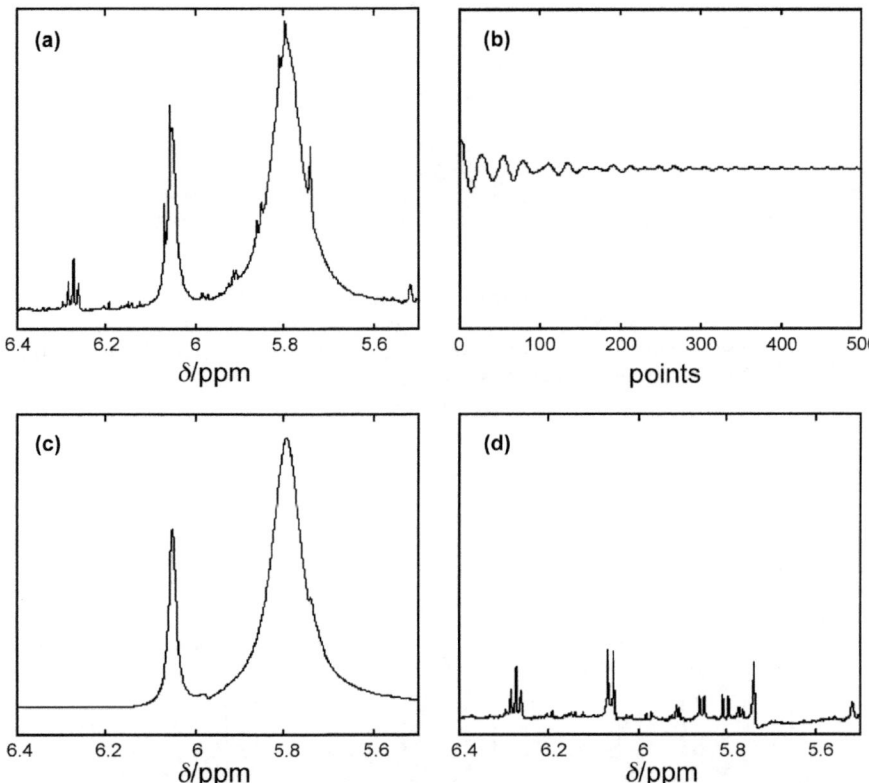

Figure 4.2   Use of 1D FDM to remove broad, intrusive peaks from urea and water in rat urine metabolite spectra.[17] The raw FT spectrum is shown in (a). In (b) the FID of only the broad peaks, as determined from the FDM parameters, is displayed. In (c) the spectrum corresponding to (b) is displayed, and in (d) this broad spectrum is subtracted from (a) to give a spectrum that shows the narrower metabolite resonances. The subtraction is imperfect, but the residual can be eliminated with standard baseline correction routines. These routines fail in the presence of the strong urea and water peaks.
Adapted from B. Dai and C. D. Eads, Efficient removal of unwanted signals in NMR spectra using the filter diagonalization method, *Magn. Reson. Chem.*, 1998, **78**, 78. Copyright © 2009 John Wiley & Sons, Ltd. Used with permission.

region of which is shown in each case in Figure 4.4. The nine sub-spectra are shown for FIDs with $N = 8192$, 4096, or 2048 complex points, and the spectra were calculated on a grid with 16 384 points across the spectral width. All peaks extracted by FDM were broadened slightly, so that they were properly digitized. (Some peaks may be very narrow, particularly those referring to noise, and so might be artificially attenuated if their center is between two frequency grid points.) The first point to note is that the ersatz FDM spectrum is comparable in every way to the FT spectrum when the data is

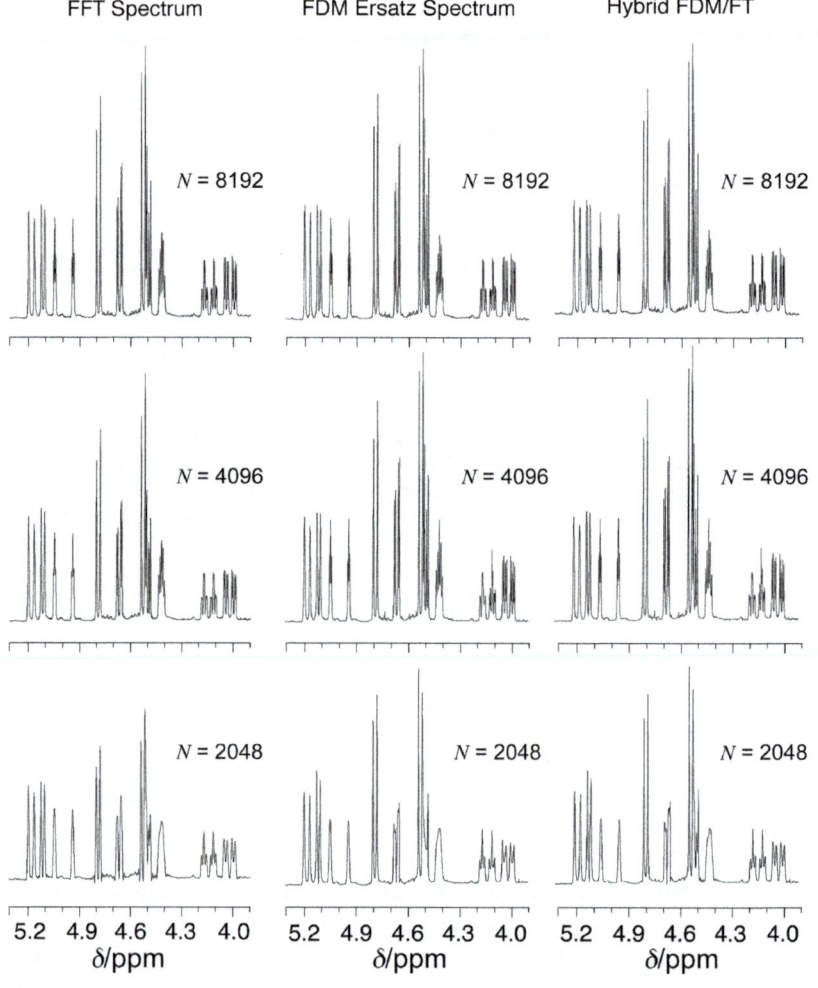

Figure 4.3 A fluorinated ribose derivative. The resonances shown in the 1.4 ppm region of the NMR spectra in Figure 4.4 arise from the protons on the ribose ring and several of the substituents. Spin–spin coupling to fluorine-19 creates extra splittings in the spectrum.

adequately sampled, $N = 8192$. That such an excellent fit can be obtained in one straightforward calculation without any iterative refinement shows the power of FDM: small peaks like carbon-13 satellites are captured accurately, and the hybrid spectrum is essentially indistinguishable from the ersatz spectrum. As the data length is lowered to $N = 4096$ there is some degradation of the performance, particularly where there are closely-spaced multiplet components. At $N = 2048$ the fine structure at $\delta = 4.44$ ppm cannot be recovered, as the local density of peaks (66 total in this case, across the plotted region) available to FDM in the window is simply insufficient. Put another way, there is not enough evidence, or information, in such a short FID to justify a host of closely spaced peaks rather than an unresolved feature. This example shows that separating two, or more, closely spaced resonances is far more difficult than capturing a narrow singlet with no near neighbors. As far as we know, no current method would be able to reliably resolve this multiplet with $N = 2048$ with the level of noise and other nearby peaks in this spectrum.

The convergence of FDM can be appreciated in the zoomed expansion of the multiplets at 4.1 ppm shown in Figure 4.5. There is incipient loss of resolution with $N = 4096$ and the FT line shapes show the influence of broadening in accordance with the time–frequency uncertainty principle and the mild apodization used to avoid frank truncation. The ersatz FDM spectrum shows narrower peaks, and hence *apparently* superior SNR, but one of the narrow doublets on the leftmost multiplet has been distorted, and another peak is apparently slightly too narrow. The hybrid spectrum largely repairs the first error, but cannot much influence the second one, because

---

Figure 4.4   A 1.4 ppm section of the 1D spectrum of **1** obtained by various data processing methods. The three FT spectra on the left show the quality of the data and change in line shape and resolution when the signal is truncated. Digital filtering and zero-filling have been used to obtain the best spectral representation. The central three spectra are ersatz spectra constructed artificially from the parameters extracted by FDM. Although not easily visible at the vertical scale of the figure, very small peaks in the baseline were identified with fidelity by FDM. The convergence onto the spectral features is somewhat different than with the FT spectra. Note that narrow spectral features, especially if relatively isolated, are located very accurately. Wide overlapping lines converge onto the correct amplitudes and widths more slowly with respect to the number of time-domain points. As follows from the ratio of the window area to the spectral width and the length of the signal, a total of 264, 132, and 66 peaks were used to construct the ersatz spectra for the cases $N = 8192$, 4096, and 2048, respectively, in a 1.4 ppm region centered on the expansion shown. Note, however, that far fewer than this number of poles have significantly large amplitudes. The far right column shows the hybrid spectrum obtained as described above.

Reprinted from *J. Magn. Reson.*, **133**, V. A. Mandelshtam, H. S. Taylor and A. J. Shaka, Application of the Filter Diagonalization Method to One- and Two-Dimensional NMR Spectra, 304. Copyright © 1998 Academic Press, with permission from Elsevier and Academic Press.

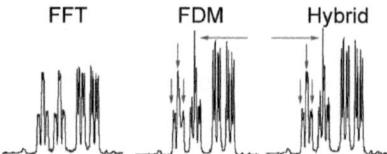

Figure 4.5 A zoomed region from the spectra of Figure 4.4 around 4.1 ppm for the case $N = 4096$. The FT spectrum, left, shows line shapes that are broadened in accordance with the time-frequency uncertainty principle. The FDM spectrum, center, enhances the resolution by determining the best-fit line width, assuming that all features are intrinsically Lorentzian. However, the basis function density, the number of peaks that are allowed to be used in the fit, is insufficient to fully resolve every line in the multiplets, giving an asymmetric appearance for the leftmost multiplet, set out by the vertical red arrows, that would be quite foreign to someone experienced in analyzing NMR spectra to determine spin–spin couplings. In the "hybrid" spectrum, in which the difference between the measured data and the FDM fit (in the time domain) is added to the FDM spectrum, the normal patterns are largely restored, although one central component, set out by a horizontal red arrow, in the second multiplet from the left is apparently slightly too narrow, and so appears to be too intense.

the FT of the noise/error residual is added to the ersatz spectrum, and it cannot have very narrow peaks as there are only the 4096 points available. Thus, even in this simple case, it is possible to see that trying to enhance the resolution can also create some new kinds of problems. In particular, the convergence of FDM hinges on the local density of peaks. This suggests that the best strategy is to disperse the peaks as widely as possible. As multidimensional NMR is the tried and true way to do this, FDM should be of most use in higher-dimensional spectra.

Noise is substantial in the NMR and especially in the multi-D NMR spectra, in which it may arise from unwanted modulation of true signal peaks, *i.e.*, "t_1-noise" or rapid-pulsing artifacts, as well as true noise from the receiver chain. It will be essential, in 2D or higher, to regularize the generalized eigenvalue problems in order to obtain stable spectra. In 1D the effect of regularization is mainly to cause all eigenvalues u_k to lie within the unit circle in the complex plane, so that they refer to decreasing exponentials. Regularization is non-linear, with smaller peaks broadening more than larger ones. Figure 4.6 shows how both noise and small peaks are affected in a simulated 1D spectrum. Note that regularization is not strictly needed in 1D spectra; it only becomes essential for multidimensional spectra.

4.3.2 2D NMR

In 2D NMR there are experiments like NOESY,[20] COSY[21] and TOCSY[22] that create peaks that align in square-like direct product patterns: at any given f_2 frequency there may be a number of responses in f_1. These degenerate peaks

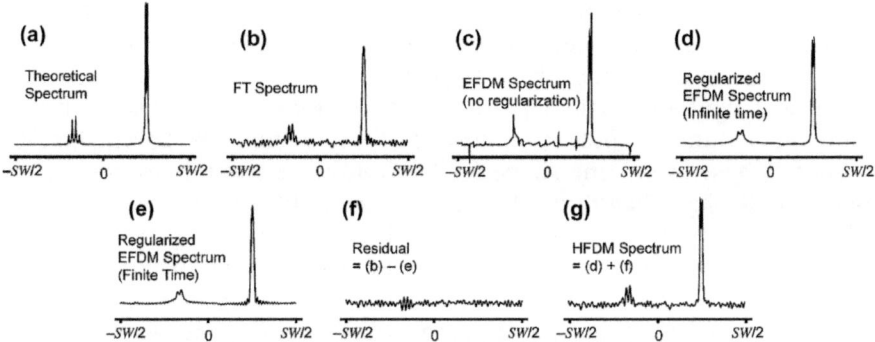

Figure 4.6 The effect of regularization on a simulated 1D NMR spectrum with an intense narrow doublet and a less intense quartet. (a) Theoretical pure Lorentzian spectrum. (b) FT of a slightly truncated FID corresponding to (a) with noise added (in the time domain). (c) Ersatz FDM spectrum with no regularization. (d) Same as (c) with regularization, but with a very dense frequency grid. (e) FT of FID corresponding to (d), but truncated at the same point in time as the FID that gave (b). (f) The residual spectrum, (b) − (e). (g) The hybrid FDM spectrum, (d) + (f). Noise is restored to the correct level, and the quartet structure, which was incorrect in the regularized ersatz spectrum, has been restored. Truncation artifacts around the base of the more intense doublet have been removed and the overall result is quite satisfactory.

Adapted with permission from C. D. Ridge and A. J. Shaka, "Ersatz" and "Hybrid" NMR Spectral Estimates Using the Filter Diagonalization Method, *J. Phys. Chem. A*, 2009, 113, 2036. Copyright © 2009 American Chemical Society.

mean that FDM cannot give a large advantage because the only way to differentiate peaks in the f_1 dimension that have degenerate frequency in f_2 is to have enough points in f_1 so that there are enough peaks to achieve a good fit. Unfortunately, in experiments like COSY a longer acquisition time in f_1 may introduce more and more 2D multiplets through long-range coupling so that the apparent number of 2D peaks keeps increasing rather than remaining fixed. This is the worst case for a parametric method, and we expect no significant advantage from using FDM in these kinds of experiments.

On the other hand, heteronuclear experiments like HSQC[23] have a fixed number of peaks, say one or two for each carbon-13 nucleus in the molecule, and the number of peaks does not change as the t_1 acquisition time is increased by taking more increments: only the resolution changes. The combination of a sparse spectrum with a fixed number of peaks and little degeneracy is absolutely ideal for FDM, and very large gains can be made for these cases.

Aside from the requirement that the number of peaks in the spectrum be small in order to get a large speedup, there are other problems in multi-dimensional NMR spectra that do not occur in 1D NMR. First, the Lorentzian line shape is not optimum for distinguishing peaks on a contour plot. The star-shaped contours, with prominent tails along the frequency axes, mean

that overlapping tails may sum up to look like a "peak" and make interpretation more difficult (see above). Usually, Lorentzian-to-Gaussian transformation, or multiplication by a shifted sine-bell function is performed to
give cleaner contours that make the contour plot easier to discern by eye.
While the natural output of multi-D FDM is a complex Lorentzian, the
so-called "phase-twist" line shape for 2D NMR, it is possible to use the parameters to display other kinds of spectra, or to modify the spectral line shape,
as discussed in the theoretical section of this chapter. When spectra are nearly
ideal, it is possible to get very impressive-looking ersatz spectra. However,
some representations are unstable when the HIP is ill-posed, and in those
cases one must be cautious, and perhaps use a more conservative formula
for the spectrum. For chemical shift assignment, where intensities are less
important, the pseudo-absorption spectrum can be employed, for example.

Many kinds of 2D spectra naturally give either a sine- or cosine-modulated
data set in t_1, depending on the chosen phase of a single pulse. If the data
has been obtained by pulsed field gradient methods, conventional processing software just combines the phase-modulated P/N-data, if present, to
obtain sin/cos amplitude-modulated data, and then proceeds as in ref. 24 to
avoid phase-twist line shapes. As discussed, a parametric method like FDM
in principle allows one to simply discard the dispersion-mode contribution
to each individual peak entry, and thereby obtain an absorption-like spectrum [*e.g.*, using eqn (4.50)] from purely phase-modulated data, such as a
single N-type data set.[25] A more conservative alternative is to calculate the
complex spectrum [eqn (4.48)], either N- or P-type, and then combine these
to obtain the desired double absorption line shape or, if intensities are not
critical, to use the pseudo-absorption spectrum, eqn (4.52), which is well
behaved. Processing either cosine- or sine-modulated data by FDM is quite
disadvantageous because each 2D spectrum has twice as many peaks in f_1. If
any of these peaks are slightly in error (see Figure 4.5) then the combination
will be imperfect. Therefore, the N/P spectra should be used, by taking appropriate combinations $\cos \pm i \sin$ if the data set is amplitude modulated,
and then the two phase-twist spectra combined to give the double absorption spectrum. The phase-sensitive FDM spectrum can also be submit-
ted, as a pseudo-FID, to any kind of Lorentzian-to-Gaussian line shape
transformation, but along the lines of conventional FT processing, and *not*
on a peak-by-peak basis as in eqn (4.51).

The best case of all is the so-called "constant-time" (CT) experiment, in
which the apparent decay with respect to t_1 is essentially zero (aside from
imperfect shimming of the magnet) because the modulation in t_1 is
achieved by rastering a refocusing pulse across a fixed constant time $2T$
that, in the case of ^{13}C-labeled proteins, might be fixed to $1/J_{CC} \approx 28$ ms. In
these experiments the resolution cannot be increased by setting the constant time interval to intermediate values: the choices are $2T \approx 0$, $2T \approx 28$ ms
or $2T \approx 56$ ms as the magnetization transfer function has nulls at values of
$(2n + 1)/2J_{CC}$, $n = 0$, 1, ... leading to very poor sensitivity. However, as T_2
relaxation occurs during the entire time $2T$ it may be that 56 ms gives poor

sensitivity for larger proteins. Very attractive, then, would be to minimize $2T$, maximizing sensitivity, and then use 2D FDM to enhance the resolution so that assignments can be made. A further important improvement is to implement a so-called doubling scheme. In principle, one could raster the refocusing pulse across the full $2T$ interval, obtaining what looks like a double-length interferogram. For 2D FT spectra this does *not* lead to any change in resolution, as it amounts to the same thing as simply implementing the usual procedure of States *et al.*[24] However, for FDM the information content of the data is doubled (and so is the basis size), because there are more total increments, and this doubling leads to a very noticeable improvement.[26] In practice, the data from the usual half-interval constant time refocusing can simply be reversed and flipped in time, with complex conjugation. In the LP literature this trick is known as mirror-image LP.[27] One must be sure the phase at $t_1 = 0$ is correct so that no discontinuities are introduced, but this requirement is easily met with a correct pulse sequence. In effect, this exploitation hinges on the fact that the line width is essentially zero in f_1, so only half of the complex frequency eigenvalue needs to be obtained or, equivalently, that the data obtained from the usual half-interval experiment seems to be twice as information-rich as it was for a decaying interferogram. While using only the measured data and trying to artificially restrict the eigenvalues to correspond to only zero-width peaks might seem like an alternative, it is not, however, as it is not numerically stable.

Figures 4.7 to 4.9 show this potential advantage in the 500 MHz CT-HSQC spectrum of the thermophilic protein MTH-1598 from the thermophilic bacterium *Methanothermobacter thermautotrophicus*.[28] The figures show the FDM spectra with the doubling scheme in the constant-time dimension and Gaussian line shapes. The 2D FT spectrum with a short constant time $2T = 7.5$ ms shows limited resolution in Figure 4.7 while 2D FDM applied to exactly the same data (and with the doubling scheme described above) gives the spectrum shown in Figure 4.8, with the boxed regions highlighting significant differences. The resolution can be improved for FT analysis by

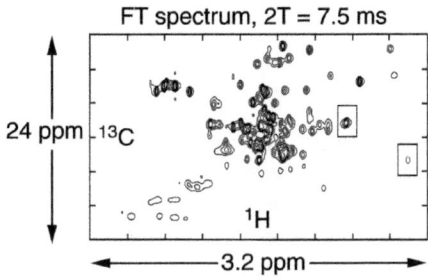

Figure 4.7 2D FT absorption mode CT-HSQC spectrum of MTH-1598 with a constant time $2T = 7.5$ ms. Sensitivity for this fairly large protein is reasonably good, but spectral resolution is limited.

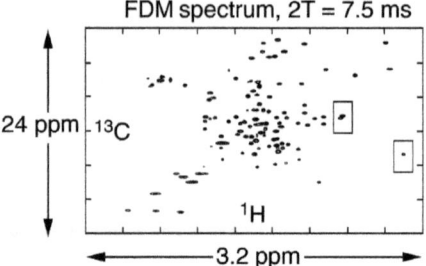

Figure 4.8 2D FDM CT-HSQC spectrum of MTH-1598 with a constant time $2T = 7.5$ ms. The same data was used as in Figure 4.7 but the apparent resolution of the 2D contour plot is much better. Note that there are two peaks in the left-hand box. Absorption-mode Gaussian line shapes have been imposed [*cf.* eqn (4.51)].

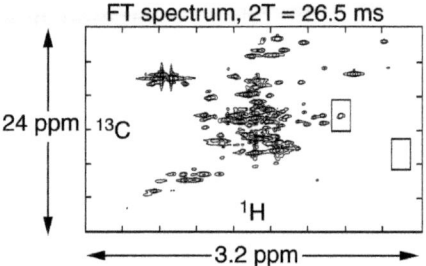

Figure 4.9 2D FT CT-HSQC spectrum of MTH-1598 with a constant time $2T = 26$ ms. The longer constant time period allows more points to be collected in f_1, improving the resolution compared with Figure 4.7, but the peak in the righthand box has been lost, as it has a short transverse relaxation time.

extending the constant time to $2T = 26.5$ ms, as shown in Figure 4.9, but unfortunately one of the faster-relaxing peaks in the righthand box is lost.

As mentioned earlier, there is little chance for rapid acquisition of crowded spectra with many hundreds if not thousands of peaks. As such, 2D NOESY spectra do not stand to benefit from 2D FDM processing, a hypothesis we have verified with some unpublished explorations.

4.3.3   3D NMR

Continuing the previous theme, the ideal 3D experiment would have two constant time dimensions, allowing doubling to be performed twice, a spectrum with relatively few peaks, and good intrinsic sensitivity. The 3D HNCO experiment can be adapted to this format, and with essentially one resonance per residue of the protein the total number of peaks to be resolved in the spectrum is not that large. In the case of the well-behaved protein ubiquitin, which tends to give excellent NMR spectra, the results are quite impressive.[29] As an example, consider the 2D integral projection of the 3D HNCO spectrum in which the columns along the ^{1}H dimension are summed so that only the ^{15}N and ^{13}C' dimensions remain. No matter how well resolved the proton dimension may be, the two indirect dimensions must display resolution in accord with the number of sampled points in these dimensions, at least for the FT spectrum. The 3D FDM spectrum behaves entirely differently, however. The 2D projection can become very sharp once the total information content of the 3D data is sufficient to describe the total number of peaks that are present. Because the true peak widths are estimated [with regularization using eqn (4.55)], and with a minimum width commensurate with the digitization of the frequency domain) the spectrum may undergo something akin to a "phase transition" in which large regions in frequency space suddenly show very high resolution. As an example, consider the results, from the pulse sequence in Figure 4.10, shown in Figure 4.11. Only 25 minutes of acquisition time (1 mM, 500 MHz, RT probe) were required to obtain the four 3D data sets, each of size $N_1 \times N_2 \times N_3 = 6 \times 8 \times 1024$ (^{13}CO, ^{15}N, ^{1}H), which were processed to give absorption line shapes in the FT spectrum, after cosine apodization to remove strong truncation artifacts in the indirect dimensions. The doubling scheme described above applied to both indirect dimensions resulted in the increase of the effective data size to $11 \times 15 \times 1024$. Since the numbers of points in both indirect dimensions were still small, the windows were implemented only along the direct dimension with six overlapping windows to cover the proton chemical shift range for the protein. The small generalized eigenvalue problems involved matrices of size 490×490. That is, the overall FDM calculation took minutes. Normally 2D planes are scrolled through sequentially to enable correct assignments in the 3D FT spectrum. However, with so few increments in both indirect dimensions, there is little resolving power. This is dramatically illustrated by Figure 4.12, which compares the integral projection onto the (C', N) plane, integrating out all the proton chemical shift information. The FT projection is essentially completely unresolved reflecting, as it must, the limited sampling time in the two displayed dimensions, while the corresponding FDM projection of the same data shows excellent definition. Of course, individual planes in the 3D FDM spectrum are sparse and contain sharp peaks throughout, making the spectrum extremely easy to analyze. Note, however, that the correct chemical shifts may lie outside of the apparent line width in some cases. The line widths can best be set by injecting a series of known signals at different positions, *i.e.* internal standards,[37] and running the 3D FDM calculation repetitively. The injected signal should mimic true signals as closely as possible, including the characteristics of the ADC and NMR receiver chain. The error statistics between the known frequencies and those extracted from the data by FDM can then be used to set a line width that is a more reliable indication of the true uncertainty. Such an approach was not undertaken in this case, however. The widths were instead set to the larger of the calculated width or an arbitrary minimum width, selected to ensure that the peaks could be

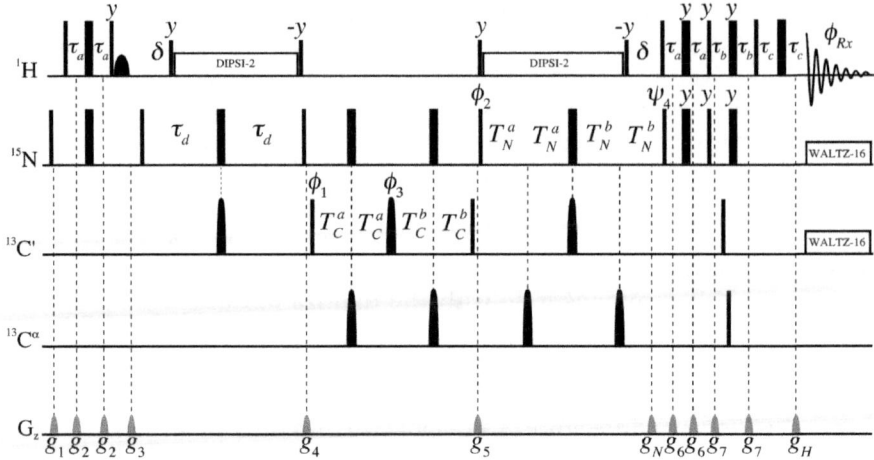

Figure 4.10 Experimental scheme for the double CT-HNCO pulse sequence. Square 90° and 180° pulses are indicated by black vertical lines, respectively, with the phases, if different from x, indicated above the pulses. The ^{1}H, ^{13}CO, and ^{15}N carriers are set to 4.75 (water), 176.0, and 120.0 ppm, respectively. All proton pulses are applied with a field strength of 24 kHz with the exception of the shaped 90° pulse, which had an E-SNOB[30] profile (2.1 ms, 270 Hz) and the DIPSI-2[31] decoupling sequence including the two surrounding 90° pulses (4.9 kHz). ^{15}N pulses employed a 6.3 kHz field. At 500 MHz, ^{13}C square pulses had a length of 64.2 μs to minimize excitation of the ^{13}C$^\alpha$ region. The shaped 180° ^{13}CO and ^{13}C$^\alpha$ pulses were applied with a G3[32] profile at a peak field strength of 7.03 kHz and had a 512 μs duration. ^{13}C$^\alpha$ pulses were applied, by phase-modulation, at 56 ppm. Decoupling was done with WALTZ-16[33] for ^{15}N (1.0 kHz field strength) and ^{13}CO (300 μs 90° pulses with SEDUCE-1[34] amplitude profile, 3.6 kHz peak power). The delays used were $\tau_a = 2.25$ ms, $\tau_b = 2.75$ ms, $\tau_c = 0.70$ ms, $\tau_c = 12.4$ ms, $\delta = 5.4$ ms, $T_C^a = (T_C - t_1/2)$, $T_C^b = (T_C + t_1/2)$, $T_N^a = (T_N - t_2/2)$, $T_N^b = (T_N + t_2/2)$ with the constant time periods $2T_C = 8.0$ ms and $2T_N = 24.8$ ms. The phase cycling was as follows: $\phi_1 = (x, -x)$, $\phi_1 = (2x, 2(-x))$, $\phi_3 = (4x, 4(-x))$, and $\phi_{Rx} = (x, -x - x, x)$. Frequency discrimination in the t_1 (^{13}C') dimension was achieved by using States-TPPI.[35] In the t_2 (^{15}N) dimension a phase-sensitive spectrum was obtained by recording a second FID for each increment of t_2 with the phase ψ_4 shifted by 180° and the amplitude of the gradient g_N simultaneously reversed.[36] For every successive t_2 increment, a 180° shift was added to ϕ_2 and ϕ_{Rx}. Durations of the pulsed field gradients (sine-bell shape) are as follows: $g_1 = (0.8$ ms, 6 G cm^{-1}), $g_2 = (0.65$ ms, 2.5 G cm^{-1}), $g_3 = (1.3$ ms, 7 G cm^{-1}), $g_4 = (1.1$ ms, 4.5 G cm^{-1}), $g_5 = (0.8$ ms, 8 G cm^{-1}), $g_6 = (0.65$ ms, 12 G cm^{-1}), $g_7 = (0.65$ ms, 4 G cm^{-1}), $g_N = (2.0$ ms, 20 G cm^{-1}), and $g_H = (0.6$ ms, 6.8 G cm^{-1}). This pulse sequence gives N/P data in the ^{15}N dimension, and sin/cos data in the ^{13}C' dimension. These were combined to produce four data sets that can be combined to give absorption-mode line shapes in the 3D FT spectrum. The FDM processing involved a single NN data set. Absorption-mode Gaussian line shapes have been imposed.

Reprinted from *J. Magn. Reson.*, **169**, J. H. Chen, D. Nietlispach, A. J. Shaka and V. A. Mandelshtam, Ultra-high resolution 3D NMR spectra from limited-size data sets, 215. Copyright © 2004 Elsevier Inc.

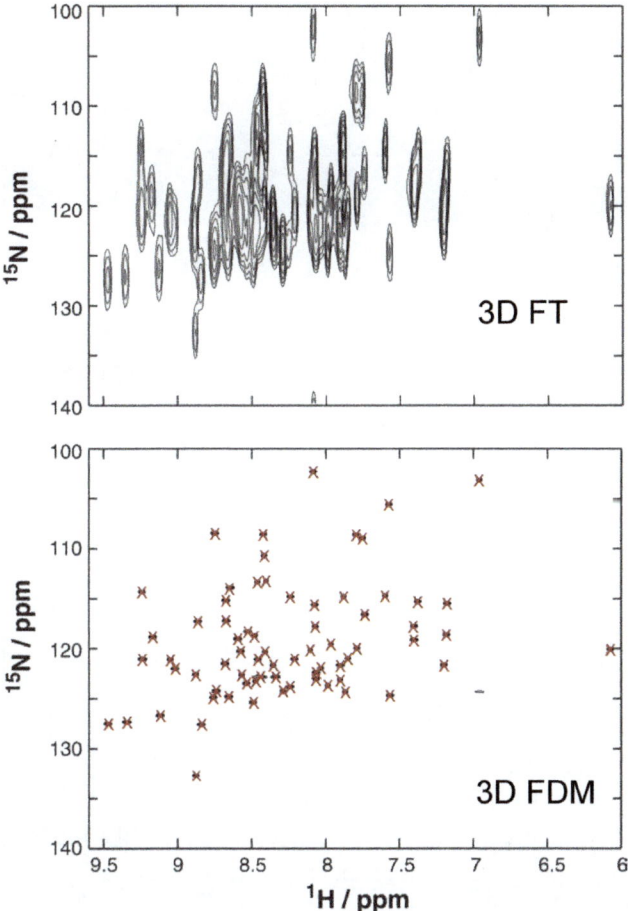

Figure 4.11 The 2D integral projections from the data obtained using the pulse sequence of Figure 4.10. The FT spectrum (top) was obtained using cosine weighting functions along all dimensions to reduce truncation artifacts. The four 3D data sets, each of size $N_1 \times N_2 \times N_3 = 6 \times 8 \times 1024$ (^{13}CO, ^{15}N, ^{1}H) were processed to give pure absorption line shapes and were zero-filled to $128 \times 128 \times 2048$ to provide high digital resolution for smooth contouring. The FDM spectrum used a single NN data set. Absorption-mode Gaussian line shapes have been imposed. The carbonyl dimension helps to improve the 3D FDM fit greatly compared to just a 2D FDM calculation, resulting in very narrow peaks in the ^{15}N–H$_N$ plane. The red crosses show the predicted positions of the peaks according to the known assignments for this ubiquitin sample, which were obtained independently from HNCA/HN(CO)CA and CBCA(CO)NH/HNCACB experiments. The unaccounted peak at ∼7 ppm, which also appears in the FT spectrum albeit with lower intensity and wider line width, arises from a poorly decoupled arginine residue side chain.
Reprinted from *J. Magn. Reson.*, **169**, J. H. Chen, D. Nietlispach, A. J. Shaka and V. A. Mandelshtam, Ultra-high resolution 3D NMR spectra from limited-size data sets, 215. Copyright © 2004 Elsevier Inc.

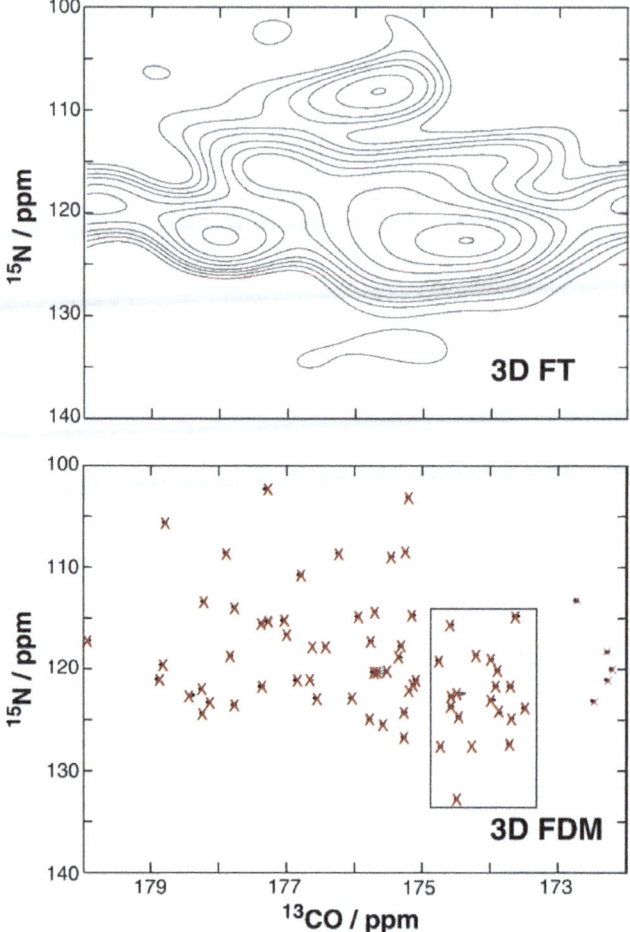

Figure 4.12   The 2D integral projections from the data obtained using the pulse
sequence of Figure 4.10. The FT spectrum (top) was obtained using
cosine weighting functions along all dimensions to reduce truncation
artifacts. The four 3D data sets, each of size $N_1 \times N_2 \times N_3 = 6 \times 8 \times 1024$
(^{13}CO, ^{15}N, ^1H), were processed to give pure absorption line shapes and
were zero-filled to $128 \times 128 \times 2048$ to provide high digital resolution for
smooth contouring. The FDM spectrum used a single NN data set.
Absorption-mode Gaussian line shapes have been imposed. The carbonyl dimension helps to improve the 3D FDM fit greatly compared to
just a 2D FDM calculation, resulting in very narrow peaks in the ^{15}N–H_N
plane. The red crosses show the predicted positions of the peaks
according to the known assignments for this ubiquitin sample,
which were obtained independently from HNCA/HN(CO)CA and
CBCA(CO)NH/HNCACB experiments.

contoured smoothly on the chosen frequency grid. Often the calculated widths are very small, making it possible for peaks to "disappear" in between frequency grid points if no precaution is taken.

4.3.4 4D NMR

As for 2D and 3D, in 4D NMR the best gains can be achieved when the SNR is high and the expected number of 4D peaks is small. In this case we anticipate that the *total* effective size (*i.e.*, that after doubling in each dimension, if any) of the 3D data that is sufficient to resolve a 3D spectrum should be about the same as that of the 4D data needed to resolve a 4D spectrum. This equivalence means that the total time to acquire the data may be roughly *the same* as that for the lower-dimensional spectrum. In FDM the added dimension, even if the resolution would be extremely poor in an FT spectrum, adds information that allows the 4-dimensional peaks to be extracted with good fidelity. As just one example, Figures 4.13–4.15 compare various 2D planes from a 2D projection of a 4D HNCOCA experiment[38] under similar conditions to those that apply to Figures 4.11 and 4.12, using an analogous pulse sequence to that in Figure 4.10, with an additional constant time C^α evolution period, kept as short as practical (<2 ms). A $4 \times 4 \times 8 \times 1024$ ($^{13}C^\alpha$, $^{13}C'$, ^{15}N, ^1H) 4D CT-HNCOCA data set was used, with the doubling method applied, to each of the three indirect dimensions, to give total size $7 \times 7 \times 15 \times 1024$. Using the full spectral width in each of the three indirect dimensions, windows in the ^1H dimension were chosen with eight basis functions, giving a series of generalized

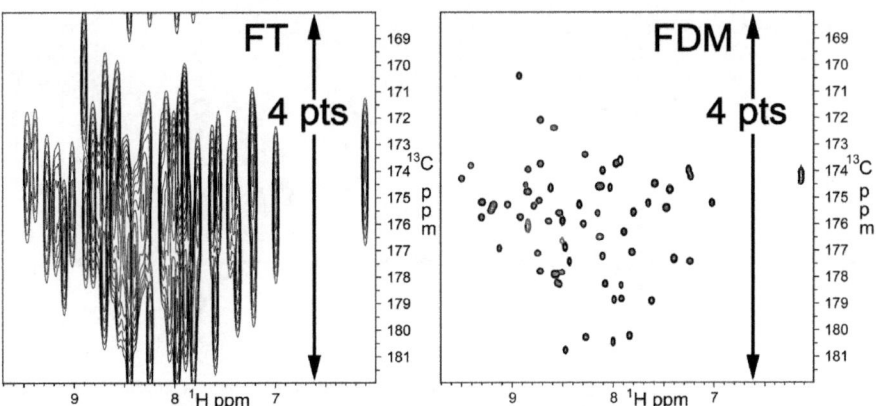

Figure 4.13   The 2D integral projections (H, C′) from the 4D HNCOCA ubiquitin data set, in which the other two dimensions have been summed over the whole spectral width. The FT spectrum (left) was obtained using cosine weighting functions along all dimensions to reduce truncation artifacts, and processed to give pure absorption-mode line shapes. The FDM spectrum used a single NNN data set. Absorption-mode Gaussian line shapes have been imposed. The gain in resolution is very large in the carbonyl dimension.

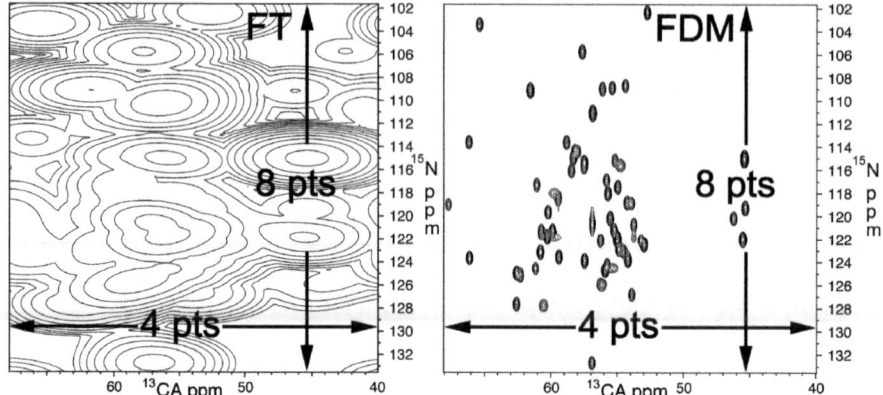

Figure 4.14   The 2D integral projections (N, C^α) from the 4D HNCOCA ubiquitin data
set; the other two dimensions have been summed over the whole
spectral width. The FT spectrum (left) was obtained using cosine
weighting functions along all dimensions to reduce truncation artifacts, and processed to give pure absorption-mode line shapes. The
FDM spectrum used a single NNN data set. Absorption-mode Gaussian
line shapes have been imposed. In this case, an essentially completely
unresolved FT spectrum resolves into sharp peaks in the FDM
spectrum.

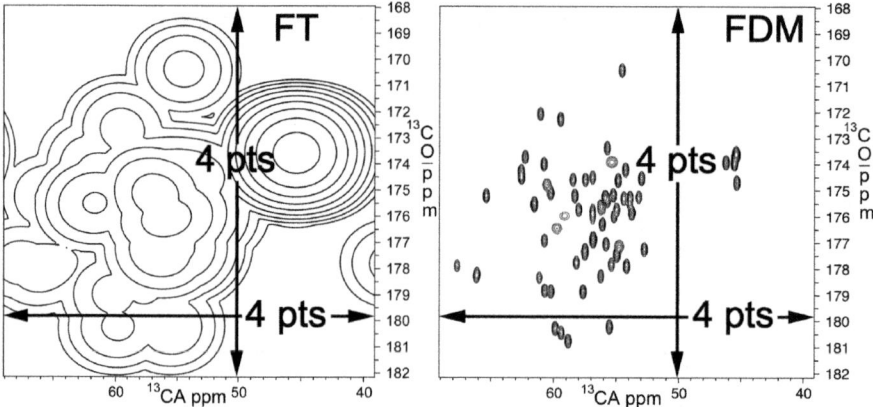

Figure 4.15   The 2D integral projections (C^α, C') from the 4D HNCOCA ubiquitin
data set; the other two dimensions have been summed over the whole
spectral width. The FT spectrum (left) was obtained using cosine
weighting functions along all dimensions to reduce truncation artifacts, and processed to give pure absorption-mode line shapes. The
FDM spectrum used a single NNN data set. Absorption-mode Gaussian
line shapes have been imposed. This dramatic example shows the high
resolving power of 4D FDM. The apparent peaks near 38 ppm and
178 ppm in the FT projection are in fact aliased tails of the peaks at
68 ppm and 178 ppm, respectively.

eigenvalue problems involving matrices of size 504×504. The FDM spectrum was calculated over a $128 \times 128 \times 128 \times 2048$ frequency grid, the latter just over the region H_N where signals were anticipated. As in the 3D case, and perhaps even more so, the difference between a dimension-by-dimension FT analysis and an integrated 4D FDM analysis is striking. The entire 4D FDM spectrum shows high resolution in all four dimensions, even though three of them have so few points that normal methods fail to give any usable information.

4.4 Conclusions

The 1D FDM is essentially a developed and well-tested technique for spectral estimation that is generally as reliable as FFT, sufficiently fast, and can often deliver spectral resolution beyond the FT uncertainty relation, if the data can be well represented by Lorentzians and is not very noisy. The algorithm is based on mapping the HIP to a generalized eigenvalue problem, thus avoiding any non-linear optimization. However, the difficulties associated with the construction of a meaningful line list for data of poor quality [*i.e.*, not described well by eqn (4.1)] exist. These difficulties have little to do with a lack of an algorithm of fitting 1D data by complex poles and amplitudes, but are rather fundamental, originating from the ill-posed nature of the HIP itself, leading to the ambiguity of the computed line list for a general data set that *a priori* does not fit any particular parametric form.

The multi-D FDM is based on mapping the multi-D HIP to a set of, in principle coupled, generalized eigenvalue problems. However, new fundamental challenges arise when trying to perform simultaneous diagonalization, which for general data cannot be accomplished within pure linear algebra. This implies that, unless the data fits well the very special form of eqn (4.31) and the SNR is excellent, direct construction of a meaningful coupled multi-D line list is nearly impossible. Fortunately, for certain classes of multi-D NMR experiments, starting with realistic data that is severely truncated in all the indirect dimensions, various spectra with superior resolution can be obtained using the independent solutions of the said generalized eigenvalue problems, thus avoiding a direct construction of a fully coupled line list. Additional challenges exist in multi-D FDM associated with the regularization of the generally ill-conditioned generalized eigenvalue problems. Some plausible regularization recipes have been identified and discussed in this chapter, which turned out to be key steps in the process of spectral estimation.

Acknowledgements

VAM would like to thank Howard Taylor, who was involved in the development of FDM in its early stages. We also acknowledge contributions to further development and implementation of FDM of our former and present group members, especially Jianhan Chen, Geoff Armstrong, Anna De

Angeles, Hasan Celik, Clark Ridge and Beau Martini, and several of our collaborators. VAM was supported by NSF grant CHE-1152845.

References

1. V. A. Mandelshtam, H. S. Taylor and A. J. Shaka, *J. Magn. Reson.*, 1998, **133**, 304.
2. M. R. Wall and D. Neuhauser, *J. Chem. Phys.*, 1995, **102**, 8011.
3. V. A. Mandelshtam and H. S. Taylor, *J. Chem. Phys.*, 1997, **107**, 6756.
4. F. Abildgaard, H. Gesmar and J. J. Led, *J. Magn. Reson.*, 1988, **79**, 78.
5. V. A. Mandelshtam, *Prog. Nucl. Magn. Reson. Spectrosc.*, 2001, **38**, 159.
6. G. H. Golub and W. Kahan, *J. SIAM. Numer. Anal., Ser. B*, 1965, **2**, 205.
7. G. H. Golub and C. Reinsch, *Numer. Math.*, 1970, **14**, 403.
8. P. Koehl, *Prog. Nucl. Magn. Reson. Spectrosc.*, 1999, **34**, 257.
9. R. Roy, A. Paulraj and T. Kailath, *IEEE Trans. Acoust., Speech, Signal Process.*, 1986, **34**, 1340.
10. Y. Hua and T. Sarkar, *IEEE Trans. Acoust., Speech, Signal Process.*, 1990, **38**, 814.
11. J. H. Chen, A. J. Shaka and V. A. Mandelshtam, *Abstr. Pap. Am. Chem. Soc.*, 2001, **222**(70), 2.
12. J. W. Pang, T. Dieckmann, J. Feigon and D. Neuhauser, *J. Chem. Phys.*, 1998, **108**, 8360.
13. V. A. Mandelshtam, N. D. Taylor, H. Hu, M. Smith and A. J. Shaka, *Chem. Phys. Lett.*, 1999, **305**, 209.
14. V. A. Mandelshtam, *J. Magn. Reson.*, 2000, **144**, 343.
15. J. H. Chen and V. A. Mandelshtam, *J. Chem. Phys.*, 2000, **112**, 4429.
16. J. H. Chen, V. A. Mandelshtam and A. J. Shaka, *J. Magn. Reson.*, 2000, **146**, 363.
17. B. Dai and C. D. Eads, *Magn. Reson. Chem.*, 1988, **78**, 78.
18. V. A. Mandelshtam, H. S. Taylor and A. J. Shaka, *J. Magn. Reson.*, 1998, **133**, 304.
19. C. D. Ridge and A. J. Shaka, *J. Phys. Chem. A*, 2009, **113**, 2036.
20. J. Jeener, B. H. Meier, P. Bachmann and R. R. Ernst, *J. Chem. Phys.*, 1979, **71**, 4546.
21. W. P. Aue, E. Bartholdi and R. R. Ernst, *J. Chem. Phys.*, 1976, **64**, 2229.
22. L. Braunschweiler and R. R. Ernst, *J. Magn. Reson.*, 1983, **53**, 521.
23. G. Bodenhausen and D. J. Ruben, *Chem. Phys. Lett.*, 1980, **69**, 185.
24. D. J. States, R. A. Haberkorn and D. J. Ruben, *J. Magn. Reson.*, 1982, **48**, 286.
25. S. Keppetipola, W. Kudlicki, B. D. Nguyen, X. Meng, K. J. Donovan and A. J. Shaka, *J. Am. Chem. Soc.*, 2006, **128**, 4508.
26. J. H. Chen, A. A. De Angelis, V. A. Mandelshtam and A. J. Shaka, *J. Magn. Reson.*, 2003, **162**, 74.
27. Z. Guang and A. Bax, *J. Magn. Reson.*, 1990, **90**, 405.

28. A. Yee, X. Chang, A. Pineda-Lucena, B. Wu, A. Semesi, B. Le, T. Ramelot, G. M. Lee, S. Bhattacharyya, P. Gutierrez, *et al.*, *Proc. Natl. Acad. Sci. U. S. A.*, 2002, **48**, 1825.

29. J. H. Chen, D. Nietlispach, A. J. Shaka and V. A. Mandelshtam, *J. Magn. Reson.*, 2004, **169**, 215.

30. E. Kupce, J. Boyd and I. D. Campbell, *J. Magn. Reson., Ser. B*, 1995, **106**, 300.

31. A. J. Shaka, C. J. Lee and A. Pines, *J. Magn. Reson.*, 1988, 77, 274.

32. L. Emsley and G. Bodenhausen, *Chem. Phys. Lett.*, 1990, **165**, 469.

33. A. J. Shaka, J. Keeler and R. Freeman, *J. Magn. Reson.*, 1983, **53**, 313.

34. M. A. McCoy and L. Mueller, *J. Am. Chem. Soc.*, 1992, **114**, 2108.

35. D. Marion, M. Ikura, R. Tschudin and A. Bax, *J. Magn. Reson.*, 1989, **85**, 393.

36. L. Kay, P. Keifer and T. Saarinen, *J. Am. Chem. Soc.*, 1992, **114**, 10663.

37. J. C. Hoch and A. S. Stern, *NMR Data Processing*, 2005, Wiley-Liss.

38. D. Yang and L. E. Kay, *J. Am. Chem. Soc.*, 1999, **121**, 2571.

39. B. G. R. de Prony, Essai éxperimental et analytique: sur les lois de la dilatabilité de fluides élastique et sur celles de la force expansive de la vapeur de l'alkool, á différentes températures, *J. Éc. Polytech.*, 1795, **1**(cahier 22), 24–76.

40. B. R. Martini, V. A. Mandelshtam, G. A. Morris, A. A. Colbourne and M. Nilsson, Filter Diagonalization Method for Processing PFG NMR Data, *J. Magn. Reson.*, 2013, **234**, 125–134.

CHAPTER 5

Acquisition and Post-processing of Reduced Dimensionality NMR Experiments

HAMID R. EGHBALNIA* AND JOHN L. MARKLEY*

University of Wisconsin-Madison, 433 Babcock Dr., Madison, WI 53706-1544, USA
*Email: heghbaln@wisc.edu; jmarkley@wisc.edu

5.1 Introduction

NMR spectroscopy investigates transitions between spin states of magnetically active nuclei in a magnetic field. The most important magnetically active nuclei for biomacromolcules are the proton (^{1}H), carbon-13 (^{13}C), nitrogen-15 (^{15}N) and phosphorus-31 (^{31}P). Each of these stable isotopes has a nuclear spin of one-half. The magnetic moment of each nucleus precesses about the external magnetic field, and other (nearby) fields influence the precession in a precise way that is dependent on the geometry and dynamics of the nearby atoms. The sum of influences on a given nuclei's spin, by neighboring spins, gives rise to NMR properties that sensitively reflect the environment of a given nuclei. Practical engineering and physical constraints place limits on achievable sensitivity and, therefore, on the experimenter's ability to identify all distinguishable signals from distinct nuclei.

Four decades after Bloch's groundbreaking demonstration of a proton NMR signal,[1] the advent of multi-nuclear, multi-dimensional Fourier transform (FT) NMR spectroscopy initiated a broad range of new applications in chemistry and biology.[2,3] Versatile pulse sequences leading to

New Developments in NMR No. 11
Fast NMR Data Acquisition: Beyond the Fourier Transform
Edited by Mehdi Mobli and Jeffrey C. Hoch
© The Royal Society of Chemistry 2017
Published by the Royal Society of Chemistry, www.rsc.org

complex "spin gymnastics" have enabled the investigation of a wealth of NMR parameters. A huge variety of NMR data collection protocols are currently at one's disposal for structural and functional investigations. Spectral resolution is an important factor in multidimensional NMR and has driven the quest for higher magnetic field strengths and the development of multidimensional NMR experiments. Higher resolution comes at a price, however. Additional dimensions lead to exponential increases in data collection time, and the number of data points required increases at polynomial rates with the spectrometer field strength.

Resolution is not the only influential factor. Sensitivity, often expressed in units of signal-to-noise (S/N), also plays a critical role. A protein sample, even at the relatively high concentration of 1 mM, may require the co-addition of eight or 16 transients at each time point in order to achieve sufficient S/N in a two- or three-dimensional NMR experiment. Cryogenic probe technology can reduce the data acquisition period by a factor as much as ten by reducing thermal noise in the probe and preamplifier circuitry.[4] The operational factor is the square-root relationship between time and sensitivity (in signal averaging): an inherent sensitivity improvement of three-fold is equivalent to a nine-fold increase in signal averaging. Cryogenic probes offer the potential for significant reduction in experiment time with few negative consequences, although caveats are associated with their use. First, a three- to four-fold sensitivity increase generally is realized only for simple 1D experiments in nonionic solutions. The RF field inhomogeneity of the ^{13}C and ^{15}N coils decreases the efficiency of the probes in multi-pulse triple resonance experiments. In practice, for a protein solution at low ionic strength, the sensitivity increase is more likely to be about two-fold. The improved sensitivity obtained from cryogenic probes is influential in more challenging systems—for example, by making it possible to lower the sample concentration by a factor of two for a molecule with low solubility.

The goal of reducing data collection times while retaining the benefits of high resolution and sensitivity remains as one of the major challenges in NMR spectroscopy.[5] While advances in experimental techniques, at the sample preparation stage, and new spectrometer technologies will continue to push for further gains, computational methodology, a significantly influential component in NMR spectroscopy, will continue to play a key role in expanding the application frontiers. As pointed out below, it can be difficult to draw sharp conceptual boundaries across stages of computation. However, for the purposes of establishing a working conceptual framework, it is useful to consider three stages: "select", "acquire", and "interpret" (Figure 5.1). The "select" stage involves the choice of NMR experiment (HNCO, HNCA, *etc.*). In the "acquire" stage, we decide on a sampling strategy—for example, uniform, non-uniform, random, radial, or adaptive.[6] The "interpret" stage contains the algorithms focused on achieving the goals of data collection—for example, projection–reconstruction, maximum entropy, geometric, or Bayesian.[7,8] In some implementations, two or more of

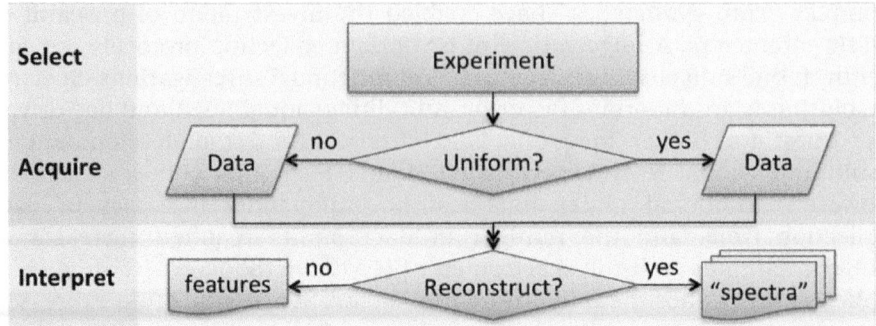

Figure 5.1   In concept, the three stages "select", "acquire", and "interpret" are distinct in that the acquire and interpret stage can vary from experiment to experiment. Furthermore, the mode of data collection (uniform or non-uniform) dictates the method of post-processing.

these stages may be coupled closely together. Nonetheless, this conceptual model can be useful in consolidating and clarifying some of the terminology and concepts in use—for example, concisely identifying strategies that differ mainly in "interpretation".

In the following section, we provide an overview of data acquisition approaches. We focus mainly on reduced dimensionality (RD) experiments,[9–11] but we briefly discuss selected examples of sampling, referred to as "non-uniform sampling", because this concept is useful in elucidating the distinction between "acquire" and "interpret" in the framework outlined in Figure 5.1. Following this overview, we expand on two conceptual approaches that are motivated by the goal of recovering features during the "interpret" stage. Next, we describe two implementations with distinct strategies—HIFI-NMR, and APSY.[12,13] Both approaches use the RD paradigm, and the feature recovery approach of both relies on a "peak recovery" paradigm for subspaces of spectral space. APSY uses a pre-specified projection regime that is straightforward, while HIFI-NMR uses a Bayesian updating procedure that tightly integrates data acquisition to data processing in real-time. The subsequent section is devoted to the discussion of robust peak recovery, which is an important aspect of robust operation for both HIFI-NMR and APSY. Maximum likelihood estimation as a prototype for statistical, and model-based, approaches for peak recovery offers the potential for improving the robustness of HIFI-NMR, APSY, or more generally any procedure that relies on a validated feature or peak recovery process.[14] The conclusion section briefly reviews the potential for future extensions.

5.2   Data Acquisition Approaches

The traditional workhorse of NMR spectroscopy is the Fourier transform. Data collected on a regular grid (regular spacing in the time domain) can be readily transformed into a representation in spectral (frequency) space in which peaks corresponding to nuclei of interest can be observed. Spectral representation of NMR data is sparse—that is, in the Fourier domain, a few peaks (non-zero values) are surrounded by a vast number of noise values (values at zero or close to zero). This fact motivates strategies for collecting data in a manner that can leverage this prior knowledge to some advantage—for example, faster data acquisition, higher sensitivity, or potentially enhanced resolution. The goal of this section is to provide a brief overview of literature relevant to the data acquisition material in this chapter. Several recent and more comprehensive reviews provide excellent coverage of the field's progress.[6,15,16]

The "accordion" experiment[17] provided the first introduction to a fast data collection approach in NMR spectroscopy. Although the goal was to reduce the time required for quantitative measurement of exchange rates that required the collection of a series of 2D spectra at different mixing times, the experiment contained conceptual elements that would later be categorized broadly as reduced dimensionality. In more general terms, reduced dimensionality (RD) methods for data collection encode multiple indirect frequencies simultaneously rather than independently, enabling the acquisition of higher-dimensional spectra in a short time and reducing the number of experiments necessary for peak assignment and structure determination. For example, in a 3D to 2D reduced dimensionality method, the chemical shift evolution of two nuclei is simultaneously monitored in a single indirect domain, thus reducing the dimensionality of NMR data collection from 3 to 2. An early example of this approach was the G-matrix Fourier transform (GFT) method.[9,18] Generalizing the G-matrix approach, the TILT experiment enables the collection of three-dimensional experiments as 2D planes by simultaneously evolving (mixing) two parameters.[11] The connection between GFT-NMR and the more general RD experiments is established by the Radon transform and the projection-slice theorem.[19] Note that the data obtained from several 2D planes in the TILT experiment form a non-uniform grid of points in 3D spectral space. The coevolution strategy as used in TILT necessitates the application of more advanced computational tools beyond the conceptually more straightforward reconstruction of the GFT approach (add/subtract). Additionally, the intensity of the peaks in the TILT experiment is proportional to the product of sine and/or cosine functions of the evolution times, which suggests that some tilt angles may provide better sensitivity than others.

The concept of tilt angle can be used to acquire data in a number of creative ways with the goal of aiding improved post-processing and signal identification. For example, radial sampling[20,21] acquires data on a regular polar-coordinate grid—a regular grid in polar coordinates. In the evolution dimensions (indirect dimensions) of the NMR experiment, data collection is parametrized by time in each dimension $\{t_i\}$, which can be experimentally controlled and specified. In more concrete terms, consider a three-dimensional experiment in which $n \times m$ time series (FIDs) are collected (along times t_1 and t_2)—a total of $n \times m \times k$, where k is the number of discrete data points in each FID. The ability to coevolve parameters t_1 and t_2 means that data for a select subset of $n \times m$ can be acquired without the need to collect the complete data set; for example, data along a plane in the three-dimensional spectral space can be selectively collected. The ability to selectively control the data collection strategy has the potential for time savings. However, the savings may come at the cost of reduced overall sensitivity and increased complexity in post-processing. Considering the advances in modern NMR spectrometers, the loss in sensitivity, in some cases, may be unnoticeable in practice.

RD methods are not alone in enabling improved data collection times. In non-uniform sampling (NUS; see also later chapters), data are collected from a subset of all uniform evolution periods at non-fixed intervals. The resulting decrease in data collection times is especially promising for larger proteins.[5,22,23] Random sampling creates a non-uniform subset at random intervals, and may be considered a subclass of NUS. Sampling (acquire) and reconstruction (post-processing) can be coupled in RD, but the coupled approach is more common in NUS. The recently released TopSpin 3.0 software package from Bruker incorporates NUS as part of a routine workflow, including optimized sampling schemes and data processing *via* multi-dimensional decomposition or compressed sensing.[24] Alternatively, processing *via* maximum entropy reconstruction[8,25] has been made more accessible to the general protein NMR community by use of the Rowland NMR Toolkit Script Generator. Another promising approach utilizes the NESTA algorithm.[26] However, it is helpful to note that while both RD and NUS methods share some of the same goals, NUS is predicated on more general assumptions that can be leveraged to obtain specialized benefits. We refer the reader to excellent reviews articles.[6,15,27,28]

5.3 Post-processing and Interpretation

The Fourier transform has been at the foundation of modern NMR spectroscopy.[3] The mathematics of Fourier methods are well developed, and modern algorithms provide fast and accurate methods for the transformation. It is, however, not without its limitations. Signal processing in NMR must ultimately balance sensitivity and resolution, and in the presence of non-uniform or sparse sampling, Fourier methods face a number of limitations. RD experiments are attractive because they have the potential to recover the needed spectral information without having to collect the entirety of uniformly sampled data. Post-processing algorithms are designed to extract spectral information from the sparsely sampled RD data. RD methods can be categorized into two groups. In the first group are methods that attempt to reconstruct the high dimensional spectra, for example, by either back-projection,[7] or the maximum entropy method.[8] Conventional NMR data analysis (for example peak picking) can be followed after reconstruction of the higher dimensional spectra. The projection–reconstruction method has the advantage of offering traditional high-dimensional spectra for further analysis, compared to other methods that provide chemical shift information instead. However, reconstruction has been the subject of intense research and has been found to have limitations[29] and to exhibit an "ill-posed nature":[30] mathematicians have shown that given any finite set of plane projection data, an infinite number of functions can be reconstructed that "fit" the given data. Specifically, perfect reconstruction of a standard 3D experimental NMR dataset from a few planes is only possible if additional conditions are carefully imposed, or, alternatively, if the quantity of data is "equivalent" to that of the standard 3D experiment—in the latter case, an equivalent amount of data is a necessary but not sufficient condition. The maximum entropy method provides the additional conditions necessary for regularizing the reconstruction problem. However, the use of maximum entropy methods come at a cost. For example, while parameter estimation using maximum entropy is conceptually straightforward, expert control over the parametric family of solutions plays a decisive role.

The second group aims at retrieving the chemical shift information (NMR peaks) directly from RD planes without reconstructing the high dimensional spectra.[12,13,31] Since the final goal in some NMR experiments is to acquire chemical shift information (for example, for backbone assignments), this approach is appealing—especially because it may be able to bypass the time-consuming process of manual peak picking.[32–35] Moreover, because the lower dimensional data can leverage the standard Fourier processing methods, the transformation portion of the procedure for feature recovery is relatively straightforward. In some algorithms, the validity of the peaks can be evaluated after the collection of each new subset of the RD experiment—providing an additional check against false discovery. The specific details of chemical shift position recovery in APSY and HIFI-NMR are given below. Chemical shift position recovery, or peak picking, is a common and important task in NMR. Maximum likelihood estimation (MLE) for spectral reconstruction belongs to the class of model-based statistical approaches that can furnish robust estimates for peak position and peak intensity.

Post-processing methods play a crucial role in NMR, and adoption and adaptation of algorithms from other domains of science continue to play a positive role in enhancing our ability for recovery of spectral information. For a statistical approach to post-processing, maximum entropy is a general method of regularization that is broadly applicable, independent of the sampling strategy.[36] Other reconstruction algorithms that have been employed to improve spectral quality include the iterative CLEAN algorithm (see also Chapter 7)[21] and Bayesian reconstruction.[37–39] Bayesian approaches are potentially powerful and robust, but their computational development remains nascent. Another approach for recovery of spectral information is to match the sampling strategy with the mathematics of processing. Radial sampling[20] and Poisson disk sampling provide examples for post-processing approaches that can, under favorable conditions, reduce artifacts,[40] which can be relatively large in the vicinity of signals. In order to obtain favorable conditions, the well-known Whittaker–Shannon Sampling theorem can provide the necessary conditions involving the convolution of NMR signals with the (discrete) Fourier transform of the Poisson-sampled disk—in practice, and for the purpose of reconstructing a fully-dimensional spectrum, achieving favorable conditions is often challenging. A more recent approach, compressed sensing (CS) theory,[41,42] offers an algebraic approach for reconstructing signals from fewer measurements than required by the Nyquist sampling rule (see also Chapter 10). CS has attracted a great deal of attention in medical imaging,[43] and computer vision,[44,45] *etc.* Application of CS to NMR spectra is meaningful because NMR spectra are inherently sparse—the discrete nature of the chemical shift groups means that most regions of a multi-dimensional NMR spectra are devoid of any signal. The choice of a sparsifying transform is pivotal in CS. Intuitively, the target signal should be sparsely represented in the transform domain. Recent work by Kazimierczuk and Orekhov[16,46] and Holland *et al.*[47] proposed the use of CS in NMR. Shrot and Frydman[48] offer an approach in which indirect domain information is spatially encoded and subsequently recovered *via* oscillating field gradients (see also Chapter 2). The use and importance of CS in NMR, is likely to grow as domain-specific methods continue to be developed.[49]

5.4   HIFI-NMR

One of the challenges in protein NMR spectroscopy is to minimize the time required for multidimensional data collection. "High-resolution Iterative Frequency Identification for NMR" (HIFI-NMR)[12] presents an original solution to this problem by combining the reduced dimensionality approach with features that take advantage of the probabilistic paradigm. For 3D triple-resonance experiments of the kind used to assign protein backbone and sidechain resonances, the probabilistic algorithm used by HIFI-NMR automatically extracts the positions (chemical shifts) of peaks with considerable time-savings compared with conventional approaches to data collection, processing, and peak picking. Numerous features and capabilities have been added to the HIFI-NMR paradigm,[31–35] but we confine our discussion to its essential concepts. We begin with a brief primer on statistical post-processing, focusing mainly on the use of Bayes' rule. This material is helpful in clarifying the statistical approach used in HIFI-NMR. The language introduced here is also useful in discussing the maximum likelihood method.

5.5   Brief Primer on Statistical Post-processing

The starting assumption in the Bayesian approach is that a set of parameters, θ, can best explain the dataset D. According to Bayes' rule, we can estimate the parameters θ using the following relationship: $p(\theta|D) = p(D|\theta) \times p(\theta)/p(D)$. From this formulation, it is immediately evident that the parameters of interest have an associated probability distribution $p(\theta)$—called the prior. Moreover, the prior scales the likelihood $p(D|\theta)$—namely, the *posterior* is equal to *likelihood×prior* divided by *evidence*. In applying the maximum likelihood estimation approach, we seek a point value for θ (a single value) that maximizes the likelihood, $p(D|\theta)$, shown in the equation above. We can denote this value as $\theta*$—this is a point estimate and not a random variable. From these equations it can be seen that MLE treats the term $p(\theta)/p(D)$ as a constant. Two equivalent interpretations of this fact are: (1) no prior belief enters the computation or (2) all parameters are equally likely. Bayesian estimation, by contrast, fully calculates (or at times approximates) the posterior distribution $p(\theta|D)$. Bayesian inference treats θ as a random variable, and, therefore, knowledge about $p(\theta)$ (the prior) is required. One trade-off is that the estimation in the Bayesian case is far more complex. In return, the outcome $p(\theta|D)$ is information-rich. HIFI-NMR uses the Bayes formula and Bayesian updating in order to find optimal reduced dimensionality projections.

In HIFI-NMR, the Bayes formula is used to ascertain the probability of observing a new peak in plane A (projection A) given that another event (peaks in the plane at projection angle B) has already been observed. We are specifically interested in answering the following question: What is the probability of observing a specific peak x, given some prior data D about the peak? Using the Bayes formula from above, we have:

$$P(x|D) = \frac{P(D|x)P(x)}{P(D)}$$

The presence of a peak x is our hypothesis, and D is the data related to the experiment. Our goal is to obtain the probability of a hypothesis (peak is present) given the evidence (earlier data) and our initial (prior) beliefs regarding that hypothesis. A key difference between Bayesian and classical statistics explains the reason for using Bayesian updating in HIFI-NMR.[50] Classical statistics is concerned with $P(D|x)$, *i.e.* if the peak is not present, what is the probability of observing the data? In HIFI-NMR, we are interested in the probability of the hypothesis (peak is present) given the data—therefore, the Bayesian approach is appropriate.

5.6   HIFI-NMR Algorithm

The HIFI-NMR algorithm (Figure 5.2) starts with the collection of NMR data from two orthogonal 2D planes. In cases where ^1H–^{15}N and ^1H–^{13}C 2D planes are common to multiple experiments, they need only be collected once. The algorithm then moves to the list of experiments specified for data collection and starts with the first experiment. For this experiment, tilted planes are selected adaptively, one-by-one, by use of probabilistic predictions. A robust statistical algorithm extracts information from each plane, and results are incorporated into the online algorithm, which maintains a

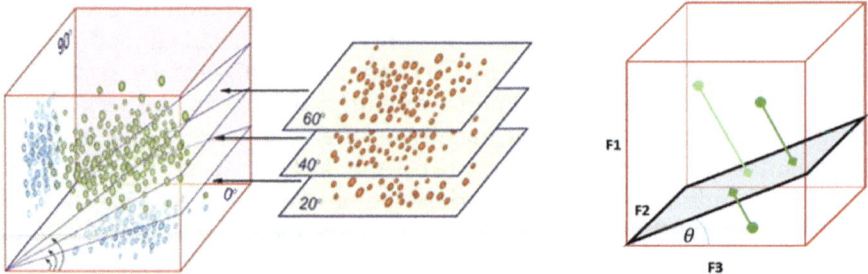

Figure 5.5 HIFI-NMR uses an iterative approach to recover the peaks in 3D spectra (green peaks) from the peaks in two orthogonal planes (blue peaks) plus a minimal number of 2D tilted planes (red peaks) collected one-by-one at optimal angles determined from analysis of the prior data collected. Projection of peaks in the 3D space along vectors orthogonal to the tilted plane at angle θ is based on the principle of simultaneously evolving chemical shifts in HIFI-NMR.

probabilistic representation of peak positions in 3D space. The updated probabilistic representation of the peak positions is used in deriving the optimal angle for the next plane to be collected. The online software predicts whether the collection of this additional data plane will improve the model of peak positions. If the answer is "yes", the next plane is collected, and the results are incorporated into the evolving model; if the answer is "no", data collection is terminated for the experiment. The algorithm then goes on to the next NMR experiment in the list to be done. Once all experiments have been run, peak lists with associated probabilities are generated directly for each experiment, without total reconstruction of the three-dimensional (3D) spectrum. These peak lists are ready for use in subsequent steps in peak assignment or structure determination.

In achieving the steps outlined above, the probabilistic algorithm in HIFI-NMR divides the 3D spectral space into discrete voxels on the basis of prior information: the number of expected peaks, the nuclei involved and their expected chemical shift ranges, and the data collection parameters. Using this discrete representation, an empirical estimate is obtained for the prior probability distribution of peak locations in multidimensional spectral space. Let $P_{pr}(X)$ denote the prior probability for the presence of a peak at position X, where the elements of the vector $X = (x_1, x_2, x_3)$ represent the voxels in *three*-dimensional space. The initial step in the HIFI approach is to collect data for the two orthogonal planes (0° and 90° planes). For experiments involving ^1H, ^{15}N, and ^{13}C dimensions (for example, HNCO, HNCACB), the 0° (^{13}C–^1H) and the 90° (^{15}N–^1H) planes are collected and peak-picked. Because the ^1H dimension is common to the two planes, the possible candidates for peaks in 3D space can be generated by considering all combinatorial possibilities for peaks in the two planes that have the same ^1H chemical shift within a given tolerance. This generates many more candidates than actual peaks, but will include most, but not all, real peaks.

These candidate peaks retain their prior probability associated to their chemical shift coordinates in spectral space generated in the earlier step. At the end of this stage, $P_0(x)$ designates the new probability distribution, which includes information gained about the probability of candidate peaks at the specified chemical shift coordinate.

The next-best tilted plane is determined by assuming that the candidate peaks have been observed on a plane at angle θ and by evaluating the information theoretic impact of this observation on the probability of the peaks. The measure of this influence is the divergence S_θ between the initial probability $P_0(x)$ (probability at the initial or 0th step) and the predicted probability after the impact of the projection at angle θ has been taken into account $Q_\theta(x)$:

$$S_\theta = -\sum Q_\theta \ln \frac{Q_\theta}{P_0}$$

where the sum is taken over all voxels in spectral space. By assuming a uniform prior over all plane selections, the maximum information approach chooses the plane with the highest "dispersion score". The dispersion score is derived by projecting the candidate peaks into the tilted planes and measuring the weighted dispersion of the resulting projection. Ambiguous candidate peaks with roughly equal probability of being "real" *versus* "noise" are assigned higher weights in the scoring function than candidate peaks with high probabilities of being either "real" or "noise".

Once the peak positions from a new tilted plane at the chosen optimal angle θ have been obtained, the projection of the current set of candidate peaks in 3D space onto that tilted plane is the map G that sends the chemical shift coordinates (x_1, x_2, x_3) to $(x_1, x_2\cos(\theta) + x_3\sin(\theta))$. HIFI-NMR first discretizes the coordinates and obtains the associated voxel labels. Next, it examines each candidate peak in turn, in light of the data from the new tilted plane and prior data, and updates its probability on the basis of Bayes' rule. To do this, HIFI-NMR compares a candidate peak x that arises from the projection $G(x)$ with the set of automatically picked peaks $a_1, a_2, a_3, \ldots, a_k$ (treated as distributions) in the latest tilted plane θ in terms of an indicator function:

$$x\_observed\_in\_\theta \equiv (\exists a_i \ni d(G(x), a_i) < \delta_i)$$

in which d represents the distance based on a normalized Euclidean measure and δ_i is the allowable statistical deviation of peak a_i. In intuitive language, measuring the event $x\_observed\_in\_\theta$ is equivalent to asking whether or not the peak x has been observed in the new tilted plane at angle θ. The use of a normalized Euclidean metric is desirable, because ^1H and the mixture of ^{15}N and ^{13}C have different tolerance values and the tolerance values vary from plane to plane.

Next, each candidate peak x is examined in turn, in light of the data from the new tilted plane and prior data, and Bayes' rule is used to update its probability $P_0(x)$ of the peak being "real":

$$P_1(x) \equiv P(x \mid x\_observed\_in\_\theta) = \frac{P(x\_observed\_in\_\theta \mid x) \cdot P_0(x)}{P(x\_observed\_in\_\theta)}$$

or

$$P_1(x) \equiv P(x \mid x\_not\_observed\_in\_\theta) = \frac{P(x\_not\_observed\_in\_\theta \mid x) \cdot P_0(x)}{P(x\_not\_observed\_in\_\theta)}$$

In the above, $x\_not\_observed\_in\_\theta$ is the complement even to t $x\_observed\_in\_\theta$. The term in the denominator, $P(x\_observed\_in\_\theta)$ [or $P(x\_not\_observed\_in\_\theta)$], represents the probability of a voxel x in plane θ to be considered a peak (or not a peak), regardless of any other consideration. The probabilities $P(x\_observed\_in\_\theta/x)$ and $P(x\_not\_observed\_in\_\theta/x)$ do not experience much angular variation and can be reasonably approximated by the empirical data.

This process of Bayesian updating is continued until no further information gain can be measured. At this stage, n peaks with probabilities above a threshold are reported as the peak list. The value of n is estimated automatically from the covalent structure of the protein, the type of experiment, and the probability distribution $P(x)$. The value of n can be adjusted in a subsequent post-processing step to include additional expert information.

Once all peaks from a given NMR experiment have been identified, the process is repeated for all NMR experiments specified by the user: for example, the set of experiments needed to achieve backbone assignments (Figure 5.3). The only user-specified parameters in HIFI-NMR is the list of experiments for which the user wishes to acquire data. No other parameters or adjustments are necessary—all other required parameters are adaptively computed by the software. The choice of "optimal" plane can be a significant factor in minimizing peak overlaps in spectra (Figure 5.4). A later enhancement of HIFI-NMR allows planes from the list of experiments provided by the user to be interleaved.[32] In this scenario, the software automatically calculates the most "informative" plane–experiment combination and then performs the necessary experiment. Moreover, a secondary "objective function"—for example, achieving backbone assignment within a range of completion—can be specified. As in the case of HIFI-NMR, no additional user-specified parameters are needed. In this sense, the user operational characteristics of the software are simple and straightforward. As expected, if the planes of experiments are allowed to be interleaved, larger time-savings and completion rates can be achieved.[34]

The statistical approach of HIFI-NMR has proven useful in another application area in which the robustness of analysis results needed to be validated—specifically, the experimental determination of NMR coupling

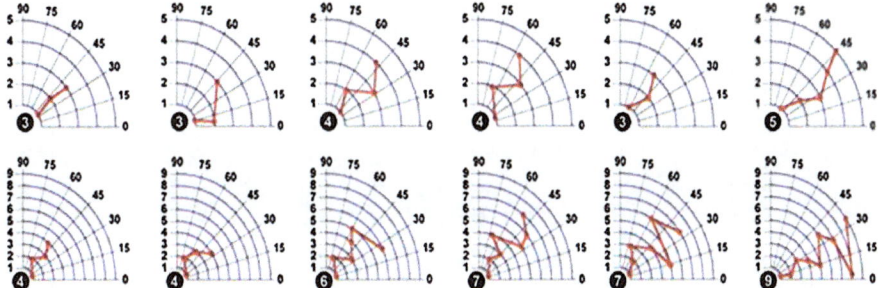

Figure 5.3   Choice of optimal angles for recovery of peaks is dependent on the specific protein under study as well as the experiment. Proteins have unique characteristics, and this is reflected in the order and number of tilt angles required for optimal data collection. In this figure the order of experiments is fixed (from left to right: HNCO, HN(CO)CA CBCA(CO)NH HNCA HN(CA)CB HNCACB). Each point on the grid specifies, in the order of selection, one data collection point at the specified tilt angle. Note the significant differences between the data collection schedules for ubiquitin (76 amino acids, top row) and flavodoxin (176 amino acids, bottom row).

constants. HIFI-C[31] represents the extension of HIFI-NMR's adaptive and intelligent data collection approach to the rapid and robust collection of coupling data. In this case, data collected from one or more optimally tilted 2D planes are used to determine couplings with high resolution and precision in less time than required by a 3D experiment. A further benefit of the approach is that data from independent planes provide a statistical measure of reliability for each measured coupling. Extensions of HIFI-NMR that integrate other steps of NMR structure determination, data processing, and data acquisition have been successfully implemented and applied to protein structure determination, further demonstrating the strength of the statistical approach. A fully integrated software package that includes HIFI-NMR within a framework that provides a complete path to structure determination has become recently available[51] and can be freely downloaded.

5.7   Automated Projection Spectroscopy

APSY (automated projection spectroscopy) combines the technique for RD NMR experiments[13,52,53] with automated peak-picking of the projected spectral data and a subsequent geometric analysis of the peak lists with the algorithm GAPRO (geometric analysis of projections). Geometrical relationships between the peak in the *N*-dimensional space and its lower dimensional projections is used by GAPRO for identifying peaks in the projections that arise from the same resonance in the *N*-dimensional frequency space (Figure 5.5). The same relationships enable the subsequent calculation of the positions of the peaks in the *N*-dimensional spectral space. Similar to HIFI-NMR, the output of an APSY experiment is thus an *N*-dimensional

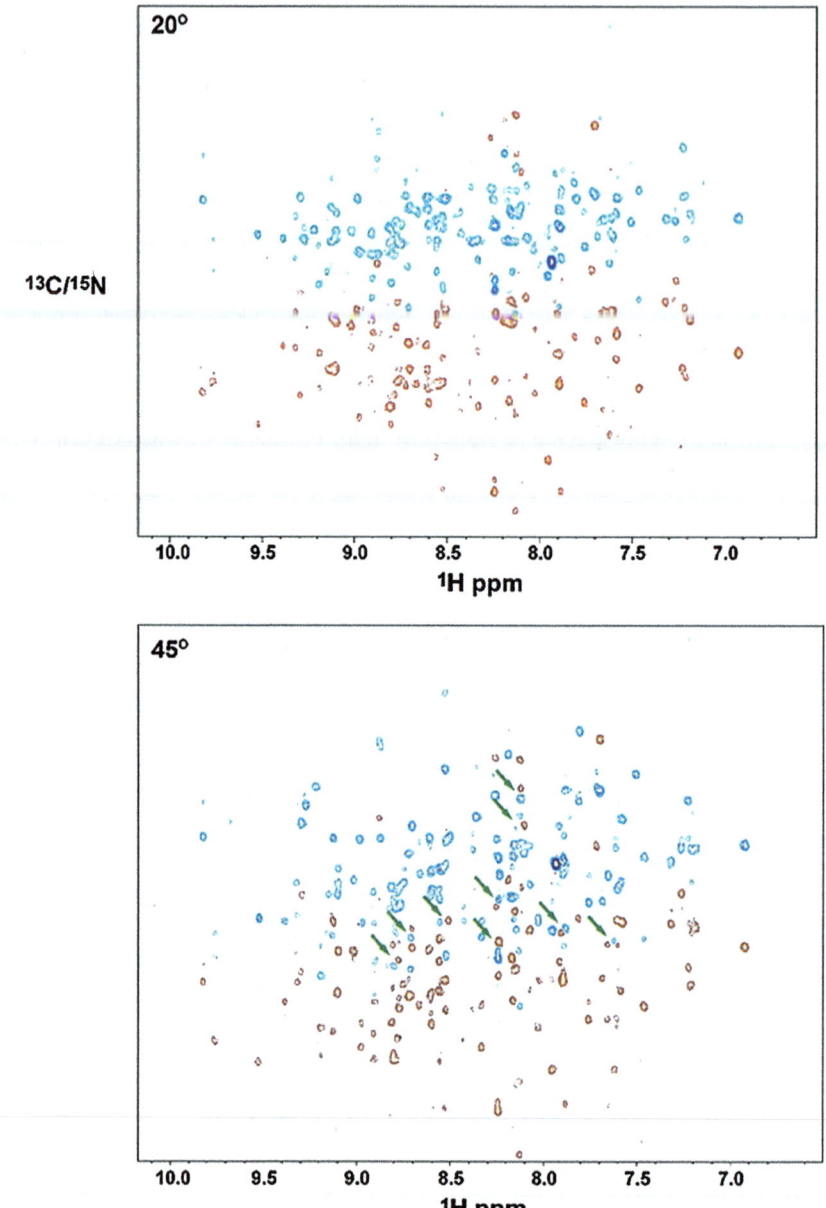

Figure 5.4 Two 2D tilted planes from the HNCACB HIFI-NMR spectrum of the mouse protein (1X03, 101 amino acids) collected as described in the text. The 2D spectrum for the optimal first plane predicted by our algorithm at 20° is compared with the 2D spectrum for 45°. The arrows indicate the positions of signals that have been cancelled because of spectral overlaps.

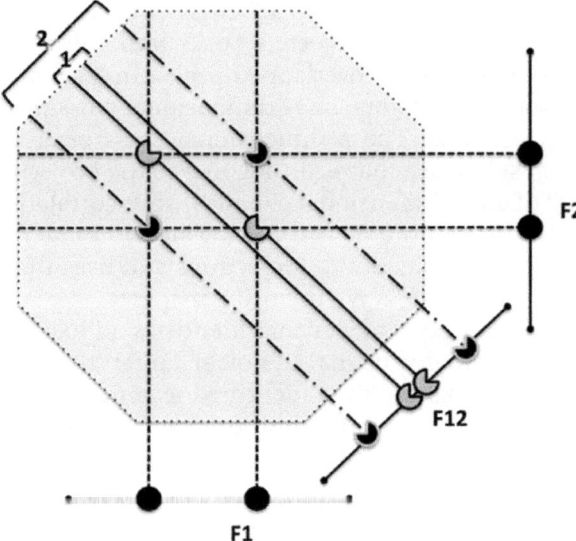

Figure 5.5   Projection onto a non-orthogonal plane can separate overlapped peaks.

chemical shift correlation list of high quality, which allows efficient and reliable subsequent use by computer algorithms. Since APSY outputs only peaks it deems as high quality, the output peak list does not maintain a probability column for the peak list, as is done in HIFI-NMR. Similar to HIFI-NMR, APSY is fully automated and operates without the need to reconstruct the high-dimensional spectrum at any point. In APSY, the experimental setup is defined before the start of the experiment and once data acquisition commences the experiment proceeds, without any adaptive steps, toward completion. For automated use, APSY provides users with guidelines that aid experimental setup. This is helpful because in a given NMR experiment (HNCO, HNCA, *etc.*), the intensity of a specific multi-dimensional NMR signal varies in each individual projection of an APSY experiment. APSY suggests an empirical formalism, linking parameters that impact the signal-to-noise ratio, in order to optimize the performance of APSY-NMR experiments. The suggested formula depends on the dimensionality of the experiment, the dimensionality of the projected space, spin type (protons, carbons, *etc.*), expected signal intensity at time zero, and projection angles, probe sensitivity, and window function applied before Fourier transform. The value of some parameters can be obtained through 1D NMR experiments or model calculations. The total number of projections must be at least *N*, the dimensionality of the experiment, and each indirect dimension needs to be evolved at least once in the set of projections. The APSY experimental protocol contains useful heuristics for setting up the experiment and improving experimental outcomes, including the choice of projection angles and the number of scans.

In order to evaluate the projection of cross peaks, APSY relies on the geometric relationship of the projected peaks and the notion that non-orthogonal planes can separate overlapped peaks. In a set of m projections along different projection vectors, an N-dimensional cross peak is projected onto m different locations. The N-dimensional cross peak is located in a lower dimensional sub-space that is orthogonal to the projection plane. The "peak subgroup" of an N-dimensional chemical shift correlation is the set of projected peaks that arise from it. The GAPRO algorithm identifies the peak subgroups in the m peak lists of the projections and uses them to calculate the coordinates of the peaks in N-dimensional space.

GAPRO implements a peak picker that identifies all local maxima of the spectrum with a sensitivity (signal-to-noise) larger than a user-defined value R_{min}. Every local maximum is identified as a peak. In particular, the GAPRO peak picker does not attempt to distinguish real peaks from spectral artifacts or random noise because the chances of the same artifact appearing in two different projections is small, therefore mitigating high false discovery rates. The implementation of this idea is achieved through a procedure GAPRO calls a secondary peak filter. To account for the imprecision in the picked peak positions owing to thermal noise, the calculation of intersections of subspaces allows a user-defined tolerance value in the direct dimension, D_{nmin}. Once the algorithm identifies the list of identified peak subgroups, the combination of subgroups are matched in order to recover the consistent peak positions in N-dimensional space. Choosing the proper number of projections in order to achieve improved data collection times while maintaining the needed sensitivity is an important factor. In APSY, this number is a user-provided parameter, and APSY provides heuristic advice for the selection of the number projections needed.

5.8 Fast Maximum Likelihood Method

APSY and HIFI-NMR rely on the detection of peaks and estimation of peak positions in several projection in order to achieve reliable results. Fast maximum likelihood reconstruction (FMLR) is an iterative optimization approach for solving the maximum likelihood estimation problem.[14] FMLR is inherently a statistical approach that achieves a "point estimate" (maximum) that is optimal for the proposed model. The model in this case is a linear combination (mixture) of basis functions that contain peak position and intensities. The mixture model is highly useful in practice—despite its shortcomings—which partly explains its popularity in many application domains.[54–57] Here, we provide a brief discussion of the mixture model to emphasize the view that the maximization algorithm performs a separate function that can be applied to any proposed model—namely, it maximizes a given cost function. This viewpoint is useful for understanding how the machinery can be applied to various NMR problems.

5.9   Mixture Models

Construction of a model is among the key steps for obtaining a statistical representation of observed data. The model used for obtaining a maximum likelihood (MLE) estimate is an example of what is commonly referred to as a "mixture model" in statistics.[56] In the absence of noise, mixture models are represented as the finite weighted sum of elementary functions. In the presence of noise, the estimation of NMR spectral parameters in a mixture model is commonly based on the assumption of white complex Gaussian noise in the signal. In general, noise in the sampled signal is not strictly white, even if the thermal noise in the receiver prior to digitization can be characterized as white Gaussian noise. However, white noise is generally considered to be a reasonable assumption. In the case of biomolecular NMR, the elementary functions are often complex (decaying) exponentials, but the same formalism can be applied, in principle, to any set of elementary functions. For a simple example, suppose we consider Gaussians as templates for the elementary functions. Then, a mixture of Gaussians is represented as:

$$p(x; \psi) = \sum_{j=1}^{k} w_j \phi(x; \mu_j, \sigma_j)$$

Here, $\phi(x; \mu_j, \sigma_j)$ denotes a Gaussian density with mean μ_j and variance σ_j. The weights $\{w_1 \ldots w_k\}$ are non-negative and sum to 1. One can also consider k, the number of components, to be an additional parameter. If we are now given a discrete set of observed data, $X = \{x_1, x_2, \ldots, x_n\}$, an optimization problem can be envisaged in which the goal is to minimize (in some norm) the difference between data and the model—that is, $\mathrm{argmin}_{\{\mu_j, \sigma_j, w_j, k\}} \| p(x; \psi) \|$. Use of mixture models is commonplace because they are considered to be intuitive, very useful, and practical. These models are described by a finite number of parameters, and the goal of estimation is to find the parameters that optimize, in a specified way, the model's description of the data. In this section we discuss some properties of these models and discuss the reasons for careful evaluation of their use. The presentation below is simplified in order to keep the technicalities manageable and to focus on the concepts.

For NMR, the specific form of the mixture model can be represented as follows:

$$p(x; \psi) = \sum_{j=1}^{k} w_j \phi(x; \mu_j, \sigma_j)$$

Where w_j are complex parameters and ϕ is the harmonic component. The form above is more general and can accommodate a variety of line shapes. For comparison, note that for Lorentzian line shapes, it is customary to write this formula as follows:

$$p_m(x; \psi) = \sum_{j=1}^{k} (d_j e^{i\theta_j} e^{-m\Delta t/\tau_j}) e^{2\pi m\Delta t \omega_j}$$

In this form, the maximum likelihood estimate is effectively an exponential fitting problem in which we seek to recover estimates for the parameter of a fit. A noteworthy observation at this point is that the MLE does not require regular sampling—it can be applied to non-uniformly sampled data. It is also worth noting that MLE mixture models face certain mathematical and numerical challenges.[58] Among these are unstable boundary values (infinite likelihoods), multimodality (leading to many local optima), non-identifiability (subspaces of the parameter space where many solutions exists), and irregularity (where convergence fails in the statistical sense). Nonetheless, with careful practice and user verification, the maximum likelihood approach to spectral deconvolution offers useful tools that can be used to for quantitative approaches. The specific numerical approach used by FMLR is iterative and uses the equivalence to the exponential fitting problem to cast the maximum likelihood estimate as an optimization problem.

5.10 FMLR Algorithm

The FMLR algorithm embodied in the software package Newton[59] uses the NMRPipe processing package[60] to perform the initial steps of NMR data processing consisting of (i) data conversion, (ii) apozidation, (iii) zero-filling, (iv) Fourier transformation, (v) phase correction, and (optionally) (vi) polynomial baseline correction. The apozidation and zero-filling parameters in steps ii–iv partially define an operator $\hat{F}$ that can be applied identically to both the acquired FID and a model FID in order to convert discrete basis functions in the time domain to discrete basis functions in the frequency domain. For each dimension of the acquired spectrum, the Newton software determines the $\hat{F}$ operator along each dimension (for example, $\hat{F}_1$ and $\hat{F}_2$ for a 2D experiment) by parsing the input NMRPipe processing scripts (see FMLR section). The theoretical basis for the approach follows the maximum likelihood method as described in the MLM primer above. The current implementation addresses practical issues pertinent to quantitation, and adds the potential for easy extension of the software in order to process higher dimensional spectra. The spectral representation of basis signals is a function that represents line shapes that are broader than a Lorentzian (power of 1), and narrower than a Gaussian—the fractional power is an estimated number between 1 and 2.

The method involves construction of a "data ensemble" consisting of three distinct spectra: (1) a data spectrum, (2) a model spectrum and (3) a residual spectrum (Figure 5.6). The data spectrum is calculated from conventional Fourier processing of the acquired FID. The processing operations define a digital transform operator along each dimension ($\hat{F}_1$ and $\hat{F}_2$, for a 2D spectrum). The data spectrum is computed by using a sequential application of the operators ($\hat{F}_1$ and $\hat{F}_2$) prior to FMLR analysis. The model spectrum is obtained by applying the same operators to the model time-domain signal in the current iteration. The relevant portion of this signal (the portion

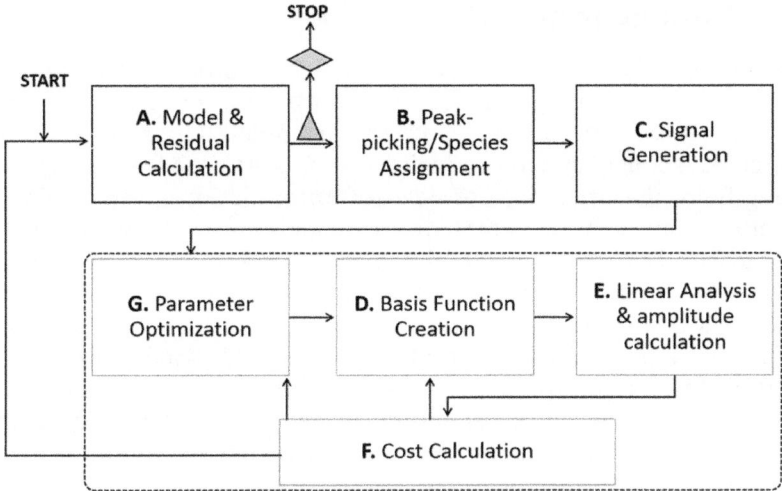

Figure 5.6 FMLR spectral deconvolution consists of steps A–G. (A) Calculation of model and residual spectra. (B) A set of most probable signals is identified by peak picking of the residual spectrum at a defined threshold. The coordinates of the peak are used to generate initial frequency values for the new signals. Parameter estimates (line widths) are obtained. (C) Signals are created for each of the peaks. A signal is constructed from assignment of initial values and constraints for each parameter. (D) Model time-domain functions are constructed from the current set of frequency and decay rate parameters for all signals. (E) The amplitudes for the current set of signals are calculated from linear least squares analysis. (F) The cost function is calculated independently for each signal group from a scalar quantity that is analogous to a "sufficient statistic"—a simplified form of the least squares statistic χ^2 with the term representing the square of the data values removed. (G) Step G is an optimization that invokes steps D–F within an inner loop. Using derivatives of this cost function with respect to each non-linear parameter, a simple gradient search algorithm is used to determine the most likely set of frequency and decay rate parameters for a fixed number of signals. This is the final step of the outer iteration loop as the optimized model resulting from this step is used as input to calculate the model and residual spectra in Step A.

coinciding with the spectral window of the data spectrum) is the portion used for further analysis. The residual spectrum is used for two purposes: (1) to estimate the noise variance and (2) to locate new signals to add to the current model by peak picking at a given threshold. The residual spectrum is the algebraic difference between the data and model spectrum. The overall algorithm consists of sequential steps A–G (see Figure 5.6). The algorithm iteratively builds the model by addition of signals from analysis of the residual. Steps D–F form an "inner loop", which represents optimization of the parameters for a fixed set of signals (step G), derived from signals identified in Steps B and C.

5.11 Conclusions and Outlook

While standard data collection methodologies in NMR are integrally linked to computation—through Fourier transform, for example—RD experiments, and more generally non-uniformly sampled data sets, strongly rely on efficient, robust, and reproducible computation. As with any rapidly evolving field, the brisk pace of new software, platform, terminology and conceptual constructs in NMR spectroscopy continues to present challenges to the scientific community. Lack of common and reproducible platforms for properly integrating the techniques into research workflows and a methodology for fully vetting the strengths and weaknesses of each approach are among key issues that need to be addressed. Improving data collection times, attaining better sensitivity, and enhanced resolution will continue to motivate the development of new approaches. Moreover, the confluence of technological improvements that drive the rapid expansion of novel and creative approaches in NMR spectroscopy will demand tighter integration with computational methodologies and more reproducible platforms. Many of the more advanced techniques and methodologies mentioned here have yet to gain favor in the community, possibly owing to issues related to ease of use, utility, and reproducibility. For example, as noted above, the conceptual separation between data collection and postprocessing has often been overlooked. Moreover, this conceptual blurring of the line between data acquisition and data processing, when carried forward to the software design and implementation level, creates the larger challenge of robustness and future reproducibility. We expect that the reality of open and platform-independent software platform will motivate the kind of future development that not only mitigates the steep learning curves but also addresses the concerns of creating reproducible platforms.

References

1. F. Bloch, Nuclear induction, *Phys. Rev.*, 1946, **70**(7–8), 460.
2. N. E. Jacobsen, *NMR Spectroscopy Explained: Simplified Theory, Applications and Examples for Organic Chemistry and Structural Biology*, 2007.
3. R. R. Ernst, G. Bodenhausen and A. Wokaun, *Principles of Nuclear Magnetic Resonance in One and Two Dimensions*, 1987.
4. P. Styles, N. F. Soffe, C. A. Scott, D. A. Cragg, F. Row, D. J. White and P. C. J. White, A high-resolution NMR probe in which the coil and preamplifier are cooled with liquid helium, *J. Magn. Reson.*, 2011, **213**(2), 347–354.
5. D. Rovnyak, J. C. Hoch, A. S. Stern and G. Wagner, Resolution and sensitivity of high field nuclear magnetic resonance spectroscopy, *J. Biomol. NMR*, 2004, **30**(1), 1–10.
6. *Novel Sampling Approaches in Higher Dimensional NMR*, ed. M. Billeter and V. Orekhov, 2012.

7. E. Kupče and R. Freeman, Projection-reconstruction technique for speeding up multidimensional NMR spectroscopy, *J. Am. Chem. Soc.*, 2004, **126**(20), 6429–6440.

8. M. Mobli, A. S. Stern and J. C. Hoch, Spectral reconstruction methods in fast NMR: reduced dimensionality, random sampling and maximum entropy, *J. Magn. Reson.*, 2006, **182**(1), 96–105.

9. S. Kim and T. Szyperski, GFT NMR, a new approach to rapidly obtain precise high-dimensional NMR spectral information, *J. Am. Chem. Soc.*, 2003, **125**(5), 1385–1393.

10. T. Szyperski, D. C. Yeh, D. K. Sukumaran, H. N. Moseley and G. T. Montelione, Reduced-dimensionality NMR spectroscopy for high-throughput protein resonance assignment, *Proc. Natl. Acad. Sci.*, 2002, **99**(12), 8009–8014.

11. E. Kupče and R. Freeman, Resolving ambiguities in two-dimensional NMR spectra: the 'TILT' experiment, *J. Magn. Reson.*, 2005, **172**(2), 329–332.

12. H. R. Eghbalnia, A. Bahrami, M. Tonelli, K. Hallenga and J. L. Markley, High-resolution iterative frequency identification for NMR as a general strategy for multidimensional data collection, *J. Am. Chem. Soc.*, 2005, **127**(36), 12528–12536.

13. S. Hiller, F. Fiorito, K. Wüthrich and G. Wider, Automated projection spectroscopy (APSY), *Proc. Natl. Acad. Sci. U. S. A.*, 2005, **102**(31), 10876–10881.

14. R. A. Chylla, K. Hu, J. J. Ellinger and J. L. Markley, Deconvolution of two-dimensional NMR spectra by fast maximum likelihood reconstruction: application to quantitative metabolomics, *Anal. Chem.*, 2011, **83**(12), 4871–4880.

15. M. W. Maciejewski, M. Mobli, A. D. Schuyler, A. S. Ster and J. C. Hoch, Data sampling in multidimensional NMR: fundamentals and strategies. *Novel Sampling Approaches in Higher Dimensional NMR*, 2011, pp. 49–77.

16. K. Kazimierczuk and V. Orekhov, Non-uniform sampling: post-Fourier era of NMR data collection and processing, *Magn. Reson. Chem.*, 2015, **53**(11), 921–926.

17. G. Bodenhausen and A. R. R. Ernst, The accordion experiment, a simple approach to three-dimensional NMR spectroscopy, *J. Magn. Reson.*, 1969, **45**(2), 363–373.

18. Y. Shen, H. S. Atreya, G. Liu and T. Szyperski, G-matrix Fourier transform NOESY-based protocol for high-quality protein structure determination, *J. Am. Chem. Soc.*, 2005, **127**(25), 9085–9099.

19. E. Kupče and R. Freeman, The Radon transform: a new scheme for fast multidimensional NMR, *Concepts Magn. Reson., Part A*, 2004, **22**(1), 4–11.

20. B. E. Coggins and P. Zhou, Sampling of the NMR time domain along concentric rings, *J. Magn. Reson.*, 2007, **184**(2), 207–221.

21. B. E. Coggins and P. Zhou, High resolution 4-D spectroscopy with sparse concentric shell sampling and FFT-CLEAN, *J. Biomol. NMR*, 2008, **42**(4), 225–239.

22. J. C. J. Barna, E. D. Laue, M. R. Mayger, J. Skilling and S. J. P. Worrall, Exponential sampling, an alternative method for sampling in two-dimensional NMR experiments, *J. Magn. Reson.*, 1987, **73**(1), 69–77.

23. D. Rovnyak, D. P. Frueh, M. Sastry, Z.-Y. J. Sun, A. S. Stern, J. C. Hoch and G. Wagner, Accelerated acquisition of high resolution triple-resonance spectra using non-uniform sampling and maximum entropy reconstruction, *J. Magn. Reson.*, 2004, **170**(1), 15–21.

24. V. Y. Orekhov, I. Ibraghimov and M. Billeter, Optimizing resolution in multidimensional NMR by three-way decomposition, *J. Biomol. NMR*, 2003, **27**(2), 165–173.

25. M. Mobli, M. W. Maciejewski, M. R. Gryk and J. C. Hoch, Automatic maximum entropy spectral reconstruction in NMR, *J. Biomol. NMR*, 2007, **39**(2), 133–139.

26. S. Sun, M. Gill, Y. Li, M. Huang and R. A. Byrd, Efficient and generalized processing of multidimensional NUS NMR data: the NESTA algorithm and comparison of regularization terms, *J. Biomol. NMR*, 2015, **62**(1), 105–117.

27. S. G. Hyberts, H. Arthanari and G. Wagner, Applications of non-uniform sampling and processing. *Novel Sampling Approaches in Higher Dimensional NMR*, 2012, pp. 125–148.

28. E. J. Candès, Compressive sampling. *Proceedings of the International Congress of Mathematcians*, 2006, vol. 3, pp. 1433–1452.

29. R. J. Gardner, *Geometric Tomography*, 1995.

30. R. B. Marr, *An overview of image reconstruction*, Proc. Internal. Symp. on Ill-posed problems, Univ. of Delaware, Newark, DE., 1979.

31. G. Cornilescu, A. Bahrami, M. Tonelli, J. L. Markley and H. R. Eghbalnia, HIFI-C: a robust and fast method for determining NMR couplings from adaptive 3D to 2D projections, *J. Biomol. NMR*, 2007, **38**(4), 341–351.

32. A. Bahrami, M. Tonelli, S. C. Sahu, K. K. Singarapu, H. R. Eghbalnia and J. L. Markley, Robust, integrated computational control of NMR experiments to achieve optimal assignment by ADAPT-NMR, *PloS One*, 2012, **7**(3), e33173.

33. W. Lee, K. Hu, M. Tonelli, A. Bahrami, E. Neuhardt, K. C. Glass and J. L. Markley, Fast automated protein NMR data collection and assignment by ADAPT-NMR on Bruker spectrometers, *J. Magn. Reson.*, 2013, **236**, 83–88.

34. W. Lee, A. Bahrami and J. L. Markley, ADAPT-NMR Enhancer: complete package for reduced dimensionality in protein NMR spectroscopy, *Bioinformatics*, 2013, **29**(4), 515–517.

35. H. Dashti, M. Tonelli and J. L. Markley, ADAPT-NMR 3.0: utilization of BEST-type triple-resonance NMR experiments to accelerate the process of data collection and assignment, *J. Biomol. NMR*, 2015, 1–6.

36. S. Sibisi, J. Skilling, R. G. Brereton, E. D. Laue and J. Staunton, Maximum entropy signal processing in practical NMR spectroscopy, *Nature*, 1984, **311**, 446–447.

37. G. L. Bretthorst, Nonuniform sampling: Bandwidth and aliasing, *Concepts Magn. Reson., Part A*, 2008, **32**(6), 417–435.

38. J. W. Yoon and S. J. Godsill, Bayesian inference for multidimensional NMR image reconstruction, Signal Processing Conference, 2006 14th European. *IEEE*, 2006, pp. 1–5.

39. J. W. Yoon, S. Godsill, E. Kupče and R. Freeman, Deterministic and statistical methods for reconstructing multidimensional NMR spectra, *Magn. Reson. Chem.*, 2006, **44**(3), 197–209.

40. S. G. Hyberts, A. G. Milbradt, A. B. Wagner, H. Arthanari and G. Wagner, Application of iterative soft thresholding for fast reconstruction of NMR data non-uniformly sampled with multidimensional Poisson Gap scheduling, *J. Biomol. NMR*, 2012, **52**(4), 315–327.

41. E. J. Candès and M. B. Wakin, An introduction to compressive sampling, *Signal Process. Mag. IEEE*, 2008, **25**(2), 21–30.

42. E. J. Candès, J. Romberg and T. Tao, Robust uncertainty principles: Exact signal reconstruction from highly incomplete frequency information, Information Theory, *IEEE Transactions on* 2006, **52**(2), 489–509.

43. M. Lustig, D. Donoho and J. M. Pauly, Sparse MRI: The application of compressed sensing for rapid MR imaging, *Magnetic resonance in medicine*, 2007, **58**(6), 1182–1195.

44. M. F. Duarte, M. A. Davenport, D. Takhar, J. N. Laska, T. Sun, K. E. Kelly and R. G. Baraniuk, Single-pixel imaging via compressive sampling, *IEEE Signal Process. Mag.*, 2008, **25**(2), 83–88.

45. R. G. Baraniuk, V. Cevher, M. F. Duarte and C. Hegde, Model-based compressive sensing, *Inf. Theory, IEEE Trans.* 2010, **56**(4), 1982–2001.

46. K. Kazimierczuk and V. Y. Orekhov, A comparison of convex and non-convex compressed sensing applied to multidimensional NMR, *J. Magn. Reson.*, 2012, **223**, 1–10.

47. D. J. Holland, M. J. Bostock, L. F. Gladden and D. Nietlispach, Fast multidimensional NMR spectroscopy using compressed sensing, *Angew. Chem.*, 2011, **123**(29), 6678–6681.

48. Y. Shrot and L. Frydman, Compressed sensing and the reconstruction of ultrafast 2D NMR data: Principles and biomolecular applications, *J. Magn. Reson.*, 2011, **209**(2), 352–358.

49. M. J. Bostock, D. J. Holland and D. Nietlispach, Compressed sensing reconstruction of undersampled 3D NOESY spectra: application to large membrane proteins, *J. Biomol. NMR*, 2012, **54**(1), 15–32.

50. H. E. Kyburg, Bayesian and non-Bayesian evidential updating, *Artifi. Intell.*, 1987, **31**(3), 271–293.

51. H. Dashti, W. Lee, M. Tonelli, C. C. Cornilescu, G. Cornilescu, F. M. Assadi-Porter, W. M. Westler, H. R. Eghbalnia and J. L. Markley, NMRFAM-SDF: a protein structure determination framework, *J. Biomol. NMR*, 2015, 1–15.

52. S. Hiller, R. Joss and G. Wider, Automated NMR assignment of protein side chain resonances using automated projection spectroscopy (APSY), *J. Am. Chem. Soc.*, 2008, **130**(36), 12073–12079.

53. S. Hiller, G. Wider and K. Wüthrich, APSY-NMR with proteins: practical aspects and backbone assignment, *J. Biomol. NMR*, 2008, **42**(3), 179–195.

54. R. A. Chylla and J. L. Markley, Theory and application of the maximum likelihood principle to NMR parameter estimation of multidimensional NMR data, *J. Biomol. NMR*, 1995, **5**(3), 245–258.

55. S. Umesh and D. W. Tufts, *Estimation of parameters of exponentially damped sinusoids using fast maximum likelihood estimation with application to NMR spectroscopy data*, *Signal Process., IEEE Trans.* 1996, **44**(9), 2245–2259.

56. M. West, Mixture models, Monte Carlo, Bayesian updating, and dynamic models, *Comput. Sci. Stat.*, 1993, 325.

57. R. Brame, D. S. Nagin and L. Wasserman, Exploring some analytical characteristics of finite mixture models, *J. Quant. Criminol.*, 2006, **22**(1), 31–59.

58. L. Wasserman, Asymptotic inference for mixture models using data-dependent priors, *J. R. Stat. Soc., Ser. B Stat. Methodol.*, 2000, 159–180.

59. R. A. Chylla, R. Van Acker, H. Kim, A. Azapira, P. Mukerjee, J. L. Markley, V. Storme, W. Boerjan and J. Ralph, Plant cell wall profiling by fast maximum likelihood reconstruction (FMLR) and region-of-interest (ROI) segmentation of solution-state 2D 1H-13C NMR spectra, *Biotechnol. Biofuels*, 2013, **6**, 45.

60. F. Delaglio, S. Grzesiek, G. W. Vuister, Z. Guang, J. Pfeifer and A. A. D. Bax, NMRPIPE: a multidimensional spectral processing system based on UNIX pipes, *J. Biomol. NMR*, 1995, **6**(3), 277–293.

CHAPTER 6

Backprojection and Related Methods

BRIAN E. COGGINS* AND PEI ZHOU*

Duke University Medical Center, Box 3711 DUMC, Durham, NC 27710, USA
*Email: bec2@duke.edu; peizhou@biochem.duke.edu

6.1 Introduction

There are many ways to carry out nonuniform sampling (NUS) in NMR. One of these, first introduced early in the FT-NMR era and still in use today, is the sampling of the time domain along radial spokes.[1] The collection of such data is straightforward, requiring only that two or more indirect dimensions be coevolved. While radially sampled data can be processed using many of the same methods discussed for randomized sampling later in this volume, the unique mathematics of radial sampling allows for a number of special possibilities. Some of these were considered in the previous chapter, which addressed the determination of higher-dimensional chemical shifts from projected positions in lower-dimensional spaces. This chapter considers another approach: the reconstruction of the full higher-dimensional spectrum from lower-dimensional projections.

Reconstruction from projections is a well-developed science of its own, going under the name *tomography*, used extensively in medical imaging and many other fields.[2,3] While tomography had been used to a limited extent in the first MRI experiments conducted by Lauterbur in the mid-1970s,[4] the idea of applying it to multidimensional NMR spectroscopy was first presented by Kupče and Freeman in 2003 as a way of reducing the

New Developments in NMR No. 11
Fast NMR Data Acquisition: Beyond the Fourier Transform
Edited by Mehdi Mobli and Jeffrey C. Hoch
© The Royal Society of Chemistry 2017
Published by the Royal Society of Chemistry, www.rsc.org

instrument time needed to record a given spectrum.[5] At first, such *projection–reconstruction* (PR) experiments were processed using intuitive, nonlinear reconstruction algorithms, which achieved extraordinary time savings in many simple cases. When the method was subsequently applied to more complex spectra, such as 4-D NOESY, it was necessary to turn to tomographic theory for more robust algorithms. The analytical inversion of the projection operator yields the most important of these, the *filtered backprojection* method—which, perhaps surprisingly, turns out to be none other than a Fourier transform in polar coordinates when applied to NMR data.

The key to reducing measurement time is being able to reconstruct full spectra accurately from a very small number of measurements. It is known today that the number of measurements needed for *quantitative* reconstruction from randomized nonuniform sampling is significantly smaller than that needed for radial sampling, and for that reason deterministic radial sampling has generally been superseded by nondeterministic random sampling.[6,7] Nonetheless, projection–reconstruction NMR played an important role in the development of NMR NUS, and both its techniques and its theoretical underpinnings may prove valuable in the future.

In this chapter, we begin by discussing the theory and practice of collecting projections, followed by the theory and practice of reconstruction. The latter leads naturally to a mathematical analysis of radial sampling itself, with implications for all approaches relying on radial data. We then survey the specific linear and nonlinear algorithms employed in NMR, and examine spectra obtained using these methods. Finally, we compare radial sampling and backprojection methods to other sampling and processing approaches.

6.2 Radial Sampling and Projections

6.2.1 Measuring Projections: The Projection-slice Theorem

Consider a 2-D time-domain FID $f(x, y)$ and its corresponding 2-D frequency-domain spectrum $F(X, Y)$, related by the Fourier transform:

$$f(x,y) \overset{\text{FT}}{\longleftrightarrow} F(X,Y). \tag{6.1}$$

A 1-D slice through this time domain, starting at the origin and extending at an angle θ from the x-axis, can be written as:

$$p_\theta(r) = f(x,y)\delta(x\sin\theta - y\cos\theta), \quad r = \sqrt{x^2 + y^2} \tag{6.2}$$

where the function:

$$\delta(u) \equiv \begin{cases} 1, & u = 0, \\ 0, & u \neq 0 \end{cases} \tag{6.3}$$

is a standard 1-D δ-function, used in the 2-D context to trace out a line impulse selecting only those values that are on the slice. The *projection-slice theorem*[8] states that the 1-D Fourier transform of this slice with respect to the radial coordinate:

$$P_\theta(R) = \int_{-\infty}^{\infty} p_\theta(r) e^{-2\pi i r R} \, dr \qquad (6.4)$$

is in fact a projection of the 2-D spectrum, integrated along lines perpendicular to the direction θ:

$$P_\theta(R) = \int_{-\infty}^{\infty} F(R\cos\theta - L\sin\theta, \; R\sin\theta + L\cos\theta) \, dL, \qquad (6.5)$$

where the variable of integration L represents a position along such a perpendicular line, or in other words:

$$f(x,y)\delta_{\text{line}}(x\sin\theta - y\cos\theta) \xleftrightarrow{\text{FT}} \int_{-\infty}^{\infty} F(R\cos\theta - L\sin\theta, R\sin\theta + L\cos\theta) \, dL,$$

$$(6.6)$$

as shown in Figure 6.1a. (Note that we allow R to assume both positive and negative frequency values, and we take θ as valid over the range of 0 to π.) This relationship arises because a line of samples in direction θ is insensitive to modulation in the perpendicular direction $\theta + \pi/2$. Since it is impossible to distinguish the frequencies of signals in this perpendicular direction, the Fourier transform shows their summation.

A slice through the time domain could also be described as a radial spoke, and a collection of such slices spaced at a regular angular interval $\Delta\theta$ would constitute a radially sampled dataset (Figure 6.1b). The Fourier transform of a single radial spoke p_θ with respect to radius yields a projection P_θ, and a radial dataset processed in this way yields a set of projections in various directions. For higher-dimensional spaces, radial samples are collected in the same way, as lines of sampling points crossing the time domain, and the space of signals that are projected onto the line at each position on the line has more dimensions. For example, if a radial spoke is measured through a 3-D time domain, its Fourier transform will show summations of planes from the full spectrum, each plane taken perpendicular to the line of the spoke (Figure 6.1c).

In the context of an NMR experiment, measuring a radial spoke would mean incrementing two or more of the indirect dimensions' evolution times simultaneously when advancing from one sampling point to the next over the course of the experiment. The precise evolution times for each indirect dimension are determined by the angle of the projection θ and the desired dwell time Δr in the radial direction. For example, in an experiment with two indirect dimensions with evolution delays x and y, the first

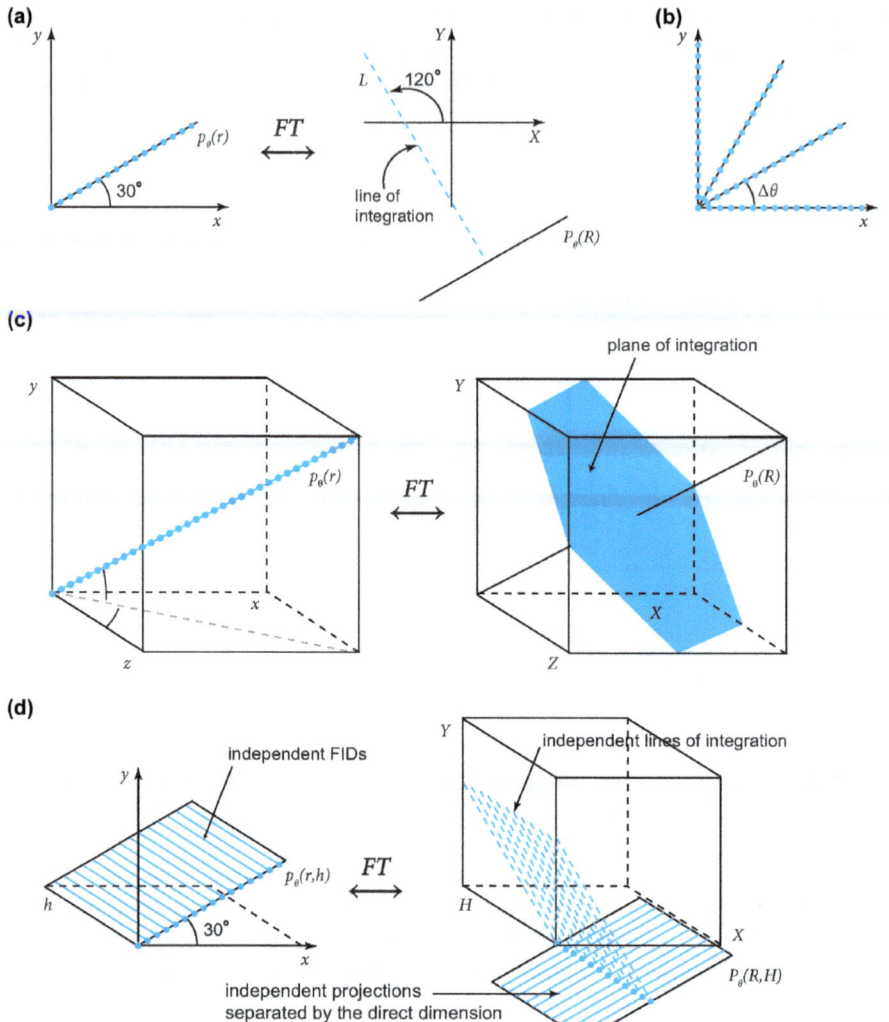

Figure 6.1   Radial sampling and projections. (a) The projection-slice theorem states
that a projection of the frequency domain at an angle θ can be obtained
by measuring a slice through the time domain at the same angle θ, and
calculating its Fourier transform. The values on the resulting projection
are line integrals of the spectrum. (b) A set of slices through the time
domain at various angles, all passing through the origin, constitutes a
radially sampled dataset. (c) The projection-slice theorem also applies in
higher-dimensional spaces. For example, the Fourier transform of a 1-D
slice through a 3-D time domain gives a projection onto a line, with
integration occurring on planes perpendicular to the projection. (d) The
direct dimension of a radial experiment is sampled conventionally,
resulting in an extra dimension in both the time domain and the
frequency domain. Each sampling position in the time domain corresponds to an FID, and each direct dimension position in the frequency
domain corresponds to an independent projection.

sampling point would allow no evolution time for either dimension $(x = y = 0)$, while the second sampling point would increment each by a scaled fraction of Δr, such that $x = \Delta r \cos \theta$ and $y = \Delta r \sin \theta$. Subsequent sampling points would be recorded at $(x = 2\Delta r \cos \theta, y = 2\Delta r \sin \theta)$, $(x = 3\Delta r \cos \theta, y = 3\Delta r \sin \theta)$, and so forth. Note that the sampling positions on the radial spoke would not, in general, align with the rectangular grid positions that would be recorded in a conventional experiment. To avoid aliasing on the projections, the radial dwell time Δr must be set shorter than the conventional dwell times Δx and Δy, as the diagonal of a rectangular spectrum is longer than the axes.[9,10]

Because the direct dimension does not participate in the coevolution and is instead sampled conventionally, it exists as an extra dimension attached to every time domain slice, frequency domain projection, and frequency domain reconstruction (Figure 6.1d). Hence measuring a sampling point at a position (x, y) on the indirect axes actually means measuring an FID in the direct dimension h with indirect evolution times x and y. In the frequency domain, each position of the direct dimension can be treated as an independent reconstruction problem. For the remainder of this chapter, we shall ignore the direct dimension.

6.2.2 Quadrature Detection and Projections

To determine the sign of the frequency of an NMR signal from its FID, that FID must be recorded in quadrature, as two parallel data streams with a 90° phase difference.[11,12] Collectively, these constitute a single stream of complex numbers. In multidimensional experiments, FIDs are recorded for all possible combinations of real and imaginary components for every dimension at every sampling position, producing a *hypercomplex* dataset. Thus in a two-dimensional dataset recording the product of a complex signal f_x in dimension x and a complex signal f_y in dimension y, each measurement $f(x, y)$ would actually consist of the four hypercomplex components:

$$f_{rr}(x,y) = \mathrm{Re}\, f_x(x) \mathrm{Re}\, f_y(y)$$

$$f_{ri}(x,y) = \mathrm{Re}\, f_x(x) \mathrm{Im}\, f_y(y)$$

$$f_{ir}(x,y) = \mathrm{Im}\, f_x(x) \mathrm{Re}\, f_y(y)$$

$$f_{ii}(x,y) = \mathrm{Im}\, f_x(x) \mathrm{Im}\, f_y(y).$$

$$(6.7)$$

Processing radial spokes to produce projections requires a 1-D FT computed over complex data, making it necessary to reorganize the hypercomplex data into complex data by taking a series of linear combinations of the hypercomplex components.[5,13–16] These linear combinations can be calculated by evaluating the product of the complex signals following conventional rules for complex numbers, for example in two dimensions as:

$$f_{+x+y}(r,\theta) = f_x(r\cos\theta)f_y(r\sin\theta)$$

$$= [\mathrm{Re}f_x(r\cos\theta) + i\,\mathrm{Im}f_x(r\cos\theta)][\mathrm{Re}f_y(r\sin\theta) + i\,\mathrm{Im}f_y(r\sin\theta)]$$

$$= [\mathrm{Re}f_x(r\cos\theta)\mathrm{Re}f_y(r\sin\theta) - \mathrm{Im}f_x(r\cos\theta)\mathrm{Im}f_y(r\sin\theta)]$$

$$+ i[\mathrm{Re}f_x(r\cos\theta)\mathrm{Im}f_y(r\sin\theta) + \mathrm{Im}f_x(r\cos\theta)\mathrm{Re}f_y(r\sin\theta)]$$

$$= [f_{rr}(r\cos\theta, r\sin\theta) - f_{ii}(r\cos\theta, r\sin\theta)]$$

$$+ i[f_{ri}(r\cos\theta, r\sin\theta) - f_{ir}(r\cos\theta, r\sin\theta)]$$

$$(6.8)$$

where the subscripts on f_{+x+y} indicate that this linear combination describes the evolution of the complex signal $f = f_x f_y$ as a function of r and θ for the $+x$ $+y$ quadrant of the time domain, that is for angles between $\theta = 0$ and $\theta = \pi/2$. Upon Fourier transformation, one obtains projections corresponding to half of the possible angle space. A second linear combination can also be formed, giving access to the rest of the angle space.[5] Relative to r, the $-x$ $+y$ quadrant of the time domain would consist of the complex conjugate of the $+x$ $+y$ quadrant—that is, the direction of precession relative to r would be reversed[1,11]—which can be calculated from the experimental data as:

$$f_{-x+y}(r,\theta) = \bar{f}_x(r\cos\theta)f_y(r\sin\theta)$$

$$= [\mathrm{Re}f_x(r\cos\theta) - i\,\mathrm{Im}f_x(r\cos\theta)][\mathrm{Re}f_y(r\sin\theta) + i\,\mathrm{Im}f_y(r\sin\theta)]$$

$$= [\mathrm{Re}f_x(r\cos\theta)\mathrm{Re}f_y(r\sin\theta) + \mathrm{Im}f_x(r\cos\theta)\mathrm{Im}f_y(r\sin\theta)]$$

$$+ i[\mathrm{Re}f_x(r\cos\theta)\mathrm{Im}f_y(r\sin\theta) - \mathrm{Im}f_x(r\cos\theta)\mathrm{Re}f_y(r\sin\theta)]$$

$$= [f_{rr}(r\cos\theta, r\sin\theta) + f_{ii}(r\cos\theta, r\sin\theta)]$$

$$+ i[f_{ri}(r\cos\theta, r\sin\theta) - f_{ir}(r\cos\theta, r\sin\theta)]$$

$$(6.9)$$

where the valid range of angles would be $\theta = \pi/2$ to π.

A comparable transformation of hypercomplex to complex data *via* linear combination of FIDs is possible for any dimensionality, by an analogous computation to that given above. The transform relationships can also be expressed concisely *via* the G-matrix formalism introduced by Kim and Szyperski.[16]

6.3 Reconstruction from Projections: Theory

The field of tomography offers a fully developed theory of reconstruction from projections, including a method that can be employed to obtain exact reconstructions in favorable cases. We shall consider this theory in due course, but we begin with the more intuitive analysis that guided the first efforts in projection–reconstruction NMR.

6.3.1 A Simple Approach: The Lattice of Possible Peak Positions

Consider a spectrum containing four peaks, together with its projections in the horizontal and vertical directions, as shown in Figure 6.2a.[5] From these two orthogonal projections, we could determine a lattice of possible peak positions (Figure 6.2b) by extending lines from each projected peak back across the spectrum; the lattice points are located where these lines intersect. We shall use the term *backprojection* to denote the extension of a line from a projected peak in a direction perpendicular to the projection itself.

The two orthogonal projections do not provide sufficient information to determine how many of the 12 lattice points in this example are actually locations of peaks. To resolve this ambiguity, an additional tilted projection could be collected (Figure 6.2d).[5] The data on the tilted projection show that only four of the lattice points describe actual peaks. Three backprojection vectors, one from each projection, intersect at each of these four lattice points (Figure 6.2e). Lattice points where fewer backprojection vectors intersect do not correspond to real signals (Figure 6.2e). This straightforward analysis inspired the first method used to reconstruct NMR spectra from projections, the *lower-value method*, described in Section 6.4.1 below, which uses a geometric construction analogous to Figure 6.2 to place peaks at lattice points where backprojection vectors from *all* projections intersect. Two additional methods, described in Sections 6.4.2 and 6.4.3, were subsequently developed based on the same ideas. Note that this basic logic is very similar to the logic employed by some of the methods described in previous chapters, including APSY[17] and HIFI,[18] where multidimensional signal frequencies are calculated based on projected peak positions in reduced dimensionality spectra.

Whether or not a single tilted projection would be able to resolve the ambiguity of the starting lattice depends on the distribution of the real peaks, the geometry of the lattice, and the choice of projection angle. Figure 6.3 shows a series of tilted projections of a spectrum containing four real signals.[19] At some projection angles, all four signals can be resolved clearly, while at other angles two or more of the signals overlap. For more complicated spectra, it is unlikely that any single projection direction would allow all such ambiguities to be resolved. As the number of signals increases, therefore, the number of tilted projections needed to resolve the ambiguities of the lattice also increases.

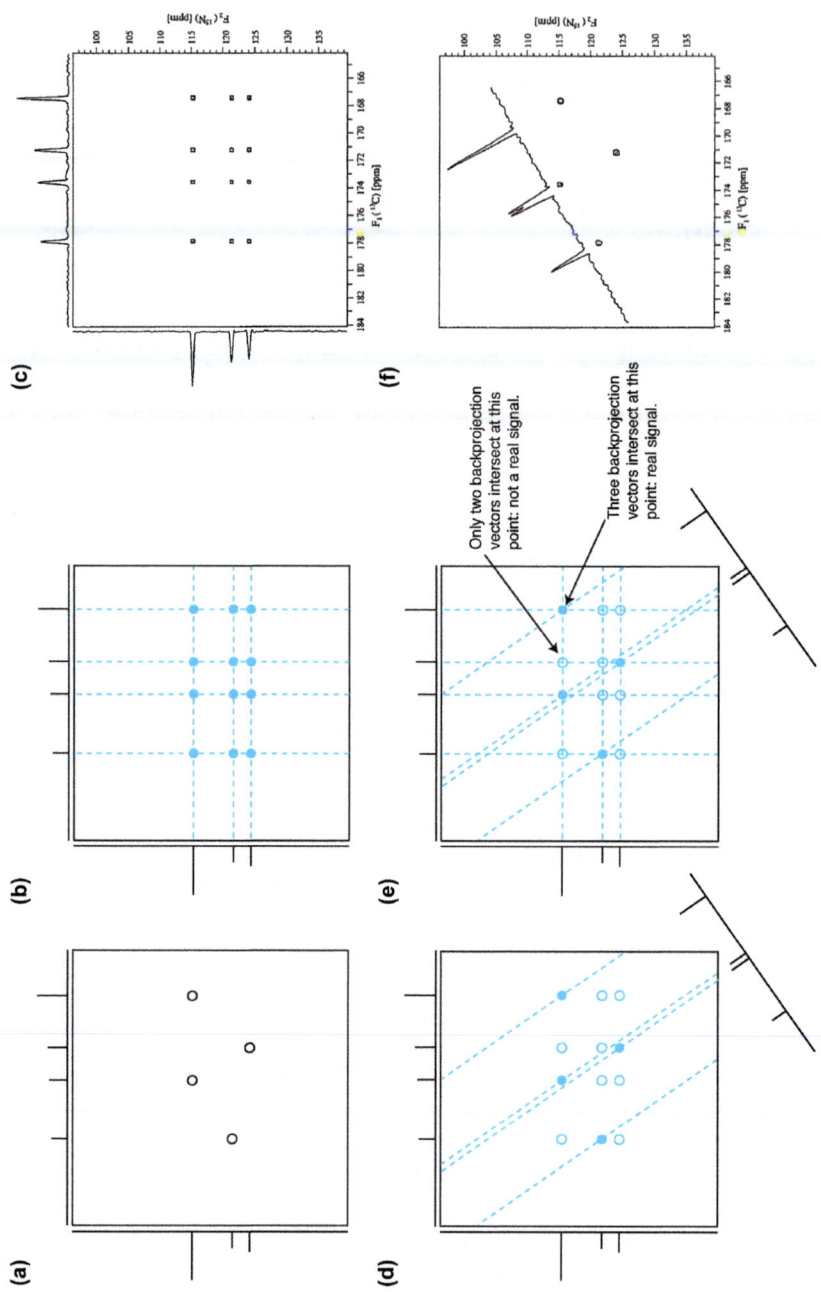

To better appreciate the nature of the problem, it may be helpful to consider it as a system of linear equations rather than geometrically. Let the known projected intensities be represented by a vector $\mathbf{I} = \{I_i: I_1, I_2, \ldots\}$ and the unknown values at the lattice points by a vector $\mathbf{L} = \{L_i: L_1, L_2, \ldots\}$. The matrix equation:

$$\mathbf{I} = \mathbf{WL} \tag{6.10}$$

would therefore describe the projection of the lattice points to produce the measured values, where $\mathbf{W}$ is a matrix of weights that depends on geometry. If only the orthogonal projections are available, this system is very much underdetermined. Depending on the geometric relationships between the projected peaks, the number of lattice points—and the degree to which the system is underdetermined—can vary widely. Tilted projections add additional known values; with enough tilted projections, it is frequently possible to solve the system uniquely from far fewer measurements than would be needed for conventional spectroscopy, but much depends on whether the rows that are added to the matrix for these additional projection angles are linearly independent of each other and of the rows corresponding to previous measurements.

Figure 6.2   Identifying possible signal locations from projections. (a) A schematic illustration of an example spectrum with four signals, together with its projections at 0° and 90°. This example is modeled on the actual experimental case shown in panels (c) and (f), reprinted from ref. 5. This was the first reported projection–reconstruction NMR experiment. (b) Using only the information on the two orthogonal projections, one can construct a lattice of possible peak positions by extending backprojection vectors (dashed lines) from the projected peaks. The possible peak positions are the points where the backprojection vectors intersect (circles). There are more lattice positions than real signals, however, and insufficient information is available from the two projections to determine which lattice positions correspond to the real signals and which do not. (c) A reconstruction of the lattice of possible peak positions for this spectrum from the experimental orthogonal projection data. (d) With an additional tilted projection, there is now sufficient information to differentiate lattice positions corresponding to real signals (filled circles) from those that do not (open circles). (e) Signals are found at those lattice points where backprojection vectors from all three projections intersect. Lattice points where only two backprojection vectors intersect cannot correspond to a real signal. (f) A reconstruction of the true spectrum from the experimental orthogonal and tilted projections. (Panels (c) and (f) reprinted from *J. Biomol. NMR*, Reconstruction of the three-dimensional NMR spectrum of a protein, from a set of plane projections, **27**, 2003, 383–387, E. Kupče and R. Freeman, (© Kluwer Academic Publishers 2003) with permission from Springer.)

6.3.2   Limitations of the Lattice Analysis and Related Reconstruction Methods

Though the analysis above, and the lower-value reconstruction method derived from it, proved effective in a number of NMR experiments, these approaches have some significant limitations.

One such limitation is that the structure of the lattice—and the extent to which the reconstruction problem is underdetermined—depends on the geometry of the spectrum as well as the projection angles. Because every reconstruction problem is different, it is not possible to generalize about the number of projections needed or the optimum projection angles to choose, and the extent of ambiguity and degeneracy faced in one reconstruction will be different from those that occur in another. One could, of course, imagine optimizing the experimental design for each spectrum, and indeed, one of the early hopes of projection–reconstruction was that projection directions could be chosen intelligently to maximize the efficiency of the data collection. The very first paper on projection–reconstruction suggested that "in principle, a program could be written to calculate undesirable choices of tilt angle—those that would catch many of the lines in *enfilade* and thus provide fewer new simultaneous equations governing the intensities."[5]

Further thoughts along these lines appeared in a subsequent paper by Kupče and Freeman, which presented the example shown in Figure 6.3, with the four crosspeaks clearly resolved for certain directions, while two or more of them become degenerate in other directions.[19] In this paper, Kupče and Freeman proposed a grid search of possible angles, calculating a scoring function to assess the number of resolvable lattice points. Subsequent work by Gledhill and Wand extended this idea to address many practical considerations, including the fact that signals are not infinitely sharp.[20] The most elaborate and successful effort to optimize the experimental strategy for the details of the specific lattice problem presented by a given spectrum is HIFI-NMR, described in the previous chapter, where possible angle selections are scored using a Bayesian probability model.[18] HIFI does not attempt a full reconstruction, but rather produces a peak list based on a probabilistic assessment of which lattice points contain signals. The success of the algorithms developed by Gledhill and Wand as well as of HIFI at selecting a small number of projection angles to maximize the available information content suggests that such approaches could have considerable value for reconstruction methods based on the lattice of possible peak positions.

In the absence of an optimized solution for a particular experiment, it is possible to estimate the lower limit of the number of tilted projections needed to resolve all hypothetically possible lattice ambiguities.[21] Assuming that signals are infinitely sharp, a simple degrees-of-freedom analysis allows one to calculate that the number of projections N_θ should be greater than or equal to $M(d-p)$, where M is the largest number of signals found at any direct-dimension position in the spectrum, d is the number of dimensions

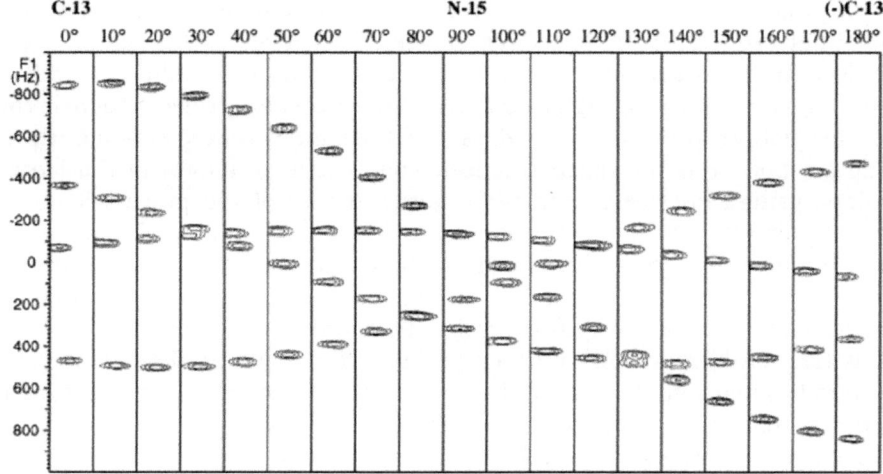

Figure 6.3   Projection angle and degeneracy. Varying the tilt angle of a time domain radial spoke changes the angle of the frequency domain projection, causing the projected positions of the signals to shift. Shown are strips from projections of the HNCO spectrum of ubiquitin at varying tilt angles from the carbon axis, ranging from 0° (along the carbon axis) at the left to 90° (along the nitrogen axis) in the middle to 180° (again along the carbon axis, but with the direction reversed) at the right. The projected positions of the four signals follow sinusoidal trajectories. Degeneracy occurs at certain angles as a result of the particular geometry of the signals in the full spectrum.
(Reprinted with permission from E. Kupče and R. Freeman, Projection–Reconstruction Technique for Speeding up Multidimensional NMR Spectroscopy, *J. Am. Chem. Soc.*, 2004, **126**, 6429–6440. Copyright 2004 American Chemical Society.)

in the overall experiment, and p is the number of dimensions in each projection. This formula can only provide an estimate since it does not take into account possible consequences from signal overlap.

It is important to remember, however, that the problem of reconstructing a spectrum is not the same as the problem of determining which lattice points contain real signals. For a reconstructed spectrum to have any additional value beyond what a peaklist could offer, that spectrum must correctly represent not only peak positions but also the many additional kinds of information found in traditional spectra, including peak intensities, lineshapes, and volumes. It is not obvious how to derive this information from the lattice formulation, and while the three major reconstruction algorithms based on this formulation (discussed in Sections 6.4.1–6.4.3) can approximate peak shapes and intensities in favorable cases, none of these algorithms can replicate peak shapes quantitatively.

An additional problem for lattice-based methods is that of sensitivity. In order to carry out the analysis described above, the projections must be collected with enough transients that each peak in the spectrum is visible on every projection. In many NMR studies, this requirement would obviate any potential reductions in measurement time expected to be achieved *via* projection–reconstruction. To overcome this limitation, an additive method was suggested, with the signals on each projection backprojected across the reconstruction space, forming ridges, and the sum or superposition of these ridges reported as the reconstruction. These ridges intersect at the lattice points, generating peaks, and the additive nature of the process leads to signal averaging and vastly improved sensitivity.[19] This method is identical to a mathematical procedure traditionally called *backprojection*.[22] Unfortunately, backprojection cannot discriminate between lattice points supported by evidence from all projections rather than only some projections, and it thus frequently generates spurious peaks at lattice points that do not correspond to real signals.[21] In addition, the ridges themselves are substantial artifacts that can interfere with the detection of weak signals.

Though the lower-value and backprojection methods can be combined to overcome some of the limitations of each, producing a method that can reconstruct sequential assignment spectra for large proteins with limited sensitivity (Section 6.4.3),[21] for more complicated spectra and for cases where quantitative reconstruction is essential—such as NOESY—it is necessary to turn to tomographic theory. This theory leads not only to a general solution for data collection and spectrum reconstruction, but also to a deeper analysis of the information content of radially sampled data—which will show that some of the problems encountered with the lattice approach are fundamental to radial sampling, regardless of the processing method used.

6.3.3 The Radon Transform and Its Inverse

Tomography formulates a solution to the problem of reconstruction in terms of integral transforms over infinitely many projections. The basic approach was first derived by Johann Radon,[23] and it has been rediscovered several times subsequently.[24–29]

The discrete projections $P_\theta(R)$ that we obtain through radial sampling of the NMR time domain followed by Fourier transformation with respect to r are mathematically related to the full spectrum $F(X, Y)$ by the integration shown in eqn (6.5). If we carried out this integration for infinitely many possible directions—that is, if we calculated infinitely many projections at infinitesimally small angular spacings over the entire range $\theta = 0$ to π—we would obtain a continuous function $P(R, \theta)$ representing a complete set of all possible projections of F. The integral transformation relating a function to its complete set of projections is known today as the *Radon transform*.

It was first shown by Radon, writing in 1917, that this transform could be inverted analytically—to reconstruct a function from its projections—using the formula:

$$F(X, Y) = -\frac{1}{2\pi^2} \int_0^\infty \frac{\mathrm{d}}{Q} \left[\int_0^{2\pi} P(X \cos\theta + Y \sin\theta + Q, \theta)\, \mathrm{d}\theta \right] \qquad (6.11)$$

where the outer integral is taken as a Stieltjes integral.[23] Rewriting the limits of integration, swapping the order of the integrals, and substituting the radial coordinate $R = X \cos \theta + Y \sin \theta$ for the Cartesian coordinates X and Y yields the form normally given today:[30]

$$F(X, Y) = \frac{1}{2\pi^2} \int_0^\pi \int_{-\infty}^\infty \frac{dQ}{R - Q} \frac{d}{dQ} P(Q, \theta) \, d\theta. \tag{6.12}$$

The outermost operation is known as *backprojection*, which we shall represent with an operator **B**, formally defined as:

$$\mathbf{B}[G(R, \theta)](X, Y) = \frac{1}{2\pi} \int_0^\pi G(X \cos \theta + Y \sin \theta, \theta) \, d\theta$$

$$= \frac{1}{2\pi} \int_0^\pi G(R, \theta) \, d\theta. \tag{6.13}$$

Backprojection was already mentioned above in the discussion of geometric analyses for reconstruction, and it has a straightforward geometric interpretation: it assigns to point (X, Y) the sum (or the average, taking into account the normalization $1/2\pi$) of that point's projections in all directions.[25,29,31] The effect of backprojection is quite literally to extend the intensity profile of each projection backwards over the frequency domain, and take the superposition of these extended profiles from the various projections.

Having defined the backprojection operation, we can now rewrite eqn (6.12) as:

$$F(X, Y) = \mathbf{B}\left[\frac{1}{\pi} \int_{-\infty}^\infty \frac{dQ}{R - Q} \frac{d}{dQ} P(Q, \theta)\right]. \tag{6.14}$$

The remaining operations have acquired the name *filtering*. They consist of taking the Hilbert transform with respect to the radial coordinate (operator **H**):

$$\mathbf{H}[G(R)] = \frac{1}{\pi} \int_{-\infty}^\infty \frac{G(R) dQ}{R - Q}, \tag{6.15}$$

of the derivative of the projections with respect to the radial coordinate (operator **D**):

$$\mathbf{D}[P(R, \theta)] = \frac{d}{dR} P(R, \theta), \tag{6.16}$$

allowing us to rewrite eqn (6.12) as:

$$F(X, Y) = \mathbf{B}[\mathbf{H}[\mathbf{D}[P(R, \theta)]]]. \tag{6.17}$$

The meaning of the filtering operation can be elucidated by moving it from the frequency domain to the time domain. First, one inserts a pair of operators $\mathbf{F}$ and $\mathbf{F}^{-1}$ representing forward and reverse one-dimensional Fourier transforms with respect to the radial coordinate, to give:

$$F(X,Y) = \mathbf{B}[\mathbf{F}[\mathbf{F}^{-1}[\mathbf{H}[\mathbf{D}[P(R,\theta)]]]]]$$

$$= \mathbf{B}\left[\int_{-\infty}^{\infty}\left[\int_{-\infty}^{\infty}\mathbf{H}[\mathbf{D}[P(R,\theta)]]\,e^{2\pi irR}\,dR\right]e^{-2\pi irR}\,dr\right]. \tag{6.18}$$

Recall[32] that the Fourier transform of the derivative of a function is:

$$\mathbf{F}[\mathbf{D}[g(r)]] = 2\pi iR\ \mathbf{F}[g(r)] \tag{6.19}$$

and that the Fourier transform of the Hilbert transform of a function is:

$$\mathbf{F}[\mathbf{H}[g(r)]] = i\,\mathrm{sgn}(R)\mathbf{F}[g(r)], \quad \mathrm{sgn}(x) = \begin{cases} 1, & x > 1, \\ -1, & x < 1 \end{cases}. \tag{6.20}$$

Applying eqn (6.19) and (6.20) to eqn (6.18), we get:

$$F(X,Y) = \mathbf{B}[\mathbf{F}[\mathbf{F}^{-1}[\mathbf{H}[\mathbf{D}[P(R,\theta)]]]]]$$

$$= \mathbf{B}[\mathbf{F}[-i\,\mathrm{sgn}(r)\times 2\pi ir\times \mathbf{F}^{-1}[P(R,\theta)]]] \tag{6.21}$$

$$= 2\pi\mathbf{B}[\mathbf{F}[\,|r|\ \mathbf{F}^{-1}[P(R,\theta)]]].$$

If we let $p(r,\theta) = \mathbf{F}^{-1}[P(R,\theta)]$ represent the time-domain slices, we obtain:

$$F(X,Y) = 2\pi\mathbf{B}[\mathbf{F}[\,|r|\ [p(r,\theta)]]$$

$$= \int_0^\pi\int_{-\infty}^{\infty} p(r,\theta)e^{-2\pi ir(X\cos\theta + Y\sin\theta)}\,|r|\,dr\,d\theta. \tag{6.22}$$

In the time domain, the filtering operation is simply a reweighting of the sampling points based on their distance from the origin.

Thus in NMR, filtering is most easily performed by applying the weighting function $|r|$ to the raw time domain data prior to FT to obtain the filtered projections:

$$\hat{P}(R,\theta) = \mathbf{F}_R[|r|\ p(r,\theta)]$$

$$= \int_{-\infty}^{\infty} p(r,\theta)e^{-2\pi irR}\,|r|\,dr. \tag{6.23}$$

Reconstruction is then carried out by backprojection in the frequency domain:

$$F(X,Y) = \mathbf{B}[\hat{P}(R,\theta)]$$

$$= \int_0^\pi \hat{P}(R,\theta)d\theta. \tag{6.24}$$

Reconstructing from projections by inversion of the Radon transform is commonly known as *filtered backprojection* (FBP), reflecting this sequence of operations.

The effect of the filtering operation on the projections can be understood through the convolution theorem of the Fourier transform.[25] Multiplying by $|r|$ in the time domain is equivalent to convolving by the Fourier transform of $|r|$ in the frequency domain:

$$|r|\, p(r, \theta) \xrightarrow{\text{FT}} \mathbf{F}_r[|r|] * P(R, \theta). \tag{6.25}$$

To find the FT of $|r|$, we can rewrite it in terms of two useful functions defined by Bracewell, the rectangle function:

$$\Pi(x) = \begin{cases} 1, & |x| < \tfrac{1}{2}, \\ 0, & |x| > \tfrac{1}{2}, \end{cases} \tag{6.26}$$

and the triangle function:

$$\Lambda(x) = \begin{cases} 1 - |x|, & |x| < 1, \\ 0, & |x| > 1, \end{cases} \tag{6.27}$$

which have Fourier transforms:

$$\Pi(x) \xrightarrow{\text{FT}} \text{sinc}(X) \quad \text{and} \quad \Lambda(x) \xrightarrow{\text{FT}} \text{sinc}^2(X) \tag{6.28}$$

Where:

$$\text{sinc}(X) = \frac{\sin \pi x}{\pi x}. \tag{6.29}$$

Since:

$$|r| = \Pi\left(\frac{r}{2r_{\text{max}}}\right) - \Lambda\left(\frac{r}{r_{\text{max}}}\right) \tag{6.30}$$

out to a radius of r_{max}, the multiplication by $|r|$ in the time domain is equivalent to convolving with a kernel:

$$K(R) = 2r_{\text{max}}\, \text{sinc}(2r_{\text{max}}R) - r_{\text{max}}\, \text{sinc}^2(r_{\text{max}}R) \tag{6.31}$$

in the frequency domain. A plot of this function is shown in Figure 6.4a; Figure 6.4b shows the result when this function is convolved with a Lorenztian signal of finite linewidth. Filtering therefore alters the lineshapes of the signals in the projections, introducing negative sidelobes. These negative sidelobes ensure that the superposition of backprojected intensity in the backprojection operation does not broaden the signals, as it otherwise would.

6.3.4 The Polar Fourier Transform and the Inverse Radon Transform

The Radon transform and the Fourier transform are closely related, as the projection-slice theorem would already suggest. A second connection becomes apparent if we convert the familiar 2-D Fourier transform:

$$F(X, Y) = \int_{-\infty}^{\infty} \int_{-\infty}^{\infty} f(x, y) e^{-2\pi i (xX + yY)} \, dx \, dy, \tag{6.32}$$

to polar coordinates by substituting $r = \sqrt{x^2 + y^2}$, $\theta = \arctan y/x$, and $dx \, dy = r \, dr \, d\theta$, yielding:[31,33-35]

$$F(X, Y) = \int_{0}^{2\pi} \int_{-\infty}^{\infty} f(r, \theta) e^{-2\pi i r (X \cos \theta + Y \sin \theta)} r \, dr \, d\theta. \tag{6.33}$$

The replacement of the area element $dx \, dy$ with $r \, dr \, d\theta$ is necessary because polar coordinates do not sample the time domain uniformly: radial spokes are closest together at the origin, and become further apart at increasing distances from the center (Figure 6.4c). Thus the sampling density diminishes according to the function $1/r$, and to obtain a correct Fourier transform it is necessary to reweight the measurements according to their distance from the origin r. If the measurements are not reweighted, earlier evolution times will be overweighted in the transform, broadening the peaks (Figure 6.4d).

If we now revise the limits of integration so that radial spokes at angles between $\theta = \pi$ and 2π are considered to be extensions into negative coordinates of the spokes at angles $\theta = 0$ to π, we have:

$$F(X, Y) = \int_{0}^{\pi} \int_{-\infty}^{\infty} p(r, \theta) e^{-2\pi i r (X \cos \theta + Y \sin \theta)} |r| \, dr \, d\theta \tag{6.34}$$

which is the same formula we derived above, starting with the inverse Radon transform, for the reconstruction of the frequency domain starting from time-domain slices [eqn (6.22)]. Thus for NMR data, we can achieve the same result with a polar Fourier transform that we would achieve by an inverse Radon transform, and in exactly the same manner: both are accomplished by filtered backprojection, with a filter function of $|r|$.

It is important to note that the inverse Radon transform and the Fourier transform in polar coordinates are *not* formally identical, even though they are practically equivalent *in the context of NMR*. The inverse Radon transform formally converts frequency-domain projections into a frequency-domain reconstruction of the spectrum:

$$P(R, \theta) \xrightarrow{\mathbf{R^{-1}}} F(X, Y) \tag{6.35}$$

where operator $\mathbf{R}^{-1}$ represents the inverse Radon transform. This is accomplished by filtering followed by backprojection, which would be:

$$P(R, \theta) \xrightarrow{\mathbf{H}_R \mathbf{D}_R} \hat{P}(R, \theta) \xrightarrow{\mathbf{B}} F(X, Y) \tag{6.36}$$

in the frequency domain or:

$$P(R, \theta) \xrightarrow{\mathbf{F}_R^{-1}} p(r, \theta) \xrightarrow{|r|} \hat{p}(r, \theta) \xrightarrow{\mathbf{F}_R} \hat{P}(R, \theta) \xrightarrow{\mathbf{B}} F(X, Y) \tag{6.37}$$

if the data are taken into the time domain temporarily for the filtering operation. The polar Fourier transform constitutes the last three operations in this sequence:

$$P(R, \theta) \underbrace{\xrightarrow{\mathbf{F}_R^{-1}} p(r, \theta) \overbrace{\xrightarrow{|r|} \hat{p}(r, \theta) \xrightarrow{\mathbf{F}_R} \hat{P}(R, \theta) \xrightarrow{\mathbf{B}} F(X, Y)}^{\text{polar Fourier transform}}}_{\text{inverse Radon transform}} \tag{6.38}$$

and takes the time-domain radial slices $p(r, \theta)$ to the frequency-domain spectrum $F(X, Y)$. In imaging applications, it is normal for the collected data to be the projections $P(R, \theta)$, and the inverse Radon transform must be performed. In NMR, the data are measured as the time-domain slices $p(r, \theta)$ to begin with, and it is sufficient to carry out a polar Fourier transform *via* the filtered projections and the backprojection operator without ever generating the unfiltered projections.

Having established this connection, the reason for the filtering operation is now clear. Both the inverse Radon transform and the polar Fourier transform work with radially sampled data. The sampling points in a radial pattern are not distributed over the time domain with a uniform density, but rather early evolution times are systematically oversampled at the expense of later evolution times. This oversampling would broaden the signals in the final spectrum if it were not corrected with the application of a reweighting function $|r|$ in the time domain. When working entirely in the frequency domain, this correction can be carried out by taking the Hilbert transform of the derivative of each projection, or alternatively by convolving with the function given in eqn (6.31), which are equivalent operations to reweighting in the time domain.

6.3.5 Reconstruction of Higher-dimensional Spectra

The above analysis extends readily to higher-dimensional datasets measured by radial sampling. Polar Fourier transforms for any dimensionality can be obtained by inserting the appropriate higher-dimensional equivalent to 2-D polar coordinates into the corresponding Cartesian Fourier transform;[33] we will represent such coordinates as a radius r and a vector of angles $\boldsymbol{\theta}$.

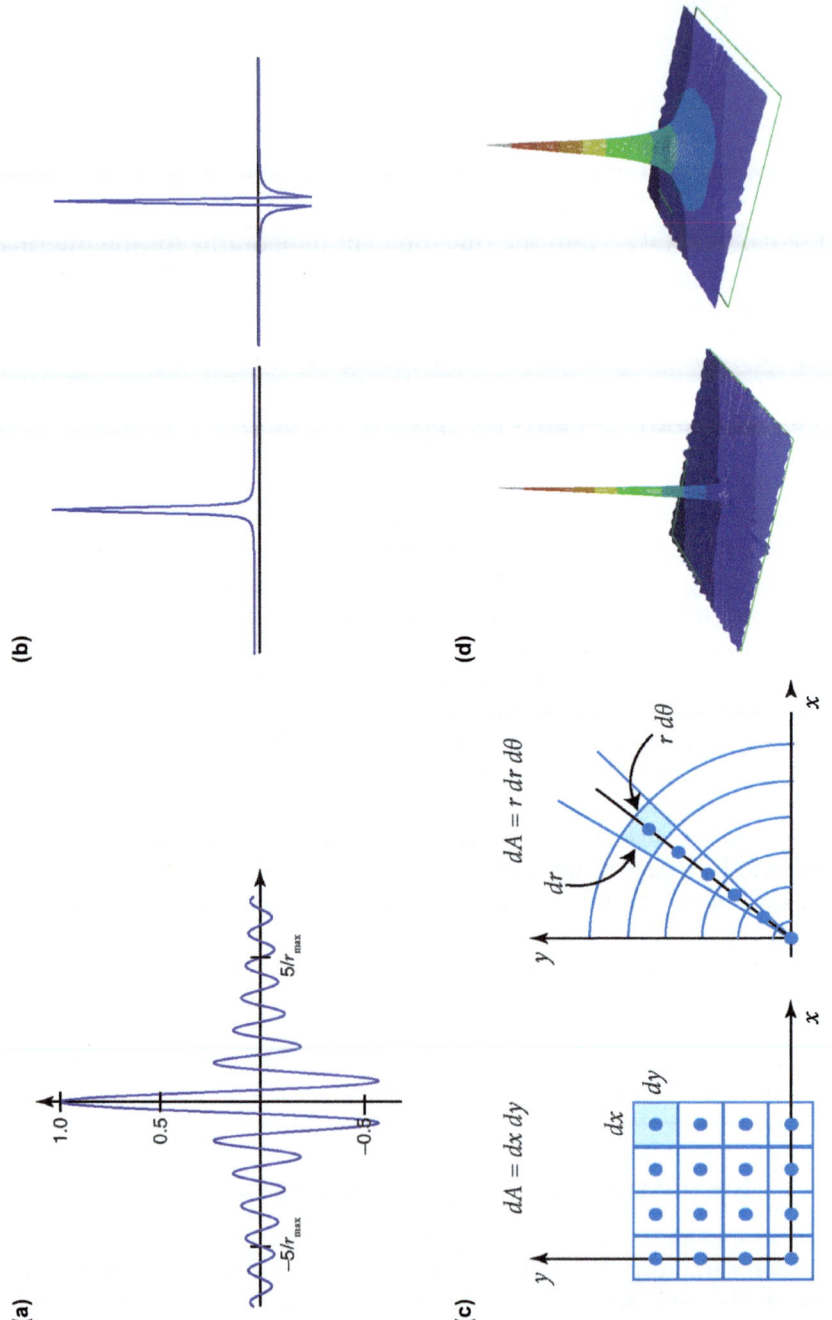

Assuming that the directions of the radial spokes were distributed uniformly in the angle space, the sampling density in the time domain will vary according to the function $1/r^{n-1}$ where n is the number of radially sampled dimensions, and the filter function to apply in the time domain will be $|r^{n-1}|$. The filtering operation would be:

$$\hat{P}(R, \boldsymbol{\theta}) = \mathbf{F}_R[\,|r^{n-1}|\; p(r, \boldsymbol{\theta})]$$

$$= \int_{-\infty}^{\infty} p(r, \boldsymbol{\theta}) e^{-2\pi i r R}\; |r^{n-1}|\; dr \qquad (6.39)$$

which would be followed by backprojection over all of the filtered projections:

$$F(\mathbf{X}) = B[\hat{P}(R, \boldsymbol{\theta})]$$

$$= \int \hat{P}(R, \boldsymbol{\theta})\, d\theta. \qquad (6.40)$$

6.3.6 The Point Response Function for Radial Sampling

If it were possible to sample the NMR time domain continuously, it would not matter whether the data were organized and processed in Cartesian coordinates or in polar coordinates. A set of infinitely many radial slices extending to infinitely long evolution times could be processed by a continuous polar Fourier transform to obtain a perfect spectrum, free of

---

Figure 6.4 Filtering in filtered backprojection. (a) A plot of the filter lineshape in the frequency domain. (b) A Lorentzian lineshape (left) and the result of convolving that lineshape with the filter lineshape (right). (c) In conventional sampling (left), the sampling points are located on a rectangular grid and have a uniform density. Because every point occupies the same area $dA = dx\, dy$, every point can be assigned the same weight. In contrast, with radial sampling the sampling points are not distributed uniformly, but rather with a density that decreases with the distance from the origin. This can be corrected by reweighting the measurements based on the area element $dA = r\, dr\, d\theta$, *i.e.* by multiplying each measurement by the filter function r. (d) Without the filter function, earlier evolution times are overrepresented in the radial pattern, and the Fourier transform shows signals that are broadened (right) by comparison with the original (left).
(Panels (b) and (d) adapted with permission from B. E. Coggins, R. A. Venters and P. Zhou, Filtered Backprojection for the Reconstruction of a High-Resolution (4,2)D CH3 − NH NOESY Spectrum on a 29 kDa Protein, *J. Am. Chem. Soc.*, 2005, **127**, 11562–11563. Copyright © 2005 American Chemical Society. Panel (c) is reprinted from *Prog. Nucl. Magn. Reson. Spectrosc.*, **57**, B. E. Coggins, R. A. Venters and P. Zhou, Radial sampling for fast NMR: Concepts and practices over three decades, 381–419, Copyright (2010), with permission from Elsevier.)

artifacts from the sampling. Likewise, a set of infinitely many frequency domain projections in all possible directions could be processed by a continuous inverse Radon transform to yield a perfect reconstruction.

Unfortunately, it is not possible to measure infinitely many radial spokes, but rather a given dataset will contain a finite number of spokes N_θ. These spokes will not extend to infinite evolution times, but rather to some finite maximum evolution time r_{max}, and only a finite number of measurements N_r at a finite spacing Δr will be recorded for each spoke. Because of these restrictions on the measurement process, the information recorded in a radial dataset with $N = N_\theta N_r$ total measurements will be different from the information recorded by conventional sampling on a rectangular grid with N grid points. This is not a consequence of the Fourier transform being computed in one coordinate system or another, but rather of the information being recorded for a very different set of positions in the time domain. It is only in the theoretical limit of infinitely many measurements that the information content would become identical. That said, it is a well-known fact that Cartesian sampling can fully describe band-limited signals as long as the Nyquist condition is met, in which case the Fourier transform yields a spectrum free of aliasing or sampling artifacts. In this section, we consider whether a similar sampling condition exists for radially sampled data, guaranteeing artifact-free spectra if the sampling is sufficient. We also consider what form aliasing artifacts from insufficient sampling would take for radial data.

The consequences of applying any type of discrete sampling to a set of continuous signals can be calculated easily *via* the *point response* or *point spread function* of the sampling process. Returning to two dimensions for simplicity, when a continuous time-domain signal $a(x, y)$ corresponding to a frequency-domain spectrum $A(X, Y)$:

$$a(x,y) \xleftrightarrow{\text{FT}} A(X, Y) \tag{6.41}$$

is measured at specific time-domain positions, it is equivalent to multiplying $a(x, y)$ by a binary mask, a function that is one at the measured points and zero otherwise, which we shall call a *sampling function*, denoted $s(x, y)$. This sampling function has a Fourier transform:

$$s(x,y) \xleftrightarrow{\text{FT}} S(X, Y) \tag{6.42}$$

which is known as the *point response* or *point spread function*. Multiplication of $a(x, y)$ and $s(x, y)$ in the time domain results in the convolution of their Fourier transforms in the frequency domain:

$$a(x,y)s(x,y) \xleftrightarrow{\text{FT}} A(X, Y) * S(X, Y). \tag{6.43}$$

Thus to understand the consequences of radial sampling, we need only to examine the point response corresponding to radial sampling, and determine whether convolving it with the true spectrum would introduce artifacts.

One way to determine this point response would be to evaluate the point response of a single spoke, and then add together the contributions from all of the spokes in the pattern.[6,31,33,35] A single spoke would consist of a line of points in the time domain in a particular direction θ. To represent a line of points in one dimension, Bracewell[32] defines the function:

$$\text{III}(u) \equiv \sum_{i=-\infty}^{\infty} \delta(u - i) \tag{6.44}$$

which describes unit impulses at unit intervals ($u = 0, \pm 1, \pm 2, \ldots$); it can be scaled on the u-axis to place the unit impulses at intervals of Δu as:

$$\left(\frac{1}{\Delta u}\right) \text{III}\left(\frac{u}{\Delta u}\right). \tag{6.45}$$

The $\text{III}(u)$ function can be used to construct a 1-D line of points extending to r_{max} in direction θ within a 2-D space:

$$l_\theta(x, y) = \left(\frac{1}{\Delta r}\right) \text{III}\left(\frac{x \cos \theta + y \sin \theta}{\Delta r}\right) \Pi\left(\frac{r}{2r_{max}}\right) \delta(x \sin \theta - y \cos \theta). \tag{6.46}$$

Using the convolution theorem and the Fourier transform pairs:

$$\text{III}(u) \xleftrightarrow{\text{FT}} \text{III}(U)$$
$$\delta(x \sin \theta - y \cos \theta) \xleftrightarrow{\text{FT}} \delta\left(X \sin \theta \left[\theta + \frac{\pi}{2}\right] - Y \cos \theta \left[\theta + \frac{\pi}{2}\right]\right) \tag{6.47}$$

together with eqn (6.46), the Fourier transform of this line of points is:

$$L_\theta(X, Y) = \Delta r \, \text{III}(\Delta r[X \cos \theta - Y \sin \theta])$$
$$* \left(2r_{max} \, \text{sinc}[2r_{max}R] - r_{max} \, \text{sinc}^2[r_{max}R]\right) \tag{6.48}$$
$$* \delta\left(X \sin\left[\theta + \frac{\pi}{2}\right] - Y \cos\left[\theta + \frac{\pi}{2}\right]\right),$$

which describes ridges of intensity running perpendicular to the original direction of sampling and spaced at intervals of $1/\Delta r$ (Figure 6.5a). The basic shape is a ridge perpendicular to the sampling direction, which arises because a line of samples in direction θ contains no information about the localization of signals in the perpendicular direction. Because the samples in direction θ are spaced at intervals of Δr, the perpendicular ridge is repeated at intervals of $1/\Delta r$. A cross-section through any individual ridge reveals a sinc-function lineshape with so-called "sinc wiggles" extending across the plane, resulting from the truncation of the sampling at r_{max} (Figure 6.5b).

(a)

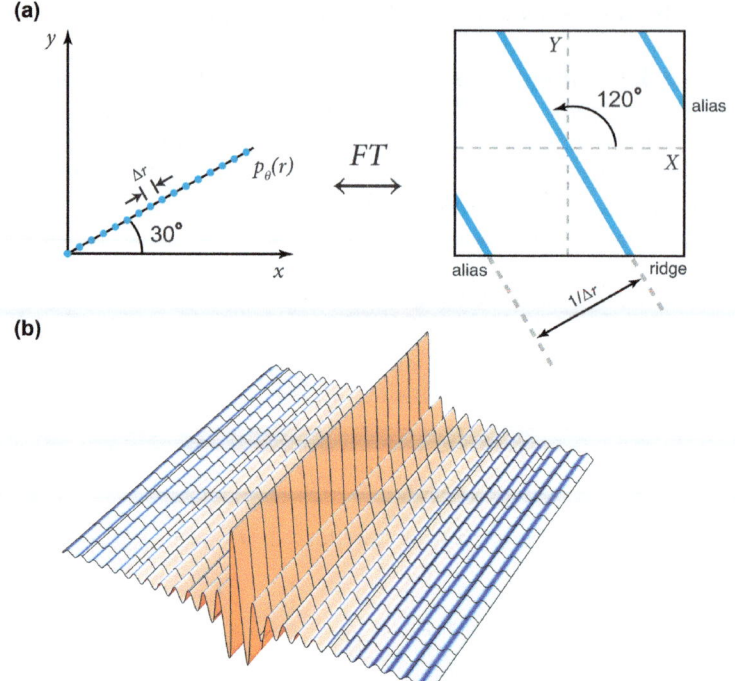

(b)

Figure 6.5 The point response of a single radial spoke. (a) The point response for a single radial spoke at an angle θ, sampled at intervals of Δr (left), consists of ridges of intensity in the direction $\theta + \pi/2$, repeated at intervals of $1/\Delta r$ (right). (b) The cross-section of each ridge has the same profile as the Fourier transform of the filter function.
(Panel (a) is reprinted from *Prog. Nucl. Magn. Reson. Spectrosc.*, **57**, B. E. Coggins, R. A. Venters and P. Zhou, Radial sampling for fast NMR: Concepts and practices over three decades, 381–419, Copyright (2010), with permission from Elsevier.)

Letting N_θ be the number of spokes after conversion from hypercomplex to complex data (*i.e.* after converting $N_\theta/2$ hypercomplex spokes over the range $\theta = 0$ to $\pi/2$ into N_θ complex spokes over the range $\theta = 0$ to π), the full sampling pattern would be:

$$s(x,y) = \sum_{j=1}^{N_\theta} l_{\theta_j}(x,y) \tag{6.49}$$

where θ_j is the direction of the jth spoke, and the point response would be the superposition of the ridges from the constituent spokes:

$$s(x,y) = \sum_{j=1}^{N_\theta} L_{\theta_j}(X,Y). \tag{6.50}$$

This function is plotted in Figure 6.6 for $N_\theta = 12$, 16, 24, 32, 48, and 64. Each sampling spoke produces its own perpendicular ridge, and these individual

(a)

(d)

(b)

(e)

(c)

(f)

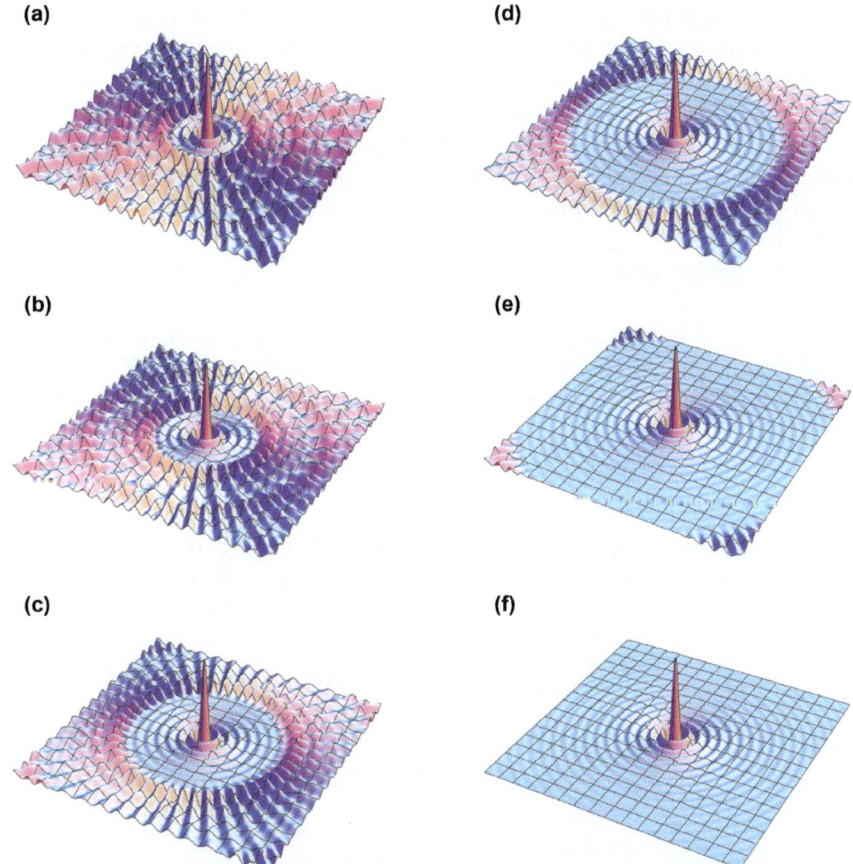

Figure 6.6 The radial sampling point response. Shown here are point responses corresponding to radial sampling patterns with varying numbers of radial spokes. Each point response was calculated by adding together N_θ copies of the ridge pattern shown in Figure 6.5b; each copy was rotated by $\Delta\theta = \pi/N_\theta$ relative to the previous, to achieve a uniform distribution in angular space. The clear zone with its circularly symmetric J_0 truncation artifacts and the higher-order J_{2N}, J_{4N}, etc. terms emerge naturally from the mutual interference of the ridges. The horizontal scaling of the pattern depends on the number of sampling points on each spoke N_r, but the area shown would approximately correspond to a region $W \times W$ where $\Delta r = 1/(2^{1/2}W)$ and $N_r = 32$ points per spoke (the exact width of the region shown is $\dfrac{80\sqrt{2}}{\pi N_r} W$ assuming $\Delta r = 1/(2^{1/2}W)$). (a) $N_\theta = 12$. A small clear zone is visible in the center. The J_{2N}, J_{4N}, and J_{6N} terms become active at progressively larger distances from the center. (b) $N_\theta = 16$. The clear zone is larger. The J_{2N} and J_{4N} terms start farther from the center than in (a), and the J_{6N} terms are only barely visible at the four corners. (c) $N_\theta = 24$. The clear zone covers half the spectrum. The J_{4N} artifacts are only visible at the corners. (d) $N_\theta = 32$. The clear zone reaches almost to the edge of the spectrum, with the J_{2N} artifacts beginning just inside the spectral boundary. (e) $N_\theta = 48$. J_{2N} artifacts are only barely visible in the corners, with the rest of the spectrum contained within the clear zone. (f) $N_\theta = 64$. The clear zone covers the entire spectrum.

ridges can easily be identified towards the edges of the point response. However, at the center of the point response, interference effects from taking the superposition of these individual ridges leads to a very different result: the negative-valued sinc sidelobes and extended sinc wiggles from one ridge cancel out some of the positive intensity from the neighboring ridges, leading to a "clear zone" in the middle of the point response where the artifacts are circularly symmetric. Within this zone, the directional artifacts from limiting sampling in θ have cancelled out.

To better understand the phenomenon of the "clear zone," consider an alternative formulation for the radial sampling pattern and its point response.[31,33,36–38] If N_r, r_{max}, and Δr are the same for all spokes in the dataset, a set of radial spokes could be described just as well as a set of concentric rings. A single ring of $2N_\theta$ sampling points at radius r_0 can be written as:

$$c_{r_0, N_\theta}(r, \theta) = \delta(r - r_0)\left(\frac{N_\theta}{\pi}\right) \mathrm{III}\left(\frac{\theta N_\theta}{\pi}\right). \tag{6.51}$$

To find the point response, we begin by rearranging the polar Fourier transform of eqn (6.34) so that the inner integration is taken over θ and so that the transform kernel is evaluated in polar coordinates:

$$F(X, Y) = \int_0^{2\pi} \int_{-0}^{\infty} f(r, \theta) e^{-2\pi i r (X\cos\theta + Y\sin\theta)} r\, dr\, d\theta$$

$$= \int_0^{\infty} \int_0^{2\pi} f(r, \theta) e^{-2\pi i r (R\cos\Theta\cos\theta + R\sin\Theta\sin\theta)} r\, d\theta\, dr \tag{6.52}$$

$$= \int_0^{\infty} \int_0^{2\pi} f(r, \theta) e^{-2\pi i r R \cos(\theta - \Theta)} r\, d\theta\, dr$$

where $\Theta = \arctan Y/X$. Using the identity:

$$e^{iu\cos\phi} = \sum_{k=-\infty}^{\infty} i^k J_k(u) e^{ik\phi}, \tag{6.53}$$

where J_k is the Bessel function of order k, and letting $u = 2\pi rR$ and $\phi = \theta - \Theta$, this becomes:[37,38]

$$F(X, Y) = \int_0^{\infty} \int_0^{2\pi} \sum_{k=-\infty}^{\infty} f(r, \theta) i^k J_k(2\pi rR) e^{ik(\Theta - \theta)} r\, d\theta\, dr$$

$$= \int_0^{\infty} \left(\sum_{k=-\infty}^{\infty} \left[\int_0^{2\pi} f(r, \theta) e^{-ik\theta}\, d\theta \right] i^k J_k(2\pi rR) e^{ik\Theta} \right) r\, dr. \tag{6.54}$$

The integral in square brackets carries out a harmonic analysis with respect to θ on a ring of values, of radius r, taken from the function $f(r, \theta)$; the resulting coefficients are used as weights for the terms of a summation, each term being a Bessel function with respect to radius and a sinusoid with respect to direction. As an example, if $f(r, \theta)$ is circularly symmetric at radius r, the harmonic analysis (inner integral) will identify that the function f is best fit at this radius by a zero-frequency (DC) wave around the ring. Only a single term of the summation will have a nonzero weight: the zero-order, circularly symmetric term. As a second example, if $f(r, \theta)$ has four-fold rotational symmetry at radius r, the harmonic analysis will identify that the function can be fit by a wave with a frequency of four cycles around the ring, along with waves at eight cycles, sixteen cycles, *etc.* The corresponding terms will be included in the summation, the fundamental term varying as a fourth-order Bessel function with respect to R and as a sinusoid with four cycles around the ring with respect to Θ, and the higher-order harmonic terms varying at integral multiples of four. Finally, the outermost integral combines the contributions from the integration and summation operations at all of the various r values. These operations produce the same result as those of eqn (6.48), where f was first transformed with respect to r and then backprojected; in this case, we first transform with respect to θ and then add together the resulting Bessel-sinusoid functions.

Having rearranged the polar Fourier transform, we now compute the transform for a single ring of sampling points at radius r_0. Inserting the angularly dependent part of eqn (6.51) into the inner integral of the polar transform, the harmonic analysis with respect to θ yields:[36–38]

$$\int_0^{2\pi} c_{r_0,N_\theta} e^{-ik\theta} \, d\theta = \int_0^{2\pi} \left(\frac{N_\theta}{\pi}\right) III\left(\frac{\theta N_\theta}{\pi}\right) e^{-ik\theta} \, d\theta$$

$$= 2N_\theta III\left(\frac{k}{2N_\theta}\right). \tag{6.55}$$

Harmonic analysis of a ring of $2N_\theta$ sampling points, represented as $III(\theta N_\theta/\pi)$, reveals that the ring can be fit by waves with $2N_\theta$ cycles or integer multiples thereof, represented as $III(k/2N_\theta)$. This in turn means that the summation in eqn (6.54) will contain only terms with orders that are integer multiples of $2N_\theta$:

$$\sum_{k=-\infty}^{\infty} \left[\int_0^{2\pi} c_{r_0,N_\theta} e^{-ik\theta} \, d\theta\right] i^k J_k(2\pi r_0 R) e^{ik\Theta} = \sum_{k=-\infty}^{\infty} 2N_\theta III\left(\frac{k}{2N_\theta}\right) i^k J_k(2\pi r_0 R) e^{ik\Theta}$$

$$= 2N_\theta \sum_{k \in K} i^k J_k(2\pi r_0 R) e^{ik\Theta},$$

$$K = \{k: 0, \pm 2N_\theta, \pm 4N_\theta, \dots\}. \tag{6.56}$$

Since we are considering only a single ring of points at $r = r_0$, the outer integral of eqn (6.54) is unnecessary, and the transform is simply:

$$C_{r_0, N_\theta} = 2N_\theta \sum_{k \in K} i^k J_k(2\pi r_0 R) e^{ik\Theta} r_0, \quad K = \{k: 0, \pm 2N_\theta, \pm 4N_\theta, \dots\}. \quad (6.57)$$

Adding together the contributions from all N_r rings of sampling points, the full point response is:

$$S(X, Y) = \sum_{j=1}^{N_r} C_{r_j, N_\theta}$$

$$= 2N_\theta \sum_{j=1}^{N_r} \sum_{k \in K} i^k J_k(2\pi r_j R) e^{ik\Theta} r_j, \quad K = \{k: 0, \pm 2N_\theta, \pm 4N_\theta, \dots\}.$$

$$(6.58)$$

Thus every ring of sampling points j at radius r_j contributes a zero-order term $J_0(2\pi r_j R)$ and a set of higher-order terms:

$$\dots ,$$

$$J_{-4N_\theta}(2\pi r_j R) e^{-4N_\theta i\Theta} r_j,$$

$$J_{-2N_\theta}(2\pi r_j R) e^{-2N_\theta i\Theta} r_j,$$

$$J_{2N_\theta}(2\pi r_j R) e^{2N_\theta i\Theta} r_j, \quad (6.59)$$

$$J_{4N_\theta}(2\pi r_j R) e^{4N_\theta i\Theta} r_j,$$

$$\dots$$

the superposition of which constitutes the point response.

To understand this point response and its implications for reconstruction from radial data, it is helpful to separate out the zero-order terms from the higher-order terms. Combining the zero-order terms from all of the rings results in the series:

$$\sum_{j=1}^{N_r} J_0(2\pi r_j R) r_j. \quad (6.60)$$

Bracewell and Thompson showed[39] that this series could be expanded as:

$$\sum_{j=1}^{N_r} J_0(2\pi r_j R) r_j = \frac{r_{max}}{2\pi R} J_1(2\pi r_{max} R) + \sum_{k=1}^{\infty} \zeta_k(R) \quad (6.61)$$

where we refer to the first term on the right as the "main term" and the series which follows as the "ringlobes." There are infinitely many ringlobes, with the kth ringlobe having the form

$$\zeta_k(u) = \int_0^\pi \xi(u\cos\phi - k/\Delta r) + \xi(u\cos\phi + k/\Delta r)\,\mathrm{d}\phi \qquad (6.62)$$

where:

$$\xi(u) = 2r_{\max}\,\text{sinc}(2r_{\max}u) - r_{\max}\,\text{sinc}^2(r_{\max}u) \qquad (6.63)$$

is the same basic lineshape that appears in the Fourier transform of the FBP filter function [eqn (6.31)] and in the point response of a single radial spoke [eqn (6.48)]. The zero-order terms are illustrated in Figure 6.7. The main term (J_1) represents the signal itself, picking up the circularly symmetric equivalent of the familiar "sinc-wiggle" truncation artifacts owing to the truncation of the sampling at $r_{\max}$ (Figure 6.7a). These ripples are observed within the clear zone of Figure 6.6, and they can be corrected by apodization in r. The same J_1 term arises in optical applications when radiation passes through a circular aperture, in which context it is known as an Airy pattern. The ringlobe terms (Figure 6.7b) are aliases resulting from the fact that the samples are collected on rings spaced at an interval of Δr. The ringlobes appear in the frequency domain spaced at intervals of $1/\Delta r$, and represent the r-dimension analog of conventional aliasing artifacts; they arise from the interaction of the alias ridges of Figure 6.5a. As long as the radial dwell time Δr is less than the reciprocal of the diagonal spectral width of the spectrum, these ringlobes will not intrude on a spectrum and can be ignored. The combination of the main and ringlobes terms yields the pattern shown in Figure 6.7c; Figure 6.7d shows how this would appear for a Lorentzian function after apodization to remove the truncation ripples.

The zero-order terms of the point response thus reflect attributes of the discrete sampling in r, including its truncation at $r_{\max}$ and its Δr periodicity. In contrast, the higher-order terms of the point response reflect the discrete sampling in θ (Figure 6.6). For this reason, they are *not* circularly symmetric, but rather have a sinusoidal shape relative to the direction Θ, with a periodicity that depends on the number of sampling points on each ring (or equivalently, on the number of radial spokes). They also have a dependence on R, and it is this dependence that gives rise to the artifact-free "clear zone," owing to the peculiar property of Bessel functions that:[38]

$$J_k(u) \approx 0, \quad u < k. \qquad (6.64)$$

Thus the higher-order terms will be nearly zero out to a frequency-domain radius that depends on both the number of sampling points and the corresponding time-domain radii of the sampling rings.[6,33] The size of the clear

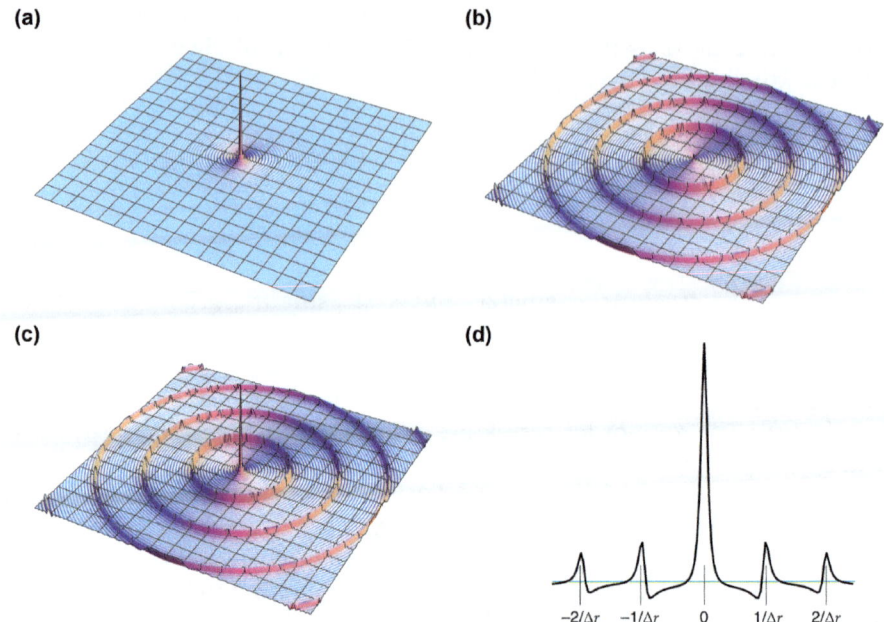

(a)   **(b)**

(c)   **(d)**

$$-2/\Delta r \quad -1/\Delta r \quad 0 \quad 1/\Delta r \quad 2/\Delta r$$

Figure 6.7   The zero-order terms of the radial point response. The J_0 terms of the radial sampling point response are circularly symmetric and arise owing to the discrete nature of the sampling with respect to r. The series of J_0 terms can be rewritten as a single J_1 term plus infinitely many ringlobes. (a) The J_1 term of the rewritten series represents the main peak and any associated truncation artifacts. (b) The ringlobes are aliases resulting from the discrete sampling in r, appearing at intervals of $1/\Delta r$. (c) The full zero-order component of the point response, combining the J_1 term and the ringlobes. (d) The convolution of a Lorentzian signal with the zero-order component of the radial point response, shown in cross-section. (Panel (d) is reprinted from *J. Magn. Reson.*, **184**, B. E. Coggins and P. Zhou, Sampling of the NMR time domain along concentric rings, 207–221, Copyright (2006), with permission from Elsevier.)

zone for a given ring j is determined by the radius $R_{cz,j}$ at which the first set of higher-order terms (orders $-2N_\theta$ and $2N_\theta$) become active:

$$J_{2N_\theta}(2\pi r_j R)e^{2N_\theta i\Theta} \approx 0, \quad R < R_{cz,j} \equiv \frac{N_\theta}{\pi r_j}. \tag{6.65}$$

Note that for a given ring, the size of the clear zone in the frequency domain is inversely proportional to the radius of the ring in the time domain. The sampling pattern as a whole consists of a set of concentric rings, all of which have $2N_\theta$ sampling points, at various r_j in the time domain; the size of the clear zone for the overall pattern R_{cz} would therefore be determined by the ring with the smallest clear zone:

$$R_{cz} = \min_j R_{cz,j} \tag{6.66}$$

which would in fact be the ring with the largest time-domain radius, r_{max}:

$$R_{cz} = R_{cz,N_\theta} = \frac{N_\theta}{\pi r_{max}}. \tag{6.67}$$

Eqn (6.67) is, in essence, a radial equivalent to the Nyquist sampling condition: it specifies the minimum number of radial samples needed to achieve artifact-free Fourier transforms at a given resolution and for a given spectral width. The Nyquist condition in conventional sampling can be derived directly from the point response for that sampling, which shows aliases at intervals of $1/\Delta t$ and clear space in between them; for a spectrum of finite width W on each axis (a *band-limited* function), if Δt is small enough, those aliases will be far enough apart that they will not intrude on the spectrum[32] (Figure 6.8a). The sampling requirement, that Δt must be less than $1/W$, arises naturally from the spacing of the aliases in the point response. The sampling requirement for radial sampling can, in a similar manner, be derived from its point response. The aliasing artifacts in radial sampling take a different form from those in conventional sampling: the artifacts cover all of the plane except for a circular region of radius R_{cz} (Figure 6.8b). For a spectrum of finite width W on each axis, as long as R_{cz} is large enough (generally, $R_{cz} \geq 2^{1/2}W$, accounting for the fact that the spectrum extends further along its diagonal than on either axis), those artifacts will not intrude on the spectrum. The criteria to determine R_{cz} are the r_{max}, the maximum evolution time or equivalently the resolution of the spectrum, and N_θ, the number of radial spokes. The size of the clear zone is directly proportional to the number of radial spokes and inversely proportional to the maximal evolution time; a given number of spokes permits the calculation of an artifact-free spectrum, but only of a limited size and at a limited resolution.

If one assumes a radial dwell time $\Delta r = 1/(2^{1/2}W)$, one can substitute into eqn (6.67) the requirement that R_{cz} must be greater than or equal to $2^{1/2}W$ and the fact that $r_{max} = N_r\Delta r$ and solve for N_θ to determine the number of radial spokes needed to ensure that an entire $W \times W$ spectrum is artifact-free. The resulting expression is remarkably simple:

$$N_\theta \geq \pi N_r, \tag{6.68}$$

and it leads to formulae for the optimum number of radial spokes and the optimum number of sampling points per spoke for a given total number of measurements $N = (N_\theta/2)N_r$:

$$N_\theta = \sqrt{2\pi N}, \quad N_r = \sqrt{\frac{2}{\pi}N}. \tag{6.69}$$

While these results show that it is possible to obtain artifact-free spectra from radial data, they unfortunately also show that it is only possible to do so with *more* measurements than would be needed for conventional sampling. One can easily see this by computing the resolution achievable in

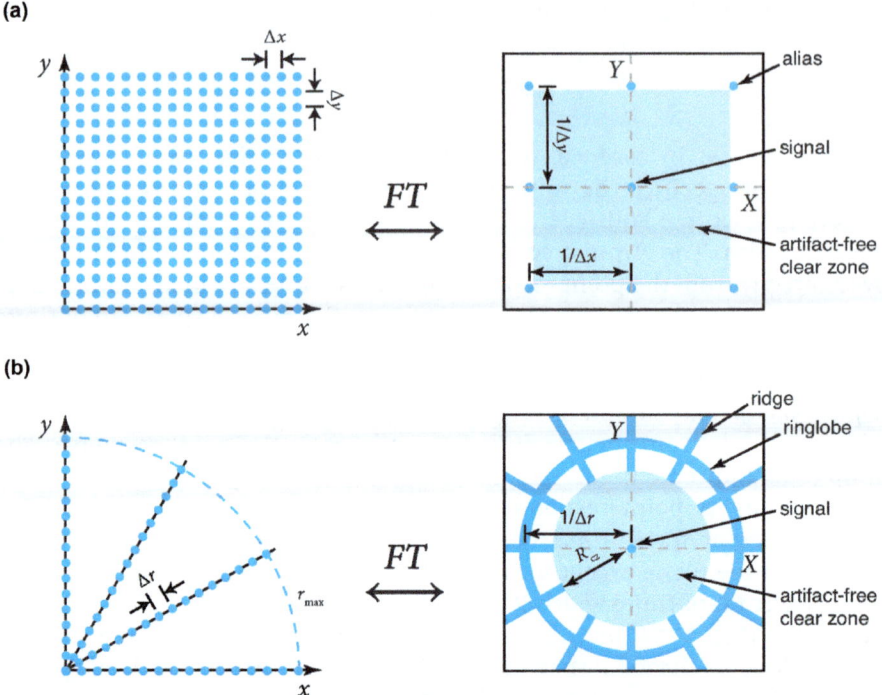

Figure 6.8   The Nyquist sampling condition and its radial equivalent can be derived from the conventional and radial point responses, respectively. (a) The Nyquist condition in conventional sampling can be derived from the point response by examining the shaded artifact-free area between alias peaks (right), which relates to the spacing of the samples in the time domain (left). (b) An equivalent condition for radial sampling can be derived from the point response by examining the size of the artifact-free clear zone (right), the shaded area inside of the first ringlobe at $1/\Delta r$ and inside of the start of the J_{2N} ridge artifacts at $R_{cz} = N_\theta/\pi r_{max}$. The size of this clear zone is determined by the number of radial spokes and the resolution.

conventional sampling $\rho_{Cartesian}$ for a spectrum of N total points and $N^{1/2}$ points per axis when the Nyquist condition ($\Delta x \leq 1/W$) is satisfied:

$$\rho_{Cartesian} = \frac{1}{x_{max}} = \frac{1}{N_x \Delta x} = \sqrt{\frac{1}{N}} W \tag{6.70}$$

where x represents one of the two axes, and comparing it to the resolution of a radially sampled spectrum ρ_{radial} of N total points with the points allocated to satisfy the radial sampling conditions in eqn (6.68) and (6.69):

$$\rho_{radial} = \frac{1}{r_{max}} = \frac{1}{N_r \Delta r} = \sqrt{\frac{\pi}{N}} W. \tag{6.71}$$

For the same number of measurements N satisfying the applicable sampling condition, the resolution achievable with radial sampling is $\pi^{1/2}$-fold worse than in conventional sampling. Alternatively, one can calculate that the number of measurements needed to achieve a resolution ρ while avoiding aliasing artifacts is:

$$N_{\text{Cartesian}} = \frac{W^2}{\rho^2} \qquad (6.72)$$

for conventional sampling and:

$$N_{\text{radial}} = \pi \frac{W^2}{\rho^2} \qquad (6.73)$$

for radial sampling, *i.e.* π-fold more measurements are needed in the radial case.

The original goal of using radial sampling rather than conventional sampling was to reduce the measurement time; it is clear from the radial sampling point response that it is not possible to do so while processing the data with a polar Fourier transform without introducing artifacts. Each peak will have its own artifact pattern, consisting of the point response centered on the peak and scaled relative to the peak height. The nature of these artifacts is as shown by the point response: each peak is surrounded by a clear zone, but outside of the clear zone a pattern of $2N_\theta$ ridges appears, with each ridge having an intensity that is approximately $1/N_\theta$ of the peak's intensity.

6.3.7 The Information Content and Ambiguity of Radially Sampled Data

The point response for radial sampling indicates the artifacts that would be produced if radial data were processed by a polar FT, or equivalently if a set of projections were used to reconstruct a spectrum by filtered back-projection. The point response also has a deeper meaning, in that it reports on the information content of the data, and on ambiguities that would affect any processing method operating on such data.

Recall that the Fourier transform is not simply a processing method, but rather a mathematical relationship describing the frequency content of a time-domain signal. Discrete sampling converts a continuous signal (the actual RF emissions from an NMR resonance) into a discontinuous signal (the measured data), and it is the discontinuous signal that a processing method must attempt to analyze. The features of the point response show how the frequency content of the continuous signal is altered when the discontinuous signal is generated by the discrete sampling process, introducing ambiguities and uncertainties. The spectrum—the frequency content—of the discontinuous signal includes the true resonance frequency as well as artifactual aliases. These aliases could be seen as alternative hypotheses for the location of the true signal, all of which fit the measured data equally well.

In conventional sampling, for example, a continuous signal at a single frequency becomes a set of identical aliases at regular intervals, reflecting the fact that samples taken at a regular interval Δx do not have enough information to distinguish frequencies that differ by factors of $1/\Delta x$; each alias position is an equally valid hypothesis about the location of the true signal, and it is only with the help of *a priori* knowledge about the frequency band of the true signal that we can resolve the ambiguity. The point response thus indicates a fundamental weakness in the measured data arising from how it was sampled, and any processing method attempting to use this information would face the same ambiguity.

The point response for radial sampling shows that this method of data collection can capture enough information to locate signals accurately within a zone of radius R_{cz}, that radius being determined by the resolution at which we seek to locate the signals and the number of spokes we have observed. Beyond that radius, aliases take the form of low-level ridges, unlike the full-size aliases of conventional Cartesian data collection. Keeping in mind the goal of reducing the measurement time, it would seem that some amount of deliberate undersampling in the radial dimension [*i.e.* setting N_θ below the value required by eqn (6.67)] might therefore be tolerable, coming only at the expense of low-level artifacts, which might be suppressed computationally.

The fact that no full-size aliases are visible in the point response does not mean that radial sampling will be free of ambiguity about signal positions, but it does mean that such ambiguity will take a different form. The ridges in the pattern provide a clue into the type of information captured by radial sampling, and also into the form of true ambiguity that can result. Recall from eqn (6.48) that the point response for a single radial spoke is a ridge running in the perpendicular direction. Thus the information from a single radial spoke in the time domain allows for a signal to be located along a line in the frequency domain. As multiple spokes are added, the locations of the signals in the frequency domain are effectively triangulated, with the signals emerging at the locations where these constraining lines intersect. This is none other than the lattice analysis from Section 6.3.1 at work again: it is as if one were tracing back from the projections to find where the original signals might lie.

The full point response shows that with enough of these constraints, the ridges in the point response cancel out everywhere other than the true signal positions, and signals can be located unambiguously. If, however, the number of constraints is low enough that ridges are visible, there is a possibility of obtaining an ambiguous spectrum, with multiple equally valid hypotheses about signal positions. True signals are found where ridges intersect, but if there are multiple signals there could be many possible intersections, some corresponding to true signals and others being spurious. It would not, however, be possible to distinguish these based on the measured data. Such problems become more serious as the number of spokes decreases, and with it both R_{cz} and the difference in relative height between the ridges and the peaks.

It is therefore possible to encounter aliases in undersampled radial spectra just as in undersampled conventional spectra, but in the radial case the positions of the aliases depend on the complex geometric inter-dependence of the measured spoke directions and the positions of the true signals. Ridge artifacts and false peaks would be seen if polar FT processing or FBP reconstruction is employed. Other reconstruction methods, such as those derived from the lattice analysis, may employ alternative computations and thereby avoid or suppress the ridge artifacts, but if the undersampling is too severe they will not be able to overcome the true ambiguity about peak positions in the measured data. For any spectrum to be processed with fil-tered backprojection, the most useful criterion for determining the per-missible degree of undersampling might well be the height of the resulting ridge artifacts and the needed dynamic range. For lattice-derived algorithms, greater undersampling may well be possible, but application of the algo-rithms described in Section 6.3.2 for selecting tilt angles is the only way to guarantee a correct reconstruction.

6.4 Reconstruction from Projections: Practice

Having described the theory of reconstruction from projections, we now consider how reconstruction has been carried out in practice in NMR, describing the various methods of reconstruction in the order of their introduction.

6.4.1 The Lower-value Algorithm

When Kupče and Freeman first introduced the idea of reconstructing a spectrum from its projections, they simultaneously introduced a simple and intuitive method for reconstruction that they termed the *lower-value method* (hereafter abbreviated "LV"; also referred to in Kupče and Freeman's later papers as the *lowest-value* or *minimum-value method*), which has the virtue of being able to reconstruct spectra from extremely small numbers of pro-jections, in many cases without introducing artifacts.[5] This method can be represented as an operator **LV** applied to the set of measured projections $\mathbf{P} = \{P_{\theta_1}, P_{\theta_2}, \ldots, P_{\theta_{N_\theta}}\}$:

$$\mathbf{LV}[\mathbf{P}](X, Y) = P_\alpha(R), \quad \alpha = \arg\min_\theta |P_\theta(R')| \tag{6.74}$$

where $R = X \cos\alpha + Y \sin\alpha$ and $R' = X \cos\theta + Y \sin\theta$. Recall the lattice an-alysis presented in Section 6.3.1, where signal positions were triangulated based on the intersections of backprojection vectors. Only those lattice points where backprojection vectors from all N_θ projections intersect could

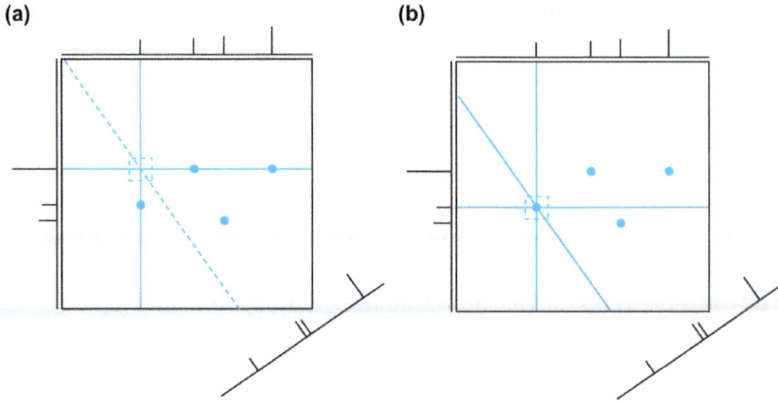

Figure 6.9 The lower-value reconstruction algorithm. (a) To determine a reconstruction value for the point at the center of the dashed box, the LV algorithm
looks to the corresponding projection values, as traced by the three lines.
On two of the projections, these values would be above the noise (solid
lines), but the third is at the noise (dashed line). The smallest of the
three values would be the noise value (dashed line), and this position in
the reconstruction would therefore correctly be assigned a noise value
rather than a signal. (b) If all three projection values are above the noise,
the reconstruction will be assigned a value above the noise. For this case
(dashed box), the three lines all lead to projection values above the noise
(solid lines), so the reconstruction correctly contains a peak at this
position.

possibly correspond to real signals, since only those lattice points are supported by data from all N_θ projections. The LV algorithm translates this
analysis into a reconstruction procedure by assessing the values at the
positions on the projections corresponding to a given point in the reconstruction, and assigning the projection value with the lowest absolute
value (Figure 6.9). Spectral positions that are not part of the lattice would
have at most only a single projection value above the noise, and many other
projection values at the noise level; taking the projection value with the
lowest magnitude would therefore result in assigning a noise value at that
position in the reconstruction. Lattice points supported by only some of the
projections would have a mix of noise values and above-noise values; taking
the lowest would result in a noise value being assigned, eliminating that
lattice position from consideration as a real peak. Only those positions
where all projection values are above the noise would be assigned reconstruction values above the noise.

The lower-value method is highly nonlinear, in a way that is difficult to
analyze theoretically; nonetheless, several conclusions can be drawn. The LV
method can successfully reconstruct a spectrum from a very small number of
projections, and the projections need not be distributed uniformly. At the
same time, if only a very small number of projections is used, false signals
can easily result, arising owing to the ambiguities discussed above in

Sections 6.3.1, 6.3.2, and 6.3.5. LV works best with point sources: the lowest-value operation does not reconstruct extended lineshapes correctly. When signals overlap on the projections, as is likely in any sufficiently complicated spectrum, the reconstructed intensities can be inaccurate.

The LV method suffers from a serious restriction in sensitivity: the sensitivity of the final reconstruction is determined by that of the weakest projection, and there is no additivity or signal-averaging across multiple projections. To use LV, therefore, it is necessary to collect each and every projection with sufficient transients so that all signals are visible above the noise on every projection. This sensitivity limitation is in some ways masked by a cosmetic consequence of the minimum operator in eqn (6.74): for points in the reconstruction that do not contain signals, the selection based on the lowest absolute value among the projections will make the thermal noise look weaker than it truly is. Thus the final reconstruction will appear to have a higher signal-to-noise ratio (SNR) than any of the input projections, even though the ability to detect signals is no better than for the projection with the weakest SNR.

6.4.2 Backprojection Without Filtering

Shortly after introducing the lower-value method, Kupče and Freeman proposed using an additive method where a given point in the reconstruction is assigned the sum of its projection values.[19] In the context of a set of discrete projections $\mathbf{P}$, this would be described by the operator:

$$\mathbf{B}[\mathbf{P}](X, Y) = \sum_{\theta} P_{\theta}(R) \tag{6.75}$$

where $R = X \cos \theta + Y \sin \theta$. This is, of course, simply a discrete version of the backprojection operator described in Section 6.3.3, which extends projected intensities across the reconstruction space and takes their superposition. This results in ridges, with peaks appearing where those ridges intersect (Figure 6.10). Like the lower-value method, it is essentially a means of translating the lattice analysis into a full-spectrum reconstruction, with the triangulation of peak positions occurring *via* the emergence of peaks where explicit backprojection ridges additively intersect. Unlike the lower-value method, adding together contributions from every projection results in full signal-averaging over all the data, the sensitivity being determined by the total measurement time for the full set of projections rather than for a single projection.

Backprojection also shows up as part of the analytical inversion of the Radon transform, but only after a filtering step. When the filtering is not included, there is no cancellation among the ridges and therefore no clear zone. Peaks are broadened owing to the overweighting of the early evolution times.

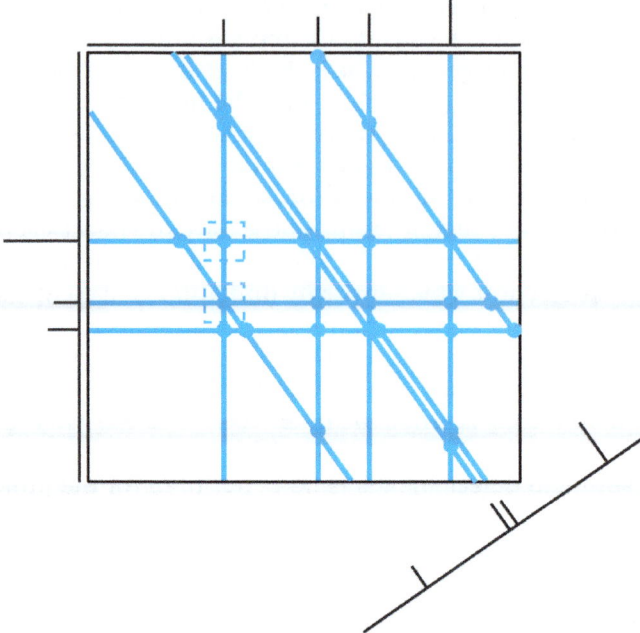

Figure 6.10 Backprojection reconstruction. Pure backprojection (without filtering) assigns to a position in the reconstruction the sum of the corresponding projection values. This results in ridges (thick lines) and peaks at every position where ridges intersect (filled circles), including positions where real signals are found (lower dashed box) as well as positions that do not correspond to real signals (upper dashed box).

A major problem in unfiltered backprojection is the occurrence of artifactual peaks where the ridges from one signal intersect with those of another. In FBP, the cancellation of ridges *via* the clear-zone phenomenon ensures that such spurious intersections are nonexistent when the sampling criterion is met, and infrequent and generally of low intensity even if the sampling criterion is not met; unfiltered backprojection, on the other hand, has no such cancellation, and typical reconstructions will generate many such spurious peaks. Kupče and Freeman suggested that a variant of the CLEAN method described in the next chapter might be useful in removing some of the artifacts that appear in backprojection reconstructions.[40]

6.4.3   The Hybrid Backprojection/Lower-value Method

The present authors proposed a hybrid backprojection/lower-value (HBLV) method that achieves a better sensitivity than the lower-value method while simultaneously suppressing ridge artifacts and, in favorable cases, spurious peaks.[21] Let $[\mathbf{P}]^k = \{\mathbf{K} : \mathbf{K} \subset \mathbf{P}, |\mathbf{K}| = k\}$ represent the set of all k-subsets of $\mathbf{P}$. The HBLV method is computed as:

$$\mathbf{HBLV}[\mathbf{P}](X, Y) = \sum_{P_\theta \in \mathbf{K}_{\min}} P_\theta(R), \quad \mathbf{K}_{\min} = \underset{\mathbf{K} \in [\mathbf{P}]^k}{\arg \min} \left| \sum_{P_\theta \in \mathbf{K}} P_\theta(R) \right| \quad (6.76)$$

where $R = X \cos \theta + Y \sin \theta$. For each position in the reconstruction, super-positions are calculated for all possible combinations $\mathbf{K}$ of k out of N_θ projections, and the superposition with the lowest absolute value $\mathbf{K}_{\min}$ is taken (Figure 6.11). By computing superpositions of k projections, signal-averaging occurs and the sensitivity is improved by $k^{1/2}$. Because the superpositions are taken over subsets of the N_θ projections, no ridge artifact will be present in all of the superpositions, and the subsequent lower-value comparison can suppress them. Spurious peaks at intersections supported by only some of

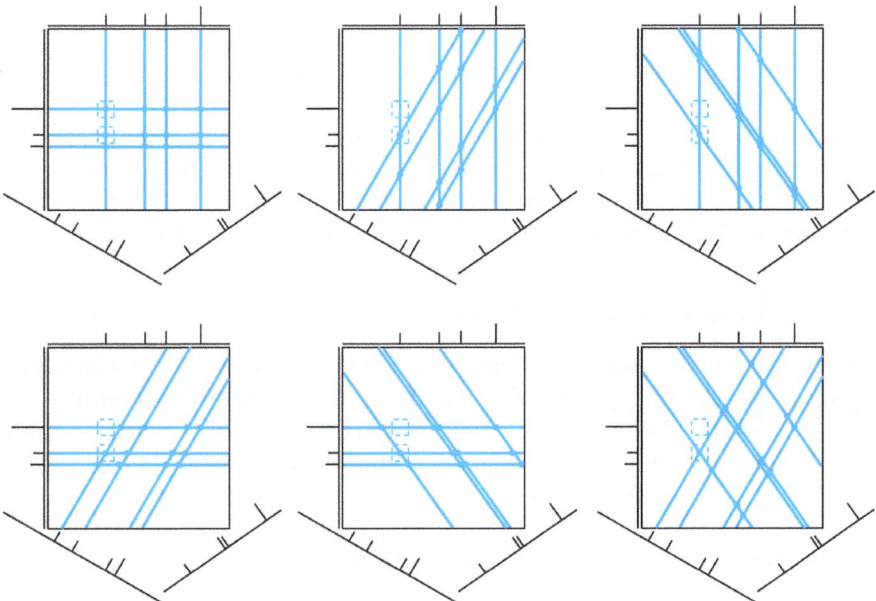

Figure 6.11 Hybrid backprojection/lower-value reconstruction. The hybrid recon-struction method requires a minimum of four projections to work successfully in this example. Using $k = 2$, the method computes back-projection reconstructions for all combinations of two out of four projections. There are six possible combinations. The method then assigns to the reconstruction the lowest corresponding value out of the six combinations. The point at the center of the upper dashed box shows a peak in one combination, ridge artifacts in four combinations, and noise in one combination; the lowest value out of the six would be the noise value, so this point would correctly contain noise rather than a peak in the final reconstruction. The point at the center of the lower dashed box contains a peak in all six combinations, so it would correctly contain a peak in the final reconstruction.

the N_θ projections are also generally removed, as they are unlikely to be present in all of the subsets, especially if the subsets are calculated over a small k. Increasing k towards N_θ boosts sensitivity at the expense of peak broadening and an increased risk of false peaks appearing in the results.

The HBLV method is computationally intensive, as:

$$_{N_\theta}C_k = \frac{N_\theta!}{k!\,(N_\theta - k)!} \tag{6.77}$$

superpositions must be calculated. A small modification can improve the performance: if any superposition has an absolute value less than a predicted noise level, further evaluation of superpositions can be discontinued. This has the added benefit of producing an apparent noise level more closely resembling the true sensitivity, without the artificial decrease in noise produced by a full lower-value comparison.

Ridge and Mandelshtam pointed out that the computational burden of HBLV can be eliminated if all signals are positive, in which case the lowest-value superposition is simply the one over the k lowest-value projections.[41] Such a simplification breaks down for spectra containing negative signals. This discussion leads naturally to a second concern that can affect all lower-value reconstruction methods, including HBLV: if negative peaks are present, positive peaks and negative peaks can cancel on one or more projections and lead to an elimination of true signals in the final reconstruction. This is a fundamental weakness of such methods.

6.4.4   Filtered Backprojection

The reconstruction methods described above all suffer from significant limitations in accuracy; most of them also suffer from limitations in sensitivity. While they were shown to be effective for reconstructing simpler spectra, they were not able to reconstruct more complicated spectra such as NOESY. To overcome those limitations, the present authors proposed using the *filtered backprojection* (FBP) method described in Section 6.3.[42] For discrete NMR data, this could be described as:

$$\mathbf{FBP}[\mathbf{P}](X, Y) = \sum_\theta \mathbf{F}_R[\,|r|\; \mathbf{F}_R^{-1}[P_\theta(R)]] \tag{6.78}$$

where $R = X\cos\theta + Y\sin\theta$, though in practice one first prepares filtered projections:

$$\hat{P}_\theta(R) = \mathbf{F}_R[\,|r|\mathbf{F}_R^{-1}[P_\theta(R)]] \tag{6.79}$$

and then carries out backprojection using those filtered projections:

$$\mathbf{FBP}[\mathbf{P}](X, Y) = \mathbf{B}[\hat{\mathbf{P}}]$$

$$= \sum_\theta \hat{P}_\theta(R). \tag{6.80}$$

As discussed above, FBP is the analytical inversion of the projection operator, and it is known to produce a correct result when the sampling meets the criteria given in eqn (6.67). When the data are undersampled, it continues to reproduce all of the real signals correctly, with accurate intensities and lineshapes, but ridge artifacts and false peaks may also appear. The intensity levels for these artifacts are predictably low as long as the number of projections is sufficiently high. Like unfiltered backprojection, filtered backprojection carries out signal-averaging over the full data, and therefore achieves a sensitivity proportional to the full measurement time of all the projections. Unlike some of the other methods, FBP requires that the projection angles be distributed uniformly over the angle space. For NMR data, FBP is equivalent to processing the radial data using a polar Fourier transform, and any artifacts it produces can be understood as aliasing artifacts reflecting true ambiguity in the data.

6.4.5 Other Proposed Approaches to Reconstruction

Several additional methods have been suggested for reconstructing NMR spectra from radially sampled data.

Reasoning from the limitations of the lower-value method, Ridge and Mandelshtam proposed assigning to each location in the reconstruction not the lowest-value among the projections, but rather the most common value, as assessed from a smoothed histogram.[41] This would have the virtue of avoiding cancellation problems when negative signals are present.

Several groups have explored iterative methods, wherein a model of the spectrum is progressively optimized by comparing backcalculated projection data to the experimental data. The algebraic reconstruction technique (ART) dates back to the earliest days of tomography, and it involves solving systems of linear equations relating the reconstruction intensities to projected intensities, as in eqn (6.10).[43] Section 6.3.1 contemplates a system of equations limited to lattice points derived from orthogonal projections, but in the original ART, all points in the spectrum were included and many more projections were needed to avoid underdetermination problems. Because of the size of the linear system, it was typically solved iteratively rather than by diagonalization methods. ART was used by Lauterbur in the first MRI experiments.[4] More recently, Yoon *et al.* proposed a least-squares method for solving such a system of equations to determine a reconstruction.[44] Yoon *et al.* also proposed several other iterative methods for building models of the spectrum, including using maximum entropy as a regularization technique (also demonstrated by Mobli *et al.*[45]) and using Monte Carlo methods incorporating Bayesian reasoning for selecting parameter values.

Multiway decomposition was successfully employed by Malmodin and Billeter.[46] In this method, the reconstruction is modeled as a collection of orthogonal 1-D "shapes," which can be convolved to generate backcalculated projection data for comparison with the experimental data.

Mueller considered the possibility of deconvolving these "shapes" directly from the projection data.[47]

6.5 Applications of Projection–Reconstruction to Protein NMR

During the period when projection–reconstruction NMR was under development, a number of practical applications to protein NMR were reported, on both small and large proteins.

Kupče and Freeman first demonstrated their idea of reconstructing a spectrum from projections by successfully calculating the 3-D HNCO spectrum of ubiquitin from three projection datasets measured at 0°, 30°, and 90° from the CO axis.[5] It was as a part of this study that they introduced the lower-value algorithm, which is able, by virtue of its nonlinearity, to produce a usable spectrum from a very small amount of data, at least in favorable cases such as this one. Example planes from this spectrum are shown in Figure 6.12. The dramatic time savings achieved in this experiment illustrated the potential advantage of undersampled radial data collection: the three projection datasets required a total measurement time of 29 min 11 s, while an equivalent conventional 3-D spectrum required 18 h 54 min.

This demonstration was quickly followed by additional 3-D examples and by extensions to more dimensions. Two months after the first publication on projection–reconstruction, Kupče and Freeman reported successful reconstructions of 3-D HNCA and HN(CO)CA spectra on ubiquitin and the larger HasA;[48] this was followed by the successful reconstruction of a 4-D HNCOCA spectrum of ubiquitin,[49] all using the lower-value reconstruction method. For the 3-D experiments, three radial spokes were collected for each dataset, while seven spokes were recorded for the 4-D experiment. Shortly thereafter, the present authors, working also with Ron Venters, used the projection–reconstruction concept to produce the first reported 5-D NMR experiment, an HACACONH spectrum for the B1 domain of protein G[9]. Data were recorded for twelve radial spokes, and reconstruction was again carried out using the lower-value method. This paper was the first to point out the connections between recording projections of an NMR spectrum and the then-recently introduced GFT method[16] of determining chemical shifts from reduced dimensionality data acquisition.

A major limitation of the lower-value method is its requirement that each projection must be collected with enough transients for all signals to be visible. As a means of overcoming that restriction, Kupče and Freeman proposed the simple *unfiltered* backprojection method described in Section 6.4.2.[19] By adding together contributions from all projections, this method achieved a signal-to-noise ratio proportional to the measurement time of the full set of projections rather than of a single projection. This method was demonstrated with a 3-D HN(CA)CO on ubiquitin (Figure 6.13). To reduce the impact of backprojection's characteristic ridge artifacts, the authors used 10 radial spokes, contributing 18 projections. This reduces the height of any single ridge artifact to 1/18 of the height of its corresponding peak. At the same time, a sensitivity boost of $18^{1/2}$ was reported over the original projections or a corresponding lower-value reconstruction.

The first application of projection–reconstruction to a truly large system was by Venters *et al.*, who showed how the concept could be used in practice for sequential assignment of 30 kDa proteins.[21] This paper reported the development of a suite of 4-D TROSY backbone assignment experiments optimized for projection–reconstruction data collection, together with a new reconstruction algorithm, the hybrid backprojection/lower-value method. As described in Section 6.4.3, this algorithm was designed to overcome the limitations of the lower-value and backprojection reconstruction methods by using the beneficial features of each to overcome the limitations of the other (Figure 6.14). Using these tools, Venters *et al.* were able to obtain 4-D TROSY-HNCACB, TROSY-HN(CO)-CACB, intra-TROSY-HNCACB, TROSY-HNCACO, TROSY-HNCOCA, TROSY-HNCO$_{i-1}$CA$_i$, HACANH, and HACA(CO)NH spectra over 12.8 days of data collection time for two proteins, the 29 kDa HCA II and the 30 kDa calbindin D$_{28K}$—a measurement time roughly equivalent to what would be needed to measure a *single* 4-D spectrum by conventional methods. For each of these 4-D experiments, eight radial spokes were recorded. At this level of sparsity, backprojection reconstruction results in significant artifacts, while recording enough transients for each spoke to make all peaks visible on all projections—as is needed for lower-value reconstruction—would have required substantially more measurement time. The hybrid algorithm successfully split the difference, producing artifact-free spectra with sufficient sensitivity to assign ~95% of residues while still allowing a significant reduction in measurement time (Figures 6.14 and 6.15). With this high-resolution 4-D data in hand, it was possible to assign 41 resonances in HCA II that could not be assigned previously from conventional 3-D data.

Sidechain assignment experiments present similar sensitivity and spectral complexity considerations to sequential assignment, and it should come as no surprise that the projection–reconstruction methodology could be applied successfully to them. Jiang *et al.* reported 4-D HC(CCO)NH and intra-HC(C)NH experiments, each reconstructed by the lower-value method from six radial spokes collected over 47.5 h (Figure 6.16).[50] In this experiment, radial sampling was carried out only over the H and C dimensions, with conventional data collection for the N dimension, reducing the complexity of the reconstruction problem. The choice of six radial spokes was based on prior conclusions about the number of projections needed to ensure a high probability of successful reconstruction, as described in Section 6.3.2. The time savings compared to conventional 4-D spectroscopy at the same resolution was ~89%.

A much more significant challenge for projection–reconstruction was its application to 4-D NOESY.[42] The complexity and sensitivity constraints of NOESY spectra, together with the requirement for quantitative accuracy in NOESY reconstructions that would be used for structure calculations,

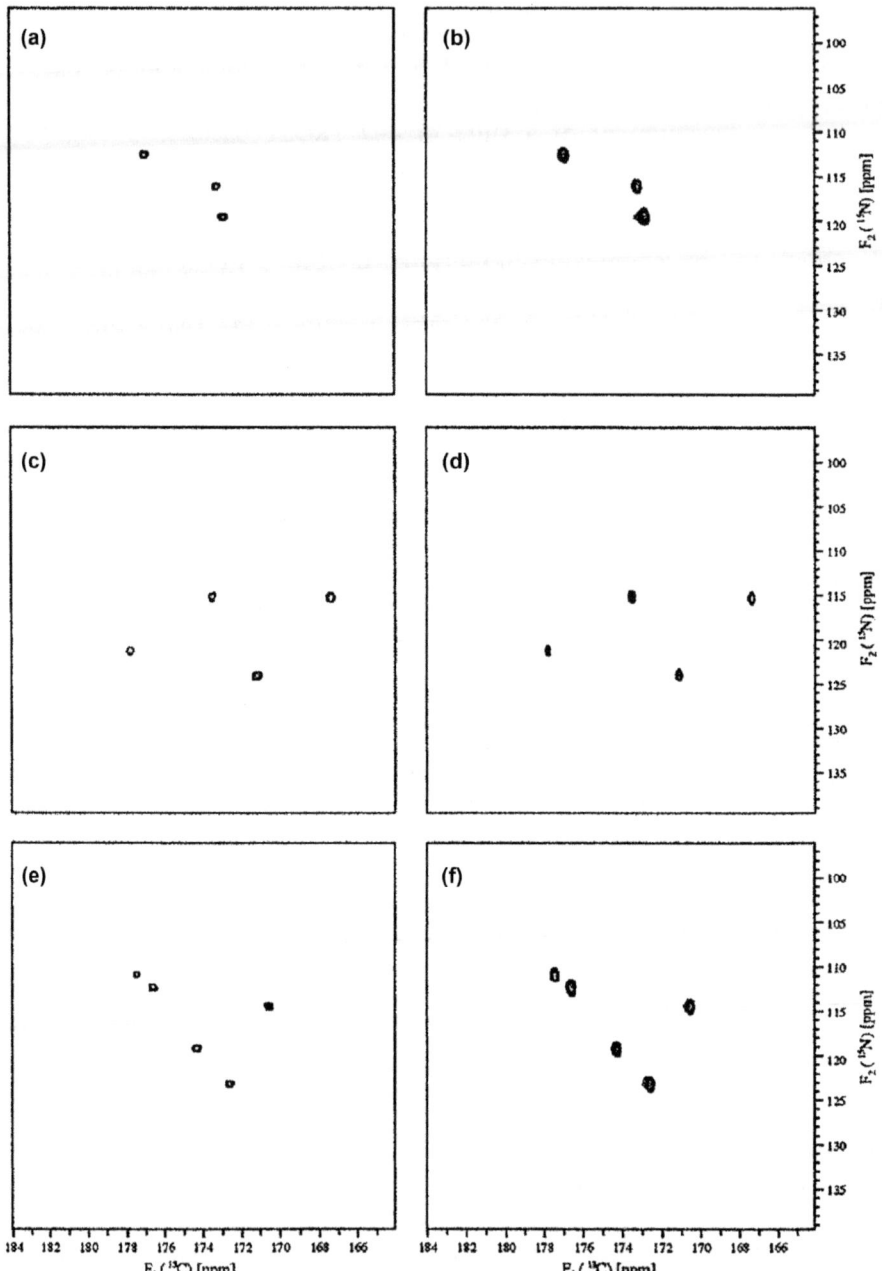

meant that alternative approaches to both data collection and reconstruction were needed. Filtered backprojection offered quantitative reconstruction of peak lineshapes and volumes, significantly reduced artifacts over unfiltered backprojection for the same number of measurements, and sensitivity proportional to the full measurement time across all radial spokes. With this method, the consequences of undersampling would be low-level ridge artifacts and potential low-level false peaks. Avoiding those pitfalls required measuring far more projections than in previous experiments—400 were used in the published example—but this still represented less than 5% of the time needed for recording a 4-D dataset at the same resolution using conventional methods. Unlike the lower-value or hybrid methods, FBP requires that projections be distributed uniformly in angle space. Following this approach, in the first reported example, the determination of a 4-D methyl/amide NOESY spectrum, the authors collected the 100 radial spokes over 88 hours on a 0.9 mM sample of the 29 kDa protein IICA II at 800 MHz. These data were processed using FBP to produce a 4-D NOESY spectrum at 128 points per axis digital resolution (Figure 6.17).

6.6 From Radial to Random

The many outstanding results obtained using projection–reconstruction between 2003 and 2007 suggested great promise, yet in only a few years the field of NMR had largely moved on, spurred by increasing awareness of the limitations of radial sampling and the emergence of generally superior approaches.

The inherent limitations of radial sampling were not immediately understood when the method was introduced, and for some time were a significant topic of discussion and investigation (see, for example, ref. 45). Lower-value methods allowed for reconstruction from very few projections, but with sometimes substantial inaccuracy in peak shapes, intensities, and volumes. Though many reconstructions were successful, there was a

---

Figure 6.12 The first reported projection–reconstruction experiment. Example C/N planes from the first reported NMR spectrum to be reconstructed from projections, a 3-D HNCO of ubiquitin, are shown in the left column, together with the corresponding planes from a conventionally sampled and processed version of the same experiment at right. The planes shown are at proton chemical shifts of (a/b) 7.28 ppm, (c/d) 8.31 ppm, (e/f) 8.77 ppm. The projections used in the reconstruction at left required 29 min 11 s of spectrometer time, while the conventional data at right required 18 h 54 min.
(Reprinted from *J. Biomol. NMR*, Reconstruction of the three-dimensional NMR spectrum of a protein, from a set of plane, **27**, 2003, 383–387, E. Kupče and R. Freeman, projections, (© Kluwer Academic Publishers 2003) with permission from Springer.)

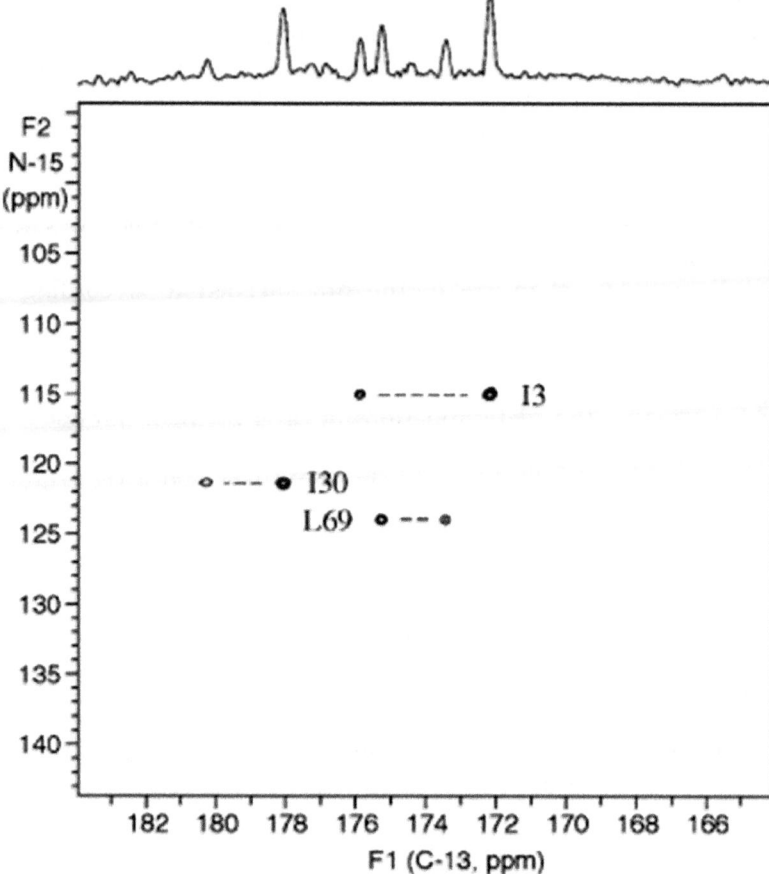

Figure 6.13 Backprojection Reconstruction. An example C/N plane at a proton coordinate of 8.3 ppm from the 3-D HN(CA)CO spectrum of ubiquitin, reconstructed from 18 projections using the backprojection method without a filter function.
(Reprinted with permission from E. Kupče and R. Freeman, Projection–Reconstruction Technique for Speeding up Multidimensional NMR Spectroscopy, *J. Am. Chem. Soc.*, 2004, **126**, 6429–6440. Copyright 2004 American Chemical Society.)

real risk of introducing false peaks or eliminating true peaks depending on factors that could not be assessed *a priori*, including the arrangement of peaks in the spectrum, the exact angles of the radial spokes, and signal-to-noise. The major alternative was backprojection, specifically filtered backprojection, which offered quantitative lineshapes and volumes, along with much greater transparency about ambiguity in the data and the risks of false peaks. Yet an analysis of the radial sampling point response makes clear that it is not possible to eliminate the risk of spurious peaks in reconstructions from radial data, regardless of reconstruction algorithm, if the data are undersampled to reduce measurement time.

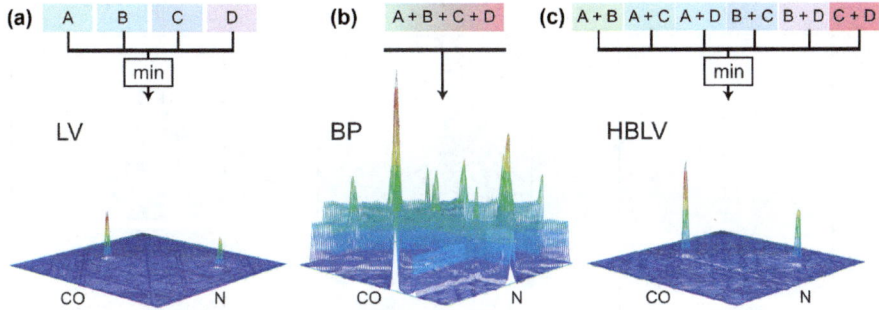

Figure 6.14 Comparison of LV, BP, and HBLV reconstruction methods. The three panels depict the same C/N plane from the 3-D HNCO of protein G B1 domain, as reconstructed using the three different reconstruction methods, with four projections as input data. (a) LV reconstruction is able to reconstruct the signals on this plane without introducing artifacts. The sensitivity is relatively low, but this is not necessarily obvious from the reconstruction, as the LV algorithm artificially reduces the noise, improving the apparent signal-to-noise, without improving the ability to detect weak signals. (b) BP reconstruction reproduces the two signals with a much higher ratio of signal to true thermal noise, but at the expense of introducing ridge artifacts and false peaks. (c) HBLV reconstruction with $k = 2$ reproduces the two signals at a higher signal-to-noise ratio, without introducing artifacts. (Reprinted with permission from R. A. Venters, B. E. Coggins, D. Kojetin, J. Cavanagh and P. Zhou, Preorganization in Highly Enantioselective Diaza-Cope Rearrangement Reaction, *J. Am. Chem. Soc.*, 2005, **127**, 8785–8795. Copyright © 2005 American Chemical Society.)

Even when that risk is very low, filtered backprojection with undersampled data will always result in low-level artifacts that can interfere with the detection of weak signals, as became especially apparent with work on NOESY data. Thus it became clear that the dream of robustly obtaining accurate, artifact-free spectra from significantly undersampled *radial* datasets simply isn't possible.

At the same time, explorations of other sampling approaches were producing more encouraging results. One interesting finding involved a generalization of radial sampling: if one views radial sampling instead as a sampling on concentric shells, one can achieve the same efficiency as conventional sampling by allowing the number of points on each shell to vary.[6] The greatest benefit, however, comes when one introduces randomization, whether in the context of concentric shells, where the orientations of the shells can be rotated randomly, or simply by selecting a random subset of grid points for measurement out of the conventional rectangular sampling grid. Undersampling according to a regular pattern leads to coherent aliasing artifacts, while undersampling in a randomized manner leads to incoherent aliasing artifacts, with the appearance of random noise. Incoherent artifacts are much to be preferred, as there is almost no chance that they can interfere constructively to produce

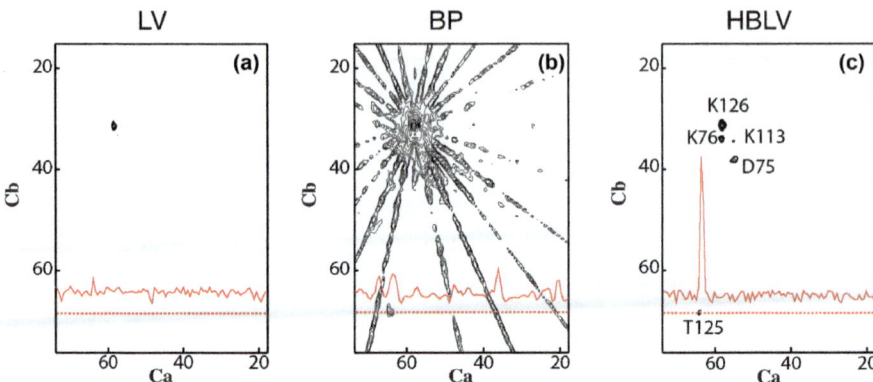

Figure 6.15 Reconstruction of 4-D sequential assignment data for large proteins. The three panels depict the same C_α/C_β plane—at the H/N position of residue K126—from the 4-D HNCACB of the 29 kDa protein HCA II, as reconstructed from 23 projections (generated from eight radial spokes) using three different reconstruction algorithms. Residue K126 partially overlaps in the H and N dimensions with both K76 and K113. (a) The LV reconstruction correctly reproduces the strong intraresidue correlation for residue K126, but all other correlations are below the detection limit and therefore are lost. (b) BP reconstruction reproduces both the intra- and the inter-residue (to T125) correlations for K126, but the inter-residue correlation is the same intensity as the ridge artifacts from the intraresidue correlation, and the interresidue peak has been broadened substantially. The ridges and the broadening of the K126 intraresidue peak make it impossible to identify other signals that may be present, such as the signals for the overlapping K76 and K113 residues. (c) HBLV reconstruction with $k = 8$ picks up the intra- and the inter-residue correlations for both K126 and K76, as well as, weakly, the intraresidue correlation for K113. No artifacts are present, and the peaks are not broadened.

(Reprinted with permission from R. A. Venters, B. E. Coggins, D. Kojetin, J. Cavanagh and P. Zhou, Preorganization in Highly Enantioselective Diaza-Cope Rearrangement Reaction, *J. Am. Chem. Soc.*, 2005, **127**, 8785–8795. Copyright © 2005 American Chemical Society.)

spurious peaks. From the standpoint of information and ambiguity, random sampling is better able to capture time domain information in all directions and at all timescales than deterministic methods based on regular patterns such as radial spokes, making it less likely than radial to have "blind spots" for certain combinations of signal frequencies. The recent development of compressed sensing (discussed in subsequent chapters) now provides a theoretical framework for understanding the advantages of incoherence in sampling. Furthermore, there are well-known algorithms for suppressing the artifacts generated by random sampling, such as CLEAN (discussed in the next chapter). The combination of random sampling together with reconstruction methods such as l_1-norm minimization or FFT followed by CLEAN has proven both more

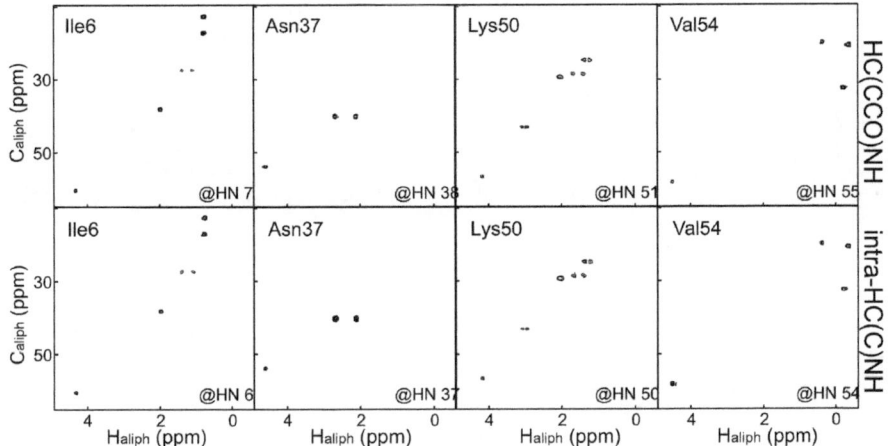

Figure 6.16 Reconstruction of 4-D sidechain assignment data. Selected 2-D H_{aliph}/C_{aliph} planes from reconstructions of a 4-D HC(CCO)NH (top row) and a 4-D intra-HC(C)NH (bottom row) on protein G B1 domain. In this experiment, the H_{aliph} and C_{aliph} dimensions were coevolved while N and HN were collected conventionally, producing 3-D projections each with two conventional axes and one tilted axis. Six radial spokes were measured, generating ten projections, and reconstruction was by the LV method.

(Reprinted from *J. Magn. Reson.*, **175**, L. Jiang, B. E. Coggins and P. Zhou, Rapid assignment of protein side chain resonances using projection–reconstruction of (4,3)D HC(CCO)NH and intra-HC(C)NH experiments, 170–176. Copyright (2005) with permission from Elsevier.)

efficient and more robust than the combination of radial sampling and tomographic reconstruction used in projection–reconstruction NMR.

The above notwithstanding, two recent publications show that radial sampling and projection–reconstruction methods continue to draw interest and may prove relevant for many years to come. Orekhov's group recently reported the reconstruction of a 4-D BEST-HNCOCA spectrum from on-grid radially sampled data, using a method that first located signals in the lattice of possible peak positions by counting the number of projections supporting the presence of a peak at each lattice point, and then computed signal intensities by solving eqn (6.10); their results confirm that this type of reconstruction is computationally faster than the newer compressed sensing methods, and that accurate results are possible when the spectrum is sufficiently sparse.[51] Kozminski's group used radial sampling to overcome a significant practical limitation with random sampling and reconstruction methods: the computational burden as the number of dimensions increases.[52] By combining radial sampling in some dimensions with random on-grid NUS in other dimensions, they could obtain 6-D and 7-D data for the backbone assignment of intrinsically disordered proteins. These studies point to the enduring appeal and utility of projection methods for extremely sparse data.

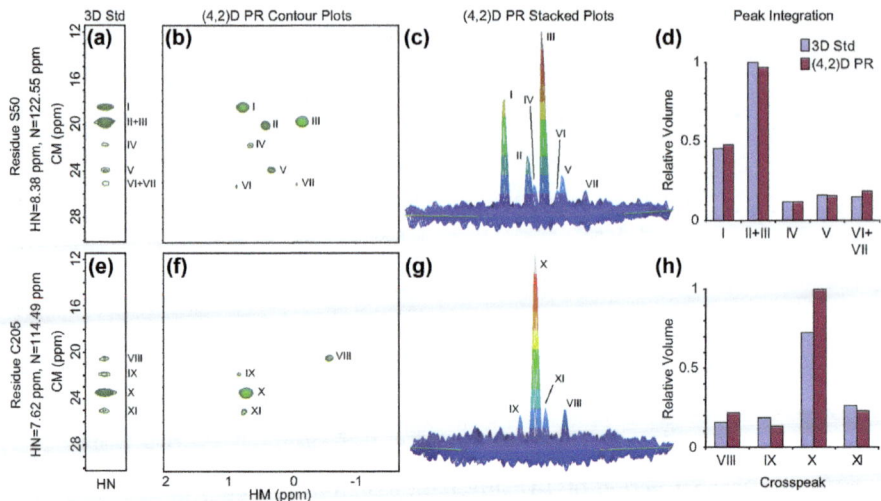

Figure 6.17 Reconstruction of 4-D NOESY data. Representative data from a filtered backprojection reconstruction of a 4-D $^{13}C/^{15}N$-separated NOESY spectrum and a conventional 3-D NOESY control spectrum at the same resolution, showing the ability of FBP reconstruction to reproduce successfully the lineshapes, volumes, and intensities of both strong and weak crosspeaks. This reconstruction was generated from 400 projections, and the reconstruction artifacts are weak enough that they do not interfere with the analysis of the spectrum. All numbered crosspeaks were successfully assigned, and the 4-D data makes it possible to separate overlapping correlations in the conventional 3-D data. (a/e) Strips from the conventional control. (b/f) The corresponding C/H planes from the reconstruction. (c/g) Stacked plots of the reconstruction, showing that it is free of artifacts. (d/h) Comparison of integrated peak volumes between the control and the reconstruction. (Reprinted with permission from B. E. Coggins, R. A. Venters and P. Zhou, Filtered Backprojection for the Reconstruction of a High-Resolution (4,2)D CH3−NH NOESY Spectrum on a 29 kDa Protein, *J. Am. Chem. Soc.*, 2005, **127**, 11562–11563. Copyright © 2005 American Chemical Society.)

6.7 Conclusions

Though radial sampling has been largely superceded by methods based on randomized sampling, the studies described in this chapter show that the projection–reconstruction approach can be used to obtain high-quality, high-resolution multidimensional spectra while significantly reducing the measurement time from that required for conventional data collection. Data are collected as a set of radial spokes, with the Fourier transform of a single radial spoke giving a projection. Reconstruction is accomplished using intuitive algorithms derived from an analysis of possible peak positions, or using the filtered backprojection algorithm derived from tomographic theory, which in the context of NMR turns out to be equivalent to a polar Fourier transform. This approach was successfully applied to sequential assignment, sidechain assignment, and NOESY experiments on proteins as large as 30 kDa.

Acknowledgements

This work was partially supported by NIH grant AI-055588 and a bridge fund from Duke University Medical Center, both to P. Z.

References

1. B. E. Coggins, R. A. Venters and P. Zhou, *Prog. Nucl. Magn. Reson. Spectrosc.*, 2010, **57**, 381–419.
2. R. N. Bracewell, *Two-Dimensional Imaging*, Prentice-Hall, Englewood Cliffs, NJ, 1995.
3. A. C. Kak and M. Slaney, *Principles of Computerized Tomographic Imaging*, IEEE Press, New York, 1999.
4. P. C. Lauterbur, *Nature*, 1973, **242**, 190–191.
5. E. Kupče and R. Freeman, *J. Biomol. NMR*, 2003, **27**, 383–387.
6. B. E. Coggins and P. Zhou, *J. Magn. Reson.*, 2007, **184**, 207–221.
7. B. E. Coggins and P. Zhou, *J. Biomol. NMR*, 2008, **42**, 225–239.
8. K. Nagayama, P. Bachmann, K. Wüthrich and R. R. Ernst, *J. Magn. Reson.*, 1978, **31**, 133–148.
9. B. E. Coggins, R. A. Venters and P. Zhou, *J. Am. Chem. Soc.*, 2004, **126**, 1000–1001.
10. T. Szyperski, G. Wider, J. H. Bushweller and K. Wuthrich, *J. Am. Chem. Soc.*, 1993, **115**, 9307–9308.
11. R. R. Ernst, G. Bodenhausen and A. Wokaun, *Principles of Nuclear Magnetic Resonance in One and Two Dimensions*, Clarendon Press, Oxford, 1987.
12. J. Cavanagh, W. J. Fairbrother, A. G. Palmer, III and N. J. Skelton, *Protein NMR Spectroscopy: Principles and Practice*, Academic Press, San Diego, 1996.
13. B. Brutscher, N. Morelle, F. Cordier and D. Marion, *J. Magn. Reson.*, 1995, **B109**, 338–342.
14. B. Bersch, E. Rossy, J. Covès and B. Brutscher, *J. Biomol. NMR*, 2003, **27**, 57–67.
15. K. Ding and A. M. Gronenborn, *J. Magn. Reson.*, 2002, **156**, 262–268.
16. S. Kim and T. Szyperski, *J. Am. Chem. Soc.*, 2003, **125**, 1385–1393.
17. S. Hiller, F. Fiorito, K. Wüthrich and G. Wider, *Proc. Natl. Acad. Sci. U. S. A.*, 2005, **102**, 10876–10881.
18. H. R. Eghbalnia, A. Bahrami, M. Tonelli, K. Hallenga and J. L. Markley, *J. Am. Chem. Soc.*, 2005, **127**, 12528–12536.
19. E. Kupče and R. Freeman, *J. Am. Chem. Soc.*, 2004, **126**, 6429–6440.
20. J. M. Gledhill and A. J. Wand, *J. Magn. Reson.*, 2008, **195**, 169–178.
21. R. A. Venters, B. E. Coggins, D. Kojetin, J. Cavanagh and P. Zhou, *J. Am. Chem. Soc.*, 2005, **127**, 8785–8795.

22. *Image Reconstruction from Projections*, ed. G. T. Herman, Springer-Verlag, Berlin, 1979.

23. J. Radon, *Ber. Verh. K. Saechs. Ges. Wiss., Math.-Phys. Kl.*, 1917, **69**, 262–277.

24. R. N. Bracewell, *Aust. J. Phys.*, 1956, **9**, 198–217.

25. R. N. Bracewell and A. C. Riddle, *Astrophys. J.*, 1967, **150**, 427–434.

26. D. J. DeRosier and A. Klug, *Nature*, 1968, **217**, 130–134.

27. A. M. Cormack, *J. Appl. Phys.*, 1963, **34**, 2722–2727.

28. R. A. Crowther, D. J. DeRosier and A. Klug, *Proc. R. Soc. London, Ser. A*, 1970, **317**, 319–340.

29. B. K. Vainshtein, *Sov. Phys. – Crystallogr.*, 1971, **15**, 781–787.

30. S. W. Rowland, in *Image Reconstruction from Projections*, ed. G. T. Herman, Springer-Verlag, Berlin, 1979.

31. P. F. C. Gilbert, *Proc. R. Soc. London, Ser. B*, 1972, **182**, 89–102.

32. R. N. Bracewell, *The Fourier Transform and Its Applications*, McGraw-Hill, Boston, 2000.

33. B. E. Coggins and P. Zhou, *J. Magn. Reson.*, 2006, **182**, 84–95.

34. D. Marion, *J. Biomol. NMR*, 2006, **36**, 45–54.

35. A. R. Thompson and R. N. Bracewell, *Astron. J.*, 1974, **79**, 11–24.

36. W. Cochran, F. H. C. Crick and V. Vand, *Acta Crystallogr.*, 1952, **5**, 581.

37. A. Klug, F. H. C. Crick and H. W. Wyckoff, *Acta Crystallogr.*, 1958, **11**, 199.

38. J. Waser, *Acta Crystallogr.*, 1955, **8**, 142.

39. R. N. Bracewell and A. R. Thompson, *Astrophys. J.*, 1973, **182**, 77–94.

40. E. Kupče and R. Freeman, *J. Magn. Reson.*, 2005, **173**, 317–321.

41. C. D. Ridge and V. A. Mandelshtam, *J. Biomol. NMR*, 2009, **43**, 151–159.

42. B. E. Coggins, R. A. Venters and P. Zhou, *J. Am. Chem. Soc.*, 2005, **127**, 11562–11563.

43. R. Gordon, R. Bender and G. T. Herman, *J. Theor. Biol.*, 1970, **29**, 471–481.

44. J. W. Yoon, S. Godsill, E. Kupče and R. Freeman, *Magn. Reson. Chem.*, 2006, **44**, 197–209.

45. M. Mobli, A. S. Stern and J. C. Hoch, *J. Magn. Reson.*, 2006, **182**, 96–105.

46. D. Malmodin and M. Billeter, *J. Am. Chem. Soc.*, 2005, **127**, 13486–13487.

47. G. A. Mueller, *J. Biomol. NMR*, 2009, **44**, 13–23.

48. E. Kupče and R. Freeman, *J. Am. Chem. Soc.*, 2003, **125**, 13958–13959.

49. E. Kupče and R. Freeman, *J. Biomol. NMR*, 2004, **28**, 391–395.

50. L. Jiang, B. E. Coggins and P. Zhou, *J. Magn. Reson.*, 2005, **175**, 170–176.

51. H. Hassanieh, M. Mayzel, L. Shi, D. Katabi and V. Y. Orekhov, *J. Biomol. NMR*, 2015, **63**, 9–19.

52. S. Zerko and W. Kozminski, *J. Biomol. NMR*, 2015, **63**, 283–290.

CLEAN

BRIAN E. COGGINS* AND PEI ZHOU*

Duke University Medical Center, Box 3711 DUMC, Durham, NC 27710, USA
*Email: bec2@duke.edu; peizhou@biochem.duke.edu

7.1   Introduction

Nonuniform sampling offers NMR spectroscopists many advantages, but not without adding a few extra challenges as well. While the information content of a nonuniform dataset, even one sparsely sampled at a very small percentage of the conventional sampling density, may be more than sufficient to answer one's spectroscopic questions—defining the frequencies, intensities, and lineshapes of the observed NMR resonances—there is no denying that this information is in a different form from the conventional FT-NMR data. The obvious question is how to extract the spectral information.

One idea is to pretend that there's nothing unusual about the data at all, and process them with a conventional discrete Fourier transform (DFT). If the randomly sampled, nonuniform data points were selected from a grid, one could fill in zero values at all the grid positions that were *not* measured, and compute the Fourier transform using any of the standard NMR processing programs. The output of this will be, in fact, the familiar NMR spectrum—albeit corrupted with artifacts owing to the limited sampling. If the sampling is random, these artifacts will also appear to be random; the more sparse the sampling, the higher the artifact level relative to the signal heights.

It is perhaps comforting to find that the FT works even on nonuniform data, and that it can, if the artifact level is low enough, provide us with a

New Developments in NMR No. 11
Fast NMR Data Acquisition: Beyond the Fourier Transform
Edited by Mehdi Mobli and Jeffrey C. Hoch
© The Royal Society of Chemistry 2017
Published by the Royal Society of Chemistry, www.rsc.org

spectrum showing all of the signals. Unfortunately, in many cases the sampling artifacts, which look like random noise and therefore decrease the apparent signal-to-noise ratio, are strong enough to obscure the weaker signals in the spectrum. To obtain a spectrum of sufficient quality in these cases, we must contend with these artifacts.

Unsurprisingly, given the long history of fundamental scientific problems recurring in one field after another, NMR spectroscopists are not the first to carry out nonuniform sampling or to contend with sampling artifacts. Radioastronomers faced very similar issues in the late 1960s and early 1970s, as new radiotelescope designs were introduced that expanded the capabilities of the field, but only at the expense of collecting nonuniformly sampled data. The solution that emerged was to remove the artifacts using an iterative algorithm known as CLEAN. Invented nearly simultaneously by Dixon and Kraus of Ohio State[1] and Jan Högbom of the Stockholm Observatory,[2,3] the method became standard practice in the world of radioastronomy and has given rise to related methods in other fields of science. Experience has shown it to be highly effective, as well as relatively fast to compute and relatively straightforward to implement.

It turns out that CLEAN is also useful in NMR, for the removal of artifacts after computing the DFT of nonuniformly sampled data. In this chapter, we recount the historical background, explain the theory and practice of the method, discuss its mathematical properties, and show examples of its successful application in biomolecular NMR. CLEAN is, in fact, a very powerful method, and in many cases it is all one needs to get a useful spectrum from nonuniformly sampled NMR data.

7.2 Historical Background: The Origins of CLEAN in Radioastronomy

Radioastronomy is the study of stars and other objects in the cosmos by observing their radiofrequency (RF) electromagnetic emissions.[4] As with optical telescopes, one seeks to observe a small area of the sky, ideally mapping the RF-emitters in a given region and then studying them individually, which requires telescopes with narrow beamwidths and fine directional control. The same physical principles apply as for optical systems, but with RF the wavelengths involved are on the same scale as the size of the telescopes, and diffractive effects, rather than ray propagation in straight lines, are therefore the norm. The telescopes themselves are simply antennas, albeit engineered with unusual requirements for directionality and sensitivity.

When a wave front from a point source in the sky arrives at an antenna surface, that wave front is seen at different points along the antenna at different times, depending on the shape of the antenna as well as the relative angle between the wave front and the antenna surface—or in other words, depending on the direction in the sky from which the wave front comes. The current induced at each position on the antenna is sinusoidal, but varies in phase depending on what part of the wave is seen at each position. Since the output from the antenna is a summation of the induced currents at these different positions, that output is in fact a weighted sum of sine and cosine waves: a Fourier transform with respect to direction in the sky. Furthermore, the antenna's sensitivity to radiation in each direction is determined by a Fourier transform of the antenna's shape. For greater directionality and better resolution, the antenna must be large; indeed, for maximal directionality, we might choose to take advantage of the size of the Earth itself. By taking repeated measurements of the same patch of sky as the Earth rotates and combining those measurements into a single dataset, it is as if we examined that patch of sky with a set of telescopes arrayed around the Earth's surface (Figure 7.1a). This *aperture synthesis* procedure allows for mapping the sky with a very narrow beam, at very high angular resolution.

The fact that an antenna's directional sensitivity is the Fourier transform of the antenna's shape means that the sampling of the sky by any finite antenna is inherently imperfect. As in NMR, the convolution theorem of the FT reveals that the multiplication of the RF distribution in the sky by the antenna's sensitivity pattern yields a convolution in the detected signal received from the antenna, a convolution of the shape of each astronomical RF source with the point spread function of the antenna. When Earth-rotation aperture synthesis is used, the problem becomes even more severe, since the sampling pattern of the antenna is now repeated at discrete points on an ellipse or arc—an instance of highly nonuniform sampling (Figure 7.1b). The resulting map of the sky is severely corrupted with sampling artifacts.

CLEAN was invented as a means to suppress these artifacts. The basic idea was first described in the literature by Dixon and Kraus, who called it a "source extraction technique" and used it to remove antenna-shape artifacts.[1] Högbom devised a related procedure almost simultaneously, which was first employed by Rogstad and Shostak a few years later in the context of aperture synthesis in a concentric ring pattern.[3,5] The method finally acquired its name a few years later, when Högbom published what is considered to be the canonical description of the procedure.[2] CLEAN has since become a standard method of the field.[6] Even in its earliest use, it produced images of very high quality[1] (Figures 7.1c and 7.2).

7.3 The CLEAN Method

7.3.1 Notation

Throughout this chapter, we use lowercase bold letters to represent time domain data, and uppercase bold letters to represent frequency domain data. The letters **a**/**A** refer to the "true" fully sampled FID and artifact-free spectrum, respectively; **s**/**S** to the sampling pattern/point-response function;

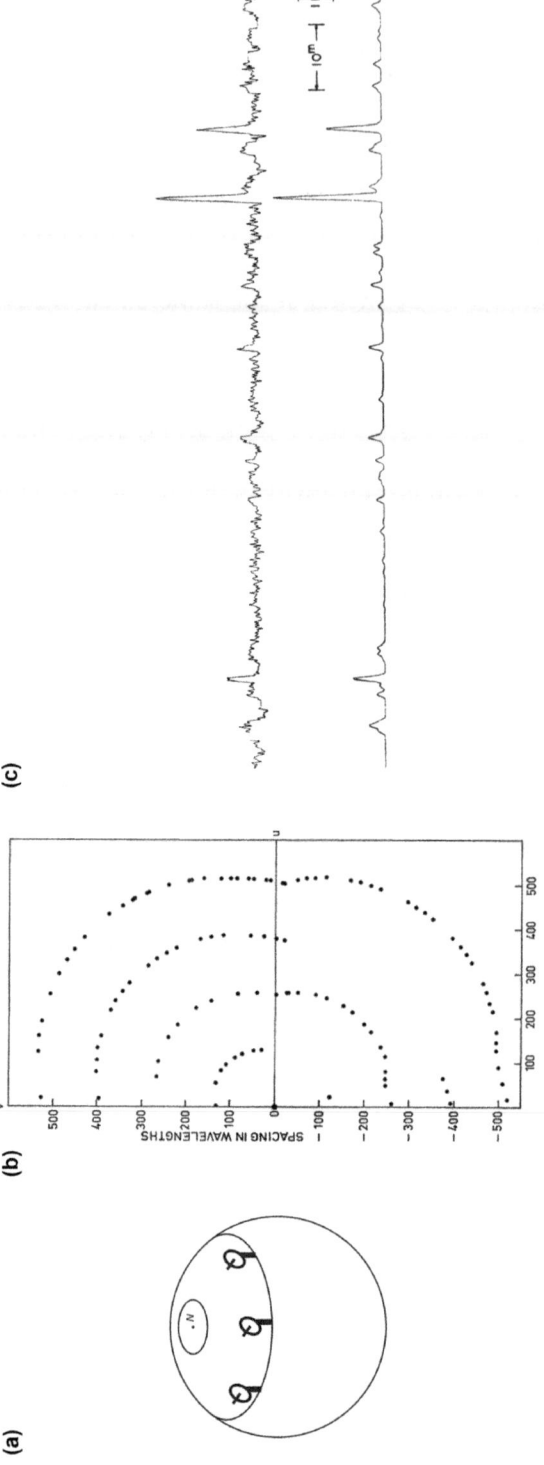

Figure 7.1 Aperture synthesis and CLEAN in astronomy. (a) The motion of a radiotelescope on the Earth's surface during the day as seen from a fixed point in space. In aperture synthesis, the RF radiation from the sky is sampled at successive times while the Earth rotates. This produces a nonuniformly sampled dataset with equivalent directionality and resolution to a much larger telescope. (Redrawn from W. N. Christiansen and J. A. Högbom, *Radiotelescopes*, Cambridge University Press, Cambridge, 1969. Copyright Cambridge University Press, 1969, 1985.) (b) The sampling pattern for the sky survey carried out by Rogstad and Shostak, the second reported study using CLEAN. Each point represents a 15 minute-long observation of the sky. (Reprinted from D. H. Rogstad and G. S. Shostak, *Astron. Astrophys.*, 1971, **13**, 99–107. Reproduced with permission © ESO.) (c) Sample 1-D cross-section from Dixon and Kraus' 1415 MHz sky survey, before (top) and after (bottom) application of CLEAN. (Reprinted from A high-sensitivity 1415-MHz survey at north declinations between 19° and 37°, R. S. Dixon and J. D. Kraus, *Astron. J.*, 1968, **73**, 381–407. Copyright 1968 American Astronomical Society. Reproduced with permission.)

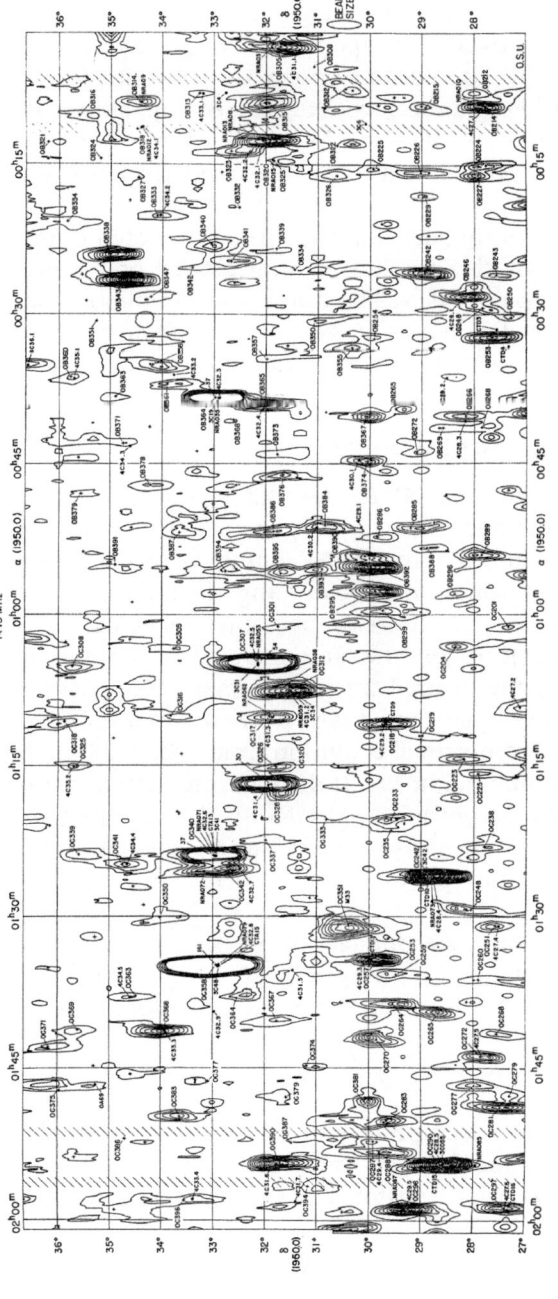

Figure 7.2 CLEAN applied to a sky survey. A panel from the 1415 MHz sky survey by Dixon and Kraus, the first known application of CLEAN, after post-processing.
(Reprinted from R. S. Dixon and J. D. Kraus, A high-sensitivity 1415-MHz survey at north declinations between $19°$ and $37°$, *Astron. J.*, 1968, 73, 381–407. Copyright 1968 American Astronomical Society. Reproduced with permission.)

d/D to the sparse NUS data/artifact-corrupted spectrum; and **c/C** to the reconstructed complete time domain/reconstructed artifact-free spectrum. In addition, **R** is a residual frequency-domain spectrum and **H** is an artifact-free point response. We frequently refer to decompositions of time-domain FIDs as the sums of sinusoidal components, using a lowercase subscript for a single component (*e.g.* a_i) and an uppercase subscript for a set of components (*e.g.* a_I). In like manner, we refer to decompositions of frequency-domain spectra as the sums of δ-functions, using a lowercase subscript for a single component (*e.g.* A_i) and an uppercase subscript for a set of components (*e.g.* A_I). The set of time-domain grid points that are measured is denoted K, while the set that are not measured is denoted U; the number of measured points is N_K, the number of unmeasured points is N_U, and the total number of grid points is N. The number of δ-function components in the true spectrum **A** is denoted as M. A vector of time-domain coordinates is **t**, while a vector of frequency-domain coordinates is ω.

7.3.2 The Problem to be Solved

In both radioastronomy and NMR, the image (sky map or spectrum, respectively) being obtained consists of generally isolated but sometimes overlapping point sources (cosmic RF emitters/nuclear resonances), of varying intensity and sometimes possessing distinctive shapes and fine structure. In both cases, the measurement process is carried out in a reciprocal domain (position on the antenna/evolution time) related by a Fourier transform to the desired imaging domain (sky position/resonance frequency). The measurement process is restricted to a finite measurement window (antenna shape/maximal evolution time) which alters the shapes of the point sources in the final image. If the measurement domain is sampled nonuniformly (aperture synthesis/NMR NUS), the resulting image is corrupted by artifacts according to a mathematical convolution between the true image and the point spread function corresponding to the sampling pattern.

Focusing on the NMR case for clarity, let **A** represent the true multi-dimensional spectrum. Representing a Fourier transform with a double-headed arrow, **A** is the Fourier transform of a time-domain FID **a**:

$$\mathbf{a} \xleftrightarrow{\text{DFT}} \mathbf{A}. \tag{7.1}$$

If **a** is sampled using a sampling pattern **s**, which is mathematically a multiplication in the time domain, the effect in the frequency domain is to convolve **A** with the Fourier transform of **s**, the point spread function **S**:

$$\mathbf{as} \xleftrightarrow{\text{DFT}} \mathbf{A} * \mathbf{S} \tag{7.2}$$

where an asterisk ($*$) is used to represent convolution. One of the basic properties of the Fourier transform is its linearity: if two time-domain signals **f** and **g** are added together, the Fourier transform will be the sum of their individual Fourier transforms **F** and **G**,

$$\mathbf{f} + \mathbf{g} \xleftrightarrow{\text{DFT}} \mathbf{F} + \mathbf{G}. \tag{7.3}$$

Taking the spectrum **A** as the sum of a set of M constituent components $\mathbf{A}_I = \{\mathbf{A}_i : i = 1, 2, \ldots, M\}$:

$$\mathbf{A} = \sum_i \mathbf{A}_i \tag{7.4}$$

(Figure 7.3) and recalling that convolution is distributive, we find that the artifact-corrupted "dirty" spectrum **D** is the sum of the convolutions of the individual components with the artifact pattern:

$$\mathbf{D} = \mathbf{A} * \mathbf{S} = \sum_i \mathbf{A}_i * \mathbf{S}. \tag{7.5}$$

Sparse sampling thus results in a replication of the point spread function **S** so that it appears once for each component, centered on that component, and scaled by the intensity of that component (Figure 7.4). Importantly, this means that the artifacts corrupting the final spectrum are completely predictable, and in principle removable, if we know the sampling pattern used to measure the data and a set of components $\mathbf{A}_I$ that collectively describe the true spectrum.

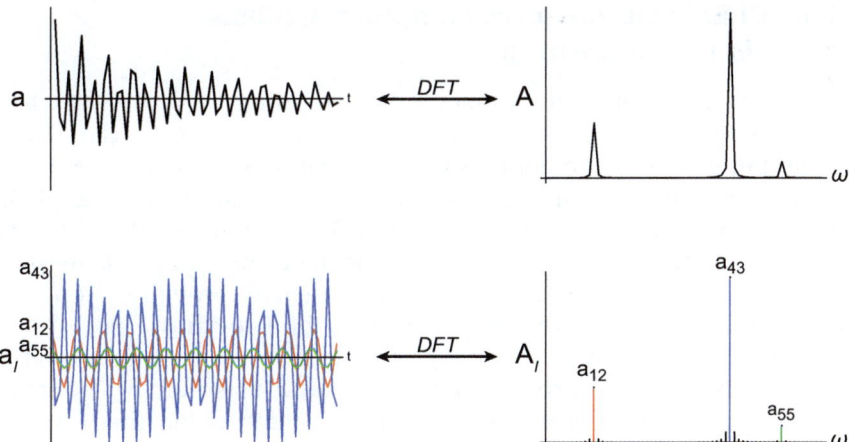

Figure 7.3 Decomposition into components. An NMR dataset may be described as the sum of complex sinusoids in the time domain or as the sum of δ-functions in the frequency domain. The spectrum shown top-right can be decomposed into the set of δ-functions shown bottom-right; the corresponding FID in the top-left is the sum of complex sinusoids, such as the three specific examples shown bottom-left.

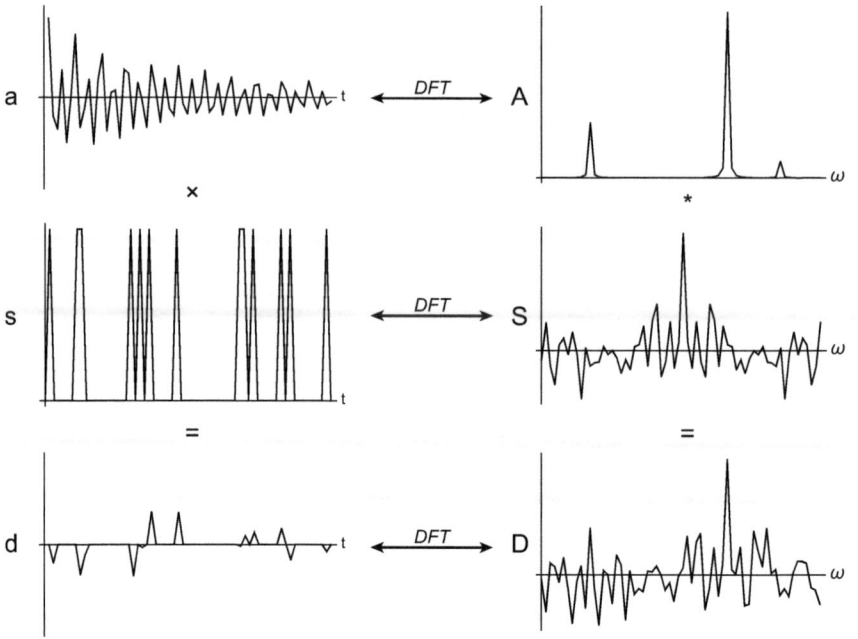

Figure 7.4 Sparse sampling and artifacts. Sparse nonuniform sampling is equivalent to multiplying a complete FID **a** by a sampling function **s** (a binary
mask) to get a set of observed data **d**. In the frequency domain, this is
equivalent to convolving the true spectrum **A** with an artifact-corrupted
point response function **S** to get an artifact-corrupted spectrum **D**.

7.3.3   CLEAN Deconvolves Sampling Artifacts *via* Decomposition

The decomposition of an artifact-corrupted "dirty" spectrum **D** into the
sum of artifact-corrupted components is the basis for CLEAN. Given **D** and **S**,
we determine a set of components $C_I = \{C_i\}$ that produce **D** when convolved
with **S**. While this decomposition could be performed at the level of NMR
peaks, it is more versatile to carry it out at the level of individual grid points in
the discrete spectrum (*i.e.* a set of δ-functions), in which case the lineshapes
and fine structure of NMR peaks will be represented through multiple components and will not need special treatment by the algorithm (Figure 7.3). The
components constituting the set C_I are therefore scaled δ-functions and
represent what we can learn from **D** about the true spectrum **A**. The artifact-
corrupted spectrum **D** normally also contains information that cannot be
modeled as convolutions of δ-functions with **S**, typically thermal noise and
in some cases signals that are too weak to model with confidence, which we
refer to as the residuals **R**. The decomposition could thus be represented as:

$$\mathbf{D} = \left(\sum_i \mathbf{C}_i * \mathbf{S} \right) + \mathbf{R}. \tag{7.6}$$

From these C_I components and the residuals $\mathbf{R}$ we can calculate an artifact-free spectrum $\mathbf{C}$ which approximates the true spectrum $\mathbf{A}$:

$$\mathbf{C} = \left(\sum_i \mathbf{C}_i * \mathbf{H} \right) + \mathbf{R} \approx \mathbf{A} \tag{7.7}$$

where $\mathbf{H}$ is a "clean" point response, free of artifacts, discussed below. The net effect of applying CLEAN is thus to deconvolve $\mathbf{S}$ from $\mathbf{D}$, achieving this through fitting a model $\mathbf{C} = \sum_i \mathbf{C}_i$ to $\mathbf{D}$ and using it to reconstruct $\mathbf{A}$.

7.3.4 Obtaining the Decomposition into Components

The strategy used in CLEAN for decomposing $\mathbf{D}$ into components is to find the strongest components first, and work towards progressively weaker ones. This has two advantages over procedures that might attempt to fit all components simultaneously. First, in a case where the true spectrum has a high dynamic range, the weak components will be obscured at first by the artifacts from the strong components, and it would be difficult or impossible to fit them. Second, there is a danger of constructive interference between the artifact patterns arising from different components, which could produce false peaks in $\mathbf{D}$. If such false peaks were incorporated into the model, they would corrupt the final clean spectrum $\mathbf{C}$. When such false peaks occur, it is owing to interference caused by strong components, and it is highly unlikely that said false peaks could have an intensity on par with those strong components. Thus if we address the strong components first, false peaks arising from artifact interference are naturally removed and are unlikely to lead the model astray.

CLEAN begins with the position in $\mathbf{D}$ of greatest intensity, located at coordinate vector:

$$\omega_{\max} = \arg\max_{\omega} |\mathbf{D}(\omega)| \tag{7.8}$$

and designate the intensity at this position as $I_{\max}$. Taking this as the strongest component, we designate a δ-function $\mathbf{C}_1$ at this position with intensity $gI_{\max}$:

$$\mathbf{C}_1(\omega) = gI_{\max}\delta(\omega - \omega_{\max}) \tag{7.9}$$

where g is a value between 0 and 1, known as the *loop gain* or simply the *gain*, and where δ is a function having a value of 1 at the origin and a value of 0 everywhere else. We then remove this component and its associated artifacts from $\mathbf{D}$ by subtraction to produce $\mathbf{D}^{(1)}$:

$$\mathbf{D}^{(1)} \leftarrow \mathbf{D} - \mathbf{C}_1 * \mathbf{S}. \tag{7.10}$$

The decomposition continues by identifying and removing, in each iteration, the next strongest component:

$$\mathbf{C}_i(\omega) = gI_{\max}^{(i-1)}\delta(\omega - \omega_{\max}^{(i-1)}) \tag{7.11}$$

where $\omega_{max}^{(i-1)}$ is the position of strongest intensity in $\mathbf{D}^{(i-1)}$, the spectrum after removing all components up to $\mathbf{C}_{i-1}$, *i.e.* the result after $i-1$ iterations, and where $I_{max}^{(i-1)}$ is the intensity at that position. The residual spectrum is then updated to:

$$\mathbf{D}^{(i)} \leftarrow \mathbf{D}^{(i-1)} - \mathbf{C}_i * \mathbf{S}. \tag{7.12}$$

Though $\mathbf{C}_i * \mathbf{S}$ could be calculated *via* multiplication in the time domain or *via* an explicit convolution integral, this quantity is in fact simply a scaled and shifted copy of $\mathbf{S}$:

$$\mathbf{C}_i(\omega) * \mathbf{S}(\omega) = g I_{max}^{(i-1)} \mathbf{S}(\omega - \omega_{max}^{(i-1)}) \tag{7.13}$$

meaning that the update in eqn (7.12) can actually be carried out by a simple and computationally efficient subtraction operation in the frequency domain. These steps are illustrated in Figure 7.5.

Note that multiple subtractions can be applied at any given position in $\mathbf{D}$, and multiple components can refer to the same position ω. Thus the single true component $\mathbf{A}_i$ at some position ω may be represented in the decomposition by a set of components $\mathbf{C}_J$ all at position ω.

The process of identifying and subtracting components is continued until it becomes impossible to distinguish any further components from the noise. At that point, the remaining contents of $\mathbf{D}$ are considered to be the residuals $\mathbf{R}$.

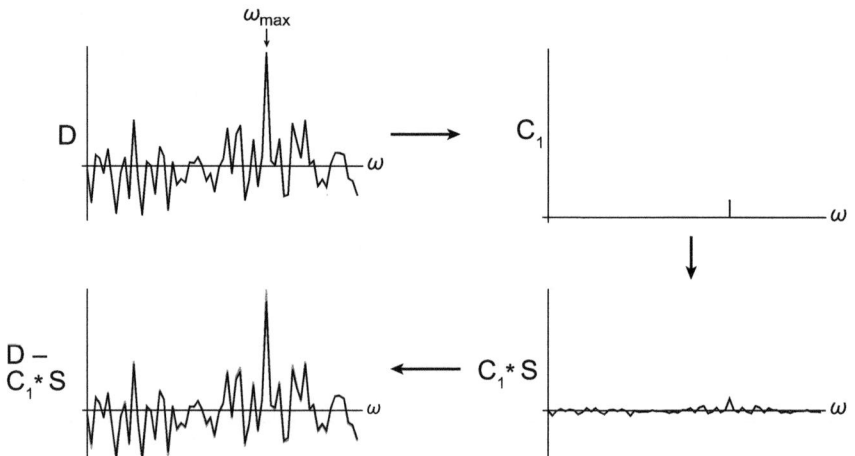

Figure 7.5 One iteration of CLEAN. First, the point of greatest intensity in the artifact-corrupted spectrum $\mathbf{D}$ is identified. A δ-function component $\mathbf{C}_1$ is added to the model at this position, with a scale equal to the intensity in $\mathbf{D}$ multiplied by a gain g (here, 10%). The scaled component $\mathbf{C}_1$ is then convolved with the point-response function $\mathbf{S}$ to give a model for the artifacts generated by that component in the artifact-corrupted spectrum $\mathbf{D}$. Finally, the modeled artifacts $\mathbf{C}_1 * \mathbf{S}$ are subtracted from $\mathbf{D}$ to give a new residual with reduced artifacts and a reduced peak height.

7.3.5 The Role of the Gain Parameter

The subtractive process described above does not remove the full component at a given position in **D** in one step, but rather over a number of steps, each time subtracting a fraction g of the remaining intensity. The purpose of this *gain* parameter is to reduce the chance of artifacts being incorporated into the components $\mathbf{C}_I$. Recall that a typical artifact pattern covers the entirety of **D**. Thus the artifacts generated by the convolution of some component $\mathbf{A}_1$ with the artifact pattern **S** will be felt at the site of some other component $\mathbf{A}_2$, and the artifacts generated from $\mathbf{A}_2$ will be felt at $\mathbf{A}_1$. The actual intensity value observed in **D** at the position ω_i of component $\mathbf{A}_i$ will therefore be:

$$\mathbf{D}(\omega_i) = \mathbf{A}_i(\omega_i) + \left(\sum_{j \neq i} [\mathbf{A}_j * \mathbf{S}](\omega_i) \right) + \varepsilon \qquad (7.14)$$

where ε represents the contribution of thermal noise. By taking the value of $g\mathbf{D}(\omega_i)$ instead of $\mathbf{D}(\omega_i)$ for the intensity of component $\mathbf{C}_i$, we avoid incorporating artifactural intensity from other positions, which can be quite significant at the start of the subtractive process. As the iterations continue, the artifacts arising from strong signals are progressively reduced throughout the spectrum, and the set of components $\mathbf{C}_I$ converges towards the real intensities $\mathbf{A}_i(\omega_i)$ rather than artifact-corrupted intensities $\mathbf{A}_i(\omega_i) + \left(\sum_{j \neq i} [\mathbf{A}_j * \mathbf{S}](\omega_i) \right)$.

Yet it is certainly possible that some interfering artifactural intensity will be incorporated as intensity errors into the components $\mathbf{C}_I$ for any $g > 0$, and one might wonder: (1) how low g needs to be to avoid trouble, (2) how serious a problem it is if g were too high, and (3) whether the algorithm might not be capable of correcting itself if artifacts wound up in the components. Taking these questions in reverse order, in many conditions the algorithm *is* in fact self-correcting: if artifactural terms are incorporated into the model at a given position, the subtraction of artifacts at that position will initially be either too aggressive or not aggressive enough, but this can trigger additional corrective iterations such that the model would ultimately converge on the correct solution in the manner of a damped oscillator.[7] However, the presence of thermal noise makes this convergence less straightforward, as it is hard for the algorithm to identify the errors from over- or under-correction when thermal noise is also present, and the additional corrective iterations may not happen. A small g allows these errors to relax away over the initial iterations, without ever being included in the model, and no subsequent corrections are needed.

Unfortunately it is not possible to name an ideal value of g from first principles, but an appropriate g can be estimated from the artifact level in **S** and the thermal noise in **D**. As a guideline, it has been found empirically that it is safest to keep g at or below 10% for 3-D NMR experiments and at or below 50% for 4-D NMR experiments.

The presence of a thermal noise term ε in eqn (7.14) is a reminder that any model fit to experimental data will be imperfect. Using a low g and stopping the decomposition when the residual signals in **D** can no longer be distinguished from the noise allows one to avoid incorporating this noise term into the CLEAN model $\mathbf{C} = \sum_i \mathbf{C}_i$, and to avoid cycles of over- or under-correction of artifacts based on noise contributions. This comes with the tradeoff that the set of components fit by the model at some position ω_i of true component $\mathbf{A}_i$ will not reach the full intensity of $\mathbf{A}_i$, and some of $\mathbf{A}_i$'s intensity will remain in the residuals **R**. As a consequence, some of the artifactual $\mathbf{A}_i * \mathbf{S}$ intensity will also remain in **R**.

7.3.6   Reconstructing the Clean Spectrum

Having decomposed **D** into components $\mathbf{C}_I$, we now construct an artifact-free spectrum, as shown in eqn (7.7), by convolving the components $\mathbf{C}_I$ with an artifact-free point response **H** and adding them to the residuals. **H** could simply be a δ-function, but it is usually better for the **H** to have some kind of shape, to avoid truncation artifacts. In fact, if quantitative accuracy is important, it is best to derive **H** from **S**. To see why, let us consider the characteristics of **S** in more detail.

The sampling function **s** could be thought of as a set of δ-functions marking the locations in the time domain that were measured; for typical on-grid NUS data, we could also imagine a multidimensional grid of the same size as our experimental time domain, with a one at every measured position and a zero at every omitted position, *i.e.* a binary mask. The Fourier transform of **s** yields the point response or point spread function **S** for the sampling process. A *point response* is so called because it describes the response generated by a measurement process from an infinitely sharp point source; *point spread function* highlights the fact that any finite measurement process will inevitably spread an infinitely sharp point to a finite width, and the point spread function quantitates this spread. The randomized sampling used in NUS NMR gives rise to a point response with a central peak surrounded by uniformly distributed, seemingly random, low-level artifacts. The convolution of the true spectrum **A** with these artifacts is what makes the "dirty" spectrum dirty, and it is our goal to remove and discard them.

For many NUS experiments, however, the central peak of **S** contains additional information that is important and useful. The shape of the central peak arises from the way the sampling points are distributed in the sampling pattern. If the sampling points are evenly distributed, the central peak will be essentially a δ-function. If, on the other hand, the sampling points are distributed with exponential weighting,[8,9] the central peak will have a Lorentzian shape reflecting the distribution of low- and high-resolution information that was measured in the experiment. By retaining this shape in the reconstructed spectrum **C**, we maximize the accuracy for the information we have obtained. Choosing an **H** with a narrower central peak would superficially seem to improve the resolution, but only by increasing the weight given to the limited number of noisier data points measured at longer evolution times. This reduces sensitivity and accuracy and in some ways defeats the purpose of the exponential weighting.

Since all signals in **D** have been convolved with **S**, all peaks in **D** will carry this shape. When the quantitative comparison of lineshapes, intensities, or volumes between the peaks of the spectrum is important, it is necessary that signals restored from C_I carry this same central peak shape to maintain quantitative accuracy. Thus it is recommended that one construct **H** by extracting the central peak from **S**, which could also be viewed as truncating **S** to exclude all but the central peak.

7.4 Mathematical Analysis of CLEAN

7.4.1 CLEAN and the NUS Inverse Problem

Obtaining spectral information from sparse, nonuniform data is often presented as an inverse problem.[10–12] In the notation we have been using throughout this chapter, one seeks a model of the time domain **c** that is consistent with the observed sparse data **d**, typically by assessing consistency in a least-squares sense. Because the data are sparsely sampled, the problem is underdetermined, and there are many solutions meeting the requirement of consistency with the data. To resolve this inherent ambiguity, a *regularization* constraint is added, a metric assessing the suitability of potential solutions, against which the model can be optimized. On grounds of parsimony—and also reflecting the nature of sampling artifacts—most regularization constraints seek to minimize the total intensity in the frequency-domain model **C**, either by minimizing a norm or by maximizing informational entropy. The inverse problem can then be solved as an iterative optimization problem, where the model **c** is adjusted so as to minimize a penalty function combining the consistency and regularization constraints, *e.g.*:

$$\min_{\mathbf{c},\mathbf{c}\leftrightarrow\mathbf{C}} ||\mathbf{c} - \mathbf{d}||^2_{l_2} + \lambda ||\mathbf{C}||_{l_1} \tag{7.15}$$

where λ is a Lagrange multiplier determining the relative weighting of the two terms, the l_2-norm for a vector **f** measured at discrete positions x_k is defined as:

$$||\mathbf{f}||_{l_2} \equiv \sqrt{\sum_k |\mathbf{f}(x_k)|^2} \tag{7.16}$$

the vertical bars indicating the complex modulus, and the l_1-norm is defined as:

$$||\mathbf{f}||_{l_1} \equiv \sqrt{\sum_k |\mathbf{f}(x_k)|}. \tag{7.17}$$

We shall refer to the two terms of eqn (7.15) as Q_C and Q_L, respectively:

$$Q_C = ||\mathbf{c} - \mathbf{d}||_{l_2}^2, \quad Q_L = ||\mathbf{C}||_{l_1} . \tag{7.18}$$

In the time domain, nonuniform sampling with a function $\mathbf{s}$ gives the observed data $\mathbf{d}$, which comprise a subset of the complete data values $\mathbf{a}$:

$$\mathbf{d(t)} = [\mathbf{a(t)} + \varepsilon\mathbf{(t)}]\mathbf{s(t)}$$

$$= \begin{cases} \mathbf{a(t)} + \varepsilon\mathbf{(t)}, & \text{if } \mathbf{s(t)} = 1(\mathbf{t} \in K) \\ 0, & \text{if } \mathbf{s(t)} = 0(\mathbf{t} \in U) \end{cases} \tag{7.19}$$

where K represents the set of sampled positions and U the set that was not sampled, and where $\varepsilon\mathbf{(t)}$ represents the thermal noise at time-domain position $\mathbf{t}$. An inverse-problems approach seeks a model $\mathbf{c}$ wherein the modeled values match the observed for those positions that were measured (set K), while the remaining positions (set U) are interpolated in the time domain so as to minimize excess intensity in the frequency domain. This process is therefore seeking to replace the missing data in set U.

Though CLEAN is not explicitly constituted as such a minimization problem, this is very close to, and in some cases identical to, what CLEAN accomplishes. CLEAN iteratively assembles a model $\mathbf{C}$ consisting of a set of δ-functions $\mathbf{C}_I$, which is equivalent to constructing a time-domain model $\mathbf{c}$ as the sum of complex sinusoids. The process of fitting $\mathbf{C}$ to $\mathbf{D}$ implicitly minimizes the difference between $\mathbf{c}$ and $\mathbf{d}$ for those data values that were measured (set K). At the same time, the replacement of artifact-corrupted components $\mathbf{C}_i * \mathbf{S}$ with artifact-free components $\mathbf{C}_i * \mathbf{H}$ is equivalent in the time domain to interpolating the missing points of $\mathbf{c}$ (set U) so as to minimize the total intensity in the final frequency-domain spectrum.

To approach this more quantitatively, consider what happens to Q_C, the first term of eqn (7.15) and the term that assesses consistency with the experimental data, in progressive iterations of CLEAN. The operation of selecting the point of strongest intensity in the artifact-corrupted spectrum $\mathbf{D}^{(i-1)}$ and setting a component $\mathbf{C}_i$ at that location is equivalent to fitting a complex sinusoid c_i in the time domain. Assuming that the point of strongest intensity corresponds to a real signal component A_i, the addition of a frequency-domain component $\mathbf{C}_i$ at a scaled fraction g of that intensity I_{max}:

$$\mathbf{C}^{(i)} \leftarrow \mathbf{C}^{(i-1)} + \mathbf{C}_i$$

$$= \mathbf{C}^{(i-1)} + gI_{max}^{(i-1)}\delta\left(\omega - \omega_{max}^{(i-1)}\right) \tag{7.20}$$

moves the equivalent time-domain model $\mathbf{c}$ a step closer to agreement with $\mathbf{d}$ for the points in set K by adding the complex sinusoid $\mathbf{c}_i(\mathbf{t}) = gI_{\max}^{(i-1)}e^{i2\pi t\omega_{\max}^{(i-1)}}$:

$$\mathbf{c}^{(i)}(\mathbf{t}) \leftarrow \mathbf{c}^{(i-1)}(\mathbf{t}) + \mathbf{c}_i(\mathbf{t})$$
$$= \mathbf{c}^{(i-1)}(\mathbf{t}) + gI_{\max}^{(i-1)}e^{i2\pi t\omega_{\max}^{(i-1)}}. \tag{7.21}$$

In the absence of noise, the values in the observed FID $\mathbf{d}$ at the measured points (set K) would be a sum of complex sinusoids [*cf.* eqn (7.4)]:

$$\mathbf{d}(\mathbf{t}_{k,k\in K}) \approx \mathbf{a}(\mathbf{t}_{k,k\in K}) = \sum_i \mathbf{a}_i(\mathbf{t}_{k,k\in K}). \tag{7.22}$$

At the start of the process, the model $\mathbf{c}$ is empty:

$$\mathbf{c}^{(0)} \leftarrow 0, \tag{7.23}$$

and the difference between the model and the observed values is at its greatest:

$$||\mathbf{c}^{(0)} - \mathbf{d}||_{l_2}^2 = \sum_{k=1}^{N_K} \sum_{i=1}^{M} |-\mathbf{a}_i(\mathbf{t}_{k,k\in K})|^2 \tag{7.24}$$

where M is the total number of components in the true FID and N_K is the number of measured data points (points in set K). Taking $g=1$ for simplicity, after the first iteration the norm is reduced, by eliminating one term in the summation over i, to:

$$||\mathbf{c}^{(1)} - \mathbf{d}||_{l_2}^2 = \sum_{k=1}^{N_K} \sum_{i=2}^{M} |-\mathbf{a}_i(\mathbf{t}_{k,k\in K})|^2. \tag{7.25}$$

After M iterations, the difference between $\mathbf{c}$ and $\mathbf{d}$ is eliminated and $Q_C = 0$. Naturally the convergence is slower when $g < 1$, when a single real component $\mathbf{a}_i$ may be modeled progressively over many iterations. Thus for any g:

$$\lim_{i\to\infty} Q_C = 0 \tag{7.26}$$

and:

$$\lim_{i\to\infty} \mathbf{c}^{(i)}(\mathbf{t}_{k,k\in K}) = \mathbf{d}(\mathbf{t}_{k,k\in K}) \approx \mathbf{a}(\mathbf{t}_{k,k\in K}) \tag{7.27}$$

where the agreement between $\mathbf{d}$ and $\mathbf{a}$ is imperfect owing to thermal noise. The key assumption in this discussion is that the components in the model $\mathbf{c}_l$ correspond only to real components $\mathbf{a}_l$. The final $\mathbf{c}$ will be correct within the measurement error *unless* at any point the fitting process was led astray by incorporating artifactual intensity into the model; this danger is reduced and in most cases eliminated by choosing a small g and by fitting components in the order of strongest to weakest. A correct solution can be guaranteed under particular conditions discussed in the sections below.

Because CLEAN begins with the sparsest possible model and adds to it only as many components as are necessary to achieve agreement with the data, it would seem intuitive that CLEAN might also minimize Q_L, the second term of eqn (7.15). For a spectrum $\mathbf{A}$ that can be decomposed as the sum of δ-functions $\mathbf{A}_I$, the smallest possible Q_L while maintaining consistency with the true spectrum:

$$\min_{\text{all possible C}} Q_L, \quad \text{subject to } \mathbf{c} = \mathbf{a} \tag{7.28}$$

would be obtained for a model $\mathbf{C}_{\text{exact}}$ containing exactly those δ-functions and nothing more:

$$\mathbf{C}_{\text{exact}} = \sum_i \mathbf{A}_i, \tag{7.29}$$

i.e. if the decomposition of $\mathbf{D}$ to $\mathbf{C}_I$ were in perfect agreement with $\mathbf{A}$. In reality, the problem we must solve is:

$$\min_{\text{all possible C}} Q_L, \quad \text{subject to } \mathbf{c}_{k,k \in K} = \mathbf{d}_{k,k \in K} \tag{7.30}$$

where we must consider the possibility that there are multiple models of differing sparsity that would agree with the data, and that CLEAN's decomposition may not always find the sparsest of those models. The problem again is the possibility of interference between overlapping artifact patterns, which could lead CLEAN into a decomposition that matches the data while containing extraneous components. It was shown by Marsh and Richardson that CLEAN does reach the sparsest solution as measured by the l_1-norm provided that certain constraints are met by both the sampling pattern and the signals in the true spectrum, in which case such interference effects are not significant enough to lead CLEAN astray.[13] These conclusions have been confirmed and extended by recent work in the compressed sensing community, showing that algorithms very similar to CLEAN do find the sparsest solution that is consistent with the data when certain requirements are met, as discussed below.

The l_1-norm is not the only possible regularization function that could be used in such a minimization problem. In principle l_0, which is defined as the number of nonzero elements in a vector, would be the ideal regularization constraint when the true spectrum is known to be sparse, but it cannot be used directly in a minimization approach as it is not a convex function.[14] It would seem that a variant of CLEAN designed to restrict the number of components used in the model might implicitly minimize the l_0, but this question has not been explored rigorously.

From this analysis, we see that CLEAN can produce similar results to the l_1-norm convex minimization method, while going about the calculation in a very different way. There are no explicit objective or penalty functions in CLEAN, and the behavior of the algorithm is not guided by such functions or their gradients. But because the action in each iteration of CLEAN is determined by the strongest element in the frequency domain difference vector $\mathbf{D} - \mathbf{C} * \mathbf{S}$, the algorithm winds up minimizing the least-squares difference between the time domain model and the data, as if it had an objective function Q_C. The path of the algorithm is always directed towards agreement of model and data, without any explicit consideration of sparsity, but a sparse model is automatically achieved owing to the sparse starting point $\mathbf{C}^{(0)} = 0$ and the minimalist approach to building up the model.

7.4.2 CLEAN as an Iterative Method for Solving a System of Linear Equations

The preceding interpretation of CLEAN as a form of iterative harmonic analysis (fitting δ-functions in the frequency domain, or equivalently fitting complex sinusoids in the time domain) was first presented by Schwarz, who was the first to analyze CLEAN quantitatively.[7] As part of his analysis, Schwarz also presented a second interpretation which is of great value in explaining some of the properties of the method.

For discrete data, convolution can be written as a matrix equation.[15] We begin with a 1-D case for simplicity. Let vector $\mathbf{S}$ of N elements represent the values of the point response at positions ω_k, $k = 1$ to N. We construct an $N \times N$ matrix $\mathbf{B}$ as follows. In the first row, circular-shift $\mathbf{S}$ so that the central peak of $\mathbf{S}$, at its midpoint $\omega_{N/2}$, is in the first element of the row:

$$\mathbf{B}_{\text{first row}} = [\mathbf{S}(\omega_{N/2}) \; \mathbf{S}(\omega_{N/2+1}) \; \cdots \; \mathbf{S}(\omega_N) \; \mathbf{S}(\omega_1) \; \cdots \; \mathbf{S}(\omega_{N/2-1})]. \quad (7.31)$$

In each subsequent row, circular shift $\mathbf{S}$ by one element to the right:

$$\mathbf{B} = \begin{bmatrix} \mathbf{S}(\omega_{N/2}) & \mathbf{S}(\omega_{N/2+1}) & \cdots & \mathbf{S}(\omega_{N/2-2}) & \mathbf{S}(\omega_{N/2-1}) \\ \mathbf{S}(\omega_{N/2-1}) & \mathbf{S}(\omega_{N/2}) & \cdots & \mathbf{S}(\omega_{N/2-3}) & \mathbf{S}(\omega_{N/2-2}) \\ \mathbf{S}(\omega_{N/2-2}) & \mathbf{S}(\omega_{N/2-1}) & \cdots & \mathbf{S}(\omega_{N/2-4}) & \mathbf{S}(\omega_{N/2-3}) \\ \vdots & \vdots & \ddots & \vdots & \vdots \\ \mathbf{S}(\omega_{N/2+1}) & \mathbf{S}(\omega_{N/2+2}) & \cdots & \mathbf{S}(\omega_{N/2-1}) & \mathbf{S}(\omega_{N/2}) \end{bmatrix}. \quad (7.32)$$

A matrix with the same element in each top-left to bottom-right diagonal is known as a Toeplitz matrix:[16]

$$\begin{bmatrix} t_0 & t_1 & t_2 & \cdots & t_{n-1} \\ t_{-1} & t_0 & t_1 & \cdots & t_{n-2} \\ t_{-2} & t_{-1} & t_0 & \cdots & t_{n-3} \\ \vdots & \cdots & \cdots & \ddots & \vdots \\ t_{-(n-1)} & \cdots & \cdots & \cdots & t_0 \end{bmatrix}, \quad (7.33)$$

and the multiplication of this Toeplitz matrix by a data vector **F** produces the convolution of **F** and the sequence of elements $\mathbf{T} = t_{-(n-1)}$ to t_{n-1}:

$$\mathbf{T} * \mathbf{F} = \begin{bmatrix} t_0 & t_1 & t_2 & \cdots & t_{n-1} \\ t_{-1} & t_0 & t_1 & \cdots & t_{n-2} \\ t_{-2} & t_{-1} & t_0 & \cdots & t_{n-3} \\ \vdots & \cdots & \cdots & \ddots & \vdots \\ t_{-(n-1)} & \cdots & \cdots & \cdots & t_0 \end{bmatrix} \begin{pmatrix} f_0 \\ f_1 \\ f_2 \\ \vdots \\ f_{n-1} \end{pmatrix}. \tag{7.34}$$

The matrix **B** above has several further properties. First, because each row is the right circular shift of the row above, it is a circulant matrix, with structure:

$$\begin{bmatrix} c_1 & c_2 & c_3 & \cdots & c_n \\ c_n & c_1 & c_2 & \cdots & c_{n-1} \\ c_{n-1} & c_n & c_1 & \cdots & c_{n-2} \\ \vdots & \cdots & \cdots & \ddots & \vdots \\ c_n & \cdots & \cdots & \cdots & c_1 \end{bmatrix}, \tag{7.35}$$

which carries out cyclic convolution when multiplied by the data vector. This arises because the convolution kernel **S** is generated from a discrete Fourier transform of the sampling pattern **s**, the output of which is always cyclic. Because **s** is defined only for positive evolution times, values of **S** for $k < N/2$ are the complex conjugates of the values for $k > N/2$:

$$\mathbf{S}(\omega_{N/2-k}) = \overline{\mathbf{S}(\omega_{N/2+k})}, \tag{7.36}$$

and **B** should be Hermitian, with the structure:

$$\begin{bmatrix} b_1 & b_2 & b_3 & \cdots & \bar{b}_3 & \bar{b}_2 \\ \bar{b}_2 & b_1 & b_2 & \cdots & \bar{b}_4 & \bar{b}_3 \\ \bar{b}_3 & \bar{b}_2 & c_1 & \cdots & \bar{b}_5 & \bar{b}_4 \\ \vdots & \cdots & \cdots & \ddots & \vdots & \vdots \\ b_3 & b_4 & b_5 & \cdots & b_1 & b_2 \\ b_2 & b_3 & b_4 & \cdots & \bar{b}_2 & b_1 \end{bmatrix}, \tag{7.37}$$

but as it is normal in NMR to discard the complex components of a spectrum, we can do the same with **S** and **B**, yielding the symmetry relationship:

$$\mathbf{S}(\omega_{N/2-k}) = \mathbf{S}(\omega_{N/2+k}) \tag{7.38}$$

and making **B** a symmetric matrix with structure:

$$\begin{bmatrix} b_1 & b_2 & b_3 & \cdots & b_3 & b_2 \\ b_2 & b_1 & b_2 & \cdots & b_4 & b_3 \\ b_3 & b_2 & b_1 & \cdots & b_5 & b_4 \\ \vdots & \cdots & \cdots & \ddots & \vdots & \vdots \\ b_3 & b_4 & b_5 & \cdots & b_1 & b_2 \\ b_2 & b_3 & b_4 & \cdots & b_2 & b_1 \end{bmatrix}. \tag{7.39}$$

It should be stressed here that this symmetry does not arise from how the points are arranged in the sampling pattern $\mathbf{s}$, but rather because of the Fourier transform relating $\mathbf{s}$ and $\mathbf{S}$.

Now let vector $\mathbf{A}$ of N elements represent the values of the true spectrum at positions ω_k, $k=1$ to N. Ignoring thermal noise, the artifact-corrupted spectrum $\mathbf{D}=\mathbf{A}*\mathbf{S}$, also a vector of N elements, can be calculated by multiplying $\mathbf{B}$ and $\mathbf{A}$:

$$\mathbf{D}=\mathbf{BA}. \tag{7.40}$$

The convolution operation is therefore a linear transformation, and the problem we are attempting to solve with CLEAN is to find a solution $\mathbf{C}\approx\mathbf{A}$ to a set of N linear equations $\mathbf{D}=\mathbf{BC}$ in the frequency domain.

Let us consider again what CLEAN does, in light of this interpretation. The first iteration of CLEAN fits a δ-function $\mathbf{C}_1$ at the position $\omega_{\max}$ where $\mathbf{D}$ has maximal intensity and adds this to an initially empty model $\mathbf{C}^{(0)}$. If we let the model $\mathbf{C}$ also be a vector with N elements, the fitting of a δ-function at position $\omega_{\max}$ is the same as setting a value to the element at index k, corresponding to $\omega_{\max}$, in vector $\mathbf{C}$. Assuming $g=1$ for simplicity and using c_k and d_k to represent the elements of $\mathbf{C}$ and $\mathbf{D}$, respectively, with index k, the change to $\mathbf{C}$ as it goes from the initially empty model $\mathbf{C}^{(0)}$ to the new model $\mathbf{C}^{(1)}$ is:

$$c_k^{(1)} \leftarrow d_k. \tag{7.41}$$

At the same time, we generate an updated $\mathbf{D}$ with the contributions from $\mathbf{C}_1*\mathbf{S}$ removed by subtraction. Every position in $\mathbf{D}$ is altered, with the new values determined by applying $\mathbf{B}$ to the updated model $\mathbf{C}^{(1)}$ and subtracting this from $\mathbf{D}$:

$$\mathbf{D}^{(1)} \leftarrow \mathbf{D} - \mathbf{BC}^{(1)}, \tag{7.42}$$

meaning that the change at any individual position k is:

$$d_k^{(1)} \leftarrow d_k - \sum_j b_{kj} c_j^{(1)} \tag{7.43}$$

where b_{kj} is the element in $\mathbf{B}$ at row k, column j. In subsequent iterations, more components are fit at various positions in the spectrum, each time locating the position k in $\mathbf{D}$ with maximum intensity, and each time resulting in a change to an element c_k according to the update rule:

$$c_k^{(i)} \leftarrow c_k^{(i-1)} + d_k^{(i-1)} \tag{7.44}$$

which adds a new component at position k on top of any existing components at the same position. The value d_k at any position k after iteration

$i-1$ will be the starting value minus the accumulated effects of the components already assigned at all positions j during the previous iterations:

$$d_k^{(i-1)} = d_k^{(0)} - \sum_j b_{kj} c_j^{(i-1)}. \tag{7.45}$$

We can combine eqn (7.44) and (7.45) to make a single update rule for how to adjust element c_k each time position k is selected for a new component:

$$c_k^{(i)} \leftarrow c_k^{(i-1)} + \left(d_k^{(0)} - \sum_j b_{kj} c_j^{(i-1)} \right), \tag{7.46}$$

or just:

$$c_k^{(i)} \leftarrow d_k^{(0)} - \sum_{j,j \neq k} b_{kj} c_j^{(i-1)}. \tag{7.47}$$

If we make one additional formal change:

$$c_k^{(i)} \leftarrow \frac{d_k^{(0)} - \sum\limits_{j,j \neq k} b_{kj} c_j^{(i-1)}}{b_{kk}} \tag{7.48}$$

where the diagonal element $b_{kk}=1$ if **S** and **B** are normalized, we now have the update rule for the Gauss–Seidel method of iteratively solving a system of linear equations.[17] The only remaining difference is in the search strategy: Gauss–Seidel builds its model **C** with components at all N positions, calculated in order of index k on repeated passes, while CLEAN builds its model **C** with components only at positions that are clearly above the noise, and taken in the order of strongest to weakest.

The Gauss–Seidel method converges to the least-squares solution for any starting model $\mathbf{C}^{(0)}$ as long as **B** is positive definite, *i.e.* all of its eigenvalues are greater than zero, and as long as **D** is in the range of **B**, meaning that **D** is in the set of possible results of applying the linear transformation **B** to some data vector. It also converges to a solution in the less strict case that **B** is positive semidefinite, *i.e.* all of its eigenvalues are greater than or equal to zero. The fact that **B** is circulant makes its decomposition into eigenvalues and eigenvectors particularly straightforward,[16] as the eigenvectors $\mathbf{V}_j = \{\mathbf{V}_j : j = 0, 1, \ldots, N-1\}$ of a circulant matrix are simply the Fourier basis vectors:

$$\mathbf{V}_j = \frac{1}{\sqrt{N}} \begin{pmatrix} 1 \\ e^{-2\pi i j \frac{1}{N}} \\ e^{-2\pi i j \frac{2}{N}} \\ \vdots \\ e^{-2\pi i j \frac{N-1}{N}} \end{pmatrix} = \frac{1}{\sqrt{N}} \begin{pmatrix} 1 \\ \omega^j \\ \omega^{2j} \\ \vdots \\ \omega^{(N-1)j} \end{pmatrix}, \quad \omega = e^{-2\pi i/N}, \tag{7.49}$$

and the corresponding eigenvalues λ_j are simply the values of the Fourier transform of the first row of the matrix:

$$\lambda_j = b_j, \quad \mathbf{b} = \mathrm{DFT}(\mathbf{B}_{\text{first row}}) \tag{7.50}$$

giving the decomposition:

$$
\mathbf{B} = \frac{1}{N}
\begin{bmatrix}
\overline{\omega^0} & \overline{\omega^0} & \overline{\omega^0} & \cdots & \overline{\omega^0} \\
\overline{\omega^0} & \overline{\omega^1} & \overline{\omega^2} & \cdots & \overline{\omega^{N-1}} \\
\overline{\omega^0} & \overline{\omega^2} & \overline{\omega^4} & \cdots & \overline{\omega^{2(N-1)}} \\
\vdots & \vdots & \vdots & \ddots & \vdots \\
\overline{\omega^0} & \overline{\omega^{N-1}} & \overline{\omega^{2(N-1)}} & \cdots & \overline{\omega^{(N-1)(N-1)}}
\end{bmatrix}
\begin{bmatrix}
\lambda_0 & 0 & \cdots & 0 \\
0 & \lambda_1 & \cdots & 0 \\
\vdots & \vdots & \ddots & \vdots \\
0 & 0 & \cdots & \lambda_{N-1}
\end{bmatrix}
$$

$$
\times
\begin{bmatrix}
\omega^0 & \omega^0 & \omega^0 & & \omega^0 \\
\omega^0 & \omega^1 & \omega^2 & \cdots & \omega^{N-1} \\
\omega^0 & \omega^2 & \omega^4 & \cdots & \omega^{2(N-1)} \\
\vdots & \vdots & \vdots & \ddots & \vdots \\
\omega^0 & \omega^{N-1} & \omega^{2(N-1)} & \cdots & \omega^{(N-1)(N-1)}
\end{bmatrix}.
\tag{7.51}
$$

The meaning of this decomposition is clear when we consider that the operator $\mathbf{B}$, which takes the true spectrum $\mathbf{A}$ and convolves it with $\mathbf{S}$, must be equivalent to an inverse DFT of the spectrum, the multiplication of the resulting time domain by the sampling function, and the FT of that time domain to give the artifact-corrupted spectrum, or in other words:

$$\mathbf{B} = \mathbf{F}\mathbf{s}_D\mathbf{F}^{-1} \tag{7.52}$$

where:

$$
\mathbf{F}^{-1} = \frac{1}{\sqrt{N}}
\begin{bmatrix}
\omega^0 & \omega^0 & \omega^0 & \cdots & \omega^0 \\
\omega^0 & \omega^1 & \omega^2 & \cdots & \omega^{N-1} \\
\omega^0 & \omega^2 & \omega^4 & \cdots & \omega^{2(N-1)} \\
\vdots & \vdots & \vdots & \ddots & \vdots \\
\omega^0 & \omega^{N-1} & \omega^{2(N-1)} & \cdots & \omega^{(N-1)(N-1)}
\end{bmatrix}
\tag{7.53}
$$

is the inverse DFT operator, $\mathbf{F}$ is the forward DFT operator (the conjugate transpose of $\mathbf{F}^{-1}$), and:

$$
\mathbf{s}_D =
\begin{bmatrix}
s(t_1) & 0 & \cdots & 0 \\
0 & s(t_2) & \cdots & 0 \\
\vdots & \vdots & \ddots & \vdots \\
0 & 0 & \cdots & s(t_N)
\end{bmatrix}
=
\begin{bmatrix}
\lambda_0 & 0 & \cdots & 0 \\
0 & \lambda_1 & \cdots & 0 \\
\vdots & \vdots & \ddots & \vdots \\
0 & 0 & \cdots & \lambda_{N-1}
\end{bmatrix}
\tag{7.54}
$$

is simply the representation of the sampling pattern as a diagonal matrix. Thus a sparse NUS sampling pattern with N_K measured points and $N_U = N - N_K$ omitted points will have N_K eigenvalues equal to one, and N_U eigenvalues equal to zero. Since all of the eigenvalues are either zero or one, the matrix $\mathbf{B}$ is positive semidefinite, and the convergence criteria are satisfied.

The case of $\mathbf{B}$ being positive definite is therefore none other than the case in which the sampling pattern is a complete grid, with the number of unknowns equal to the number of knowns, and in that case there would be a single solution $\mathbf{C} \approx \mathbf{A}$ to the matrix equation $\mathbf{D} = \mathbf{BC}$, which CLEAN is guaranteed to find. The case we are interested in, however, is the positive semidefinite case, as occurs whenever we carry out sparse sampling. In this case the rank of $\mathbf{B}$ is N_K, its determinant is zero, there are fewer knowns than unknowns, and the solution to $\mathbf{D} = \mathbf{BC}$ is not, in general, unique. Whether or not it is unique in any specific case depends on the details of the true spectrum $\mathbf{A}$ and the point response $\mathbf{S}$.

The eigendecomposition of $\mathbf{B}$ provides insight into the ambiguity of the possible solutions to $\mathbf{D} = \mathbf{BC}$ in the undersampled case, showing how this ambiguity relates to the sampling pattern. An eigenvector of $\mathbf{B}$ is a vector $\mathbf{V}$ such that:

$$\mathbf{BV} = \lambda\mathbf{V}. \tag{7.55}$$

In the context of NMR, this would mean a frequency-domain spectrum $\mathbf{V}$ that is not altered in any way by convolution with $\mathbf{S}$, except to be scaled vertically by the factor λ. Clearly $\mathbf{V}$ must be an unusual signal, as we expect convolution with $\mathbf{S}$ both to alter the lineshape and to introduce artifacts throughout the spectrum. As we saw in the decomposition above, the N independent eigenvectors $\mathbf{V}_J$ of $\mathbf{B}$ are in fact the Fourier basis functions, or in other words complex sinusoids with frequencies of j cycles over N points, and their eigenvalues correspond to specific values (1 or 0) of the sampling function. We therefore have a set of N_K complex sinusoids $\mathbf{V}_K$ at frequencies k/N such that:

$$\mathbf{BV}_k = \mathbf{V}_k, \quad \text{if } k \in K \tag{7.56}$$

and a second set of N_U complex sinusoids $\mathbf{V}_U$ such that:

$$\mathbf{BV}_u = 0, \quad \text{if } u \in U. \tag{7.57}$$

A signal that exactly matched one of the $\mathbf{V}_K$ eigenvectors would be unaffected by sparse sampling because its time domain equivalent falls entirely on a measured sampling point (Figure 7.6a). A signal that exactly matched one of the $\mathbf{V}_U$ eigenvectors would be completely lost by the sparse sampling as its time domain equivalent falls entirely on a sampling point that was omitted from the sampling pattern (Figure 7.6b).

Fortunately, typical NMR signals extend to most or all of the time domain and fall on both measured and omitted sampling points (Figure 7.6d). These signals would be represented in the frequency domain as the sum of the eigenvectors in both sets, measured (K) and unmeasured (U). The effect of sparse sampling is to omit all data corresponding to eigenvectors in set U, such that the spectrum must be calculated from the set K data alone. The omission of some of the sinusoids from the Fourier basis introduces artifacts that distort the appearance of the frequency domain spectrum, but the signals are not lost since they are still represented on K.

This analysis shows that the sampling process introduces ambiguity because it causes particular basis vectors to be omitted from the Fourier transform. The infinitely many possible solutions to $\mathbf{D} = \mathbf{BC}$ differ from one another in their inclusion or not of the missing sinusoidal components in set U. A DFT of the sparse data presents one solution to $\mathbf{D} = \mathbf{BC}$, in which it is assumed that these sinusoidal components should all be zero. This most likely does not correctly represent the true signal, and the goal of CLEAN is to find a better solution from the set that satisfy $\mathbf{D} = \mathbf{BC}$.

The sparse sampling process is therefore insensitive to any features in the spectrum that are only represented by basis vectors in set U. Any signal with a time-domain representation falling solely on points in U would be completely lost, and one could imagine a pathological case like Figure 7.6c in which a complex frequency domain disappears owing to the sampling. However, it is highly unlikely that we would ever encounter an NMR FID consisting of spikes at specific evolution times, which might be lost if a spike fell at a point in U; instead, NMR FIDs contain sinusoids spanning the whole time domain, which can still be recognized as sinusoids and fitted accurately when many points have been omitted. Moreover, CLEAN explicitly assumes frequency domain sparseness, introducing components only when they are needed to match the data, meaning that in the time domain it explicitly fits sinusoids filling the whole domain, so it is well-suited both for the type of data encountered in NMR and for the particular ambiguities introduced by sparse sampling. The fact that NUS sparse sampling patterns are normally random further increases the chances of a successful fitting, as there are no particular families of signals or ranges of frequencies that would be selectively disfavored. These intuitive notions have been formalized recently by compressed sensing theory, as described in the next section, putting rigorous bounds on the probability of failure when an NMR signal is sparsely sampled in the time domain.

Though the discussion above is framed in terms of a one-dimensional spectrum, a similar formalism can be constructed for higher-dimensional cases, and the same analysis applies.[7] For each position in the sampling pattern, there is an eigenvalue that is either one or zero, and a corresponding eigenvector consisting of a multidimensional complex sinusoid, with frequencies matching the indices of the sampling point in the various dimensions. The convergence and uniqueness criteria are the same.

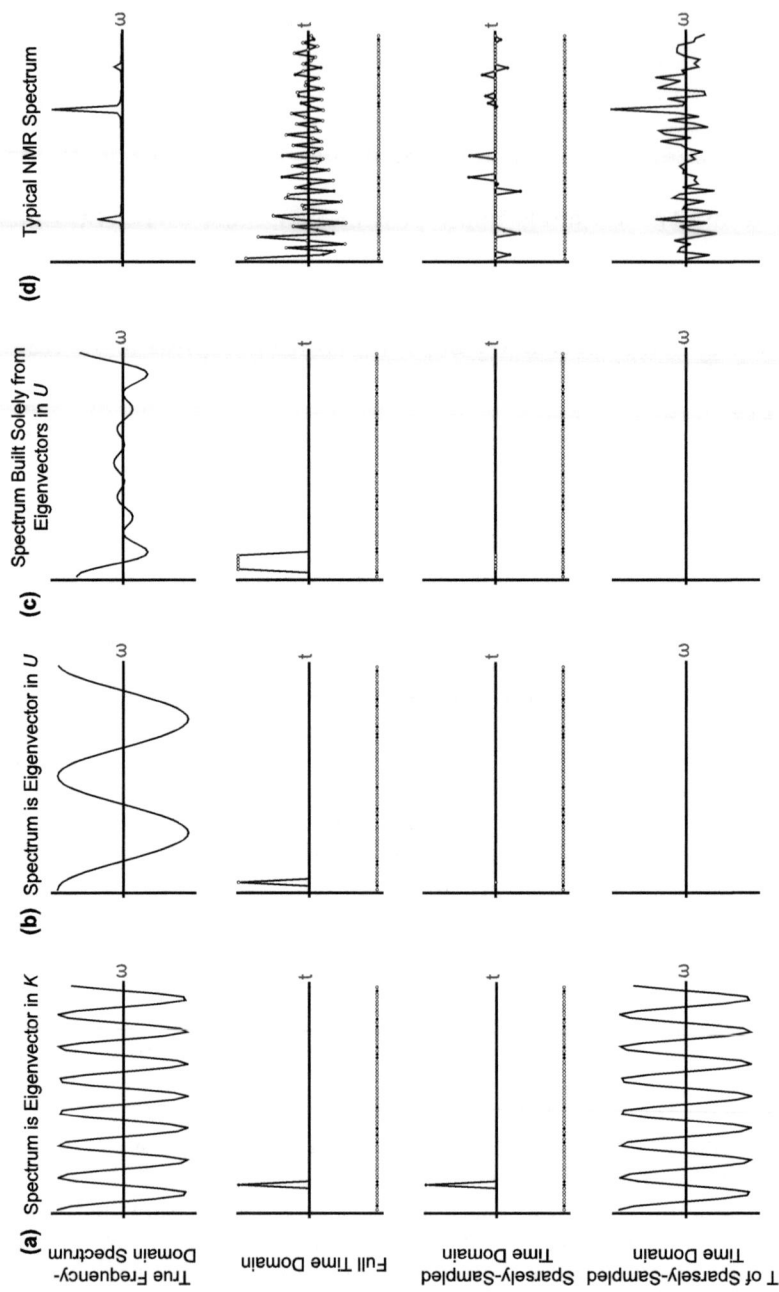

(a) Spectrum is Eigenvector in *K*

(b) Spectrum is Eigenvector in *U*

(c) Spectrum Built Solely from Eigenvectors in *U*

(d) Typical NMR Spectrum

True Frequency-Domain Spectrum

Full Time Domain

Sparsely-Sampled Time Domain

FT of Sparsely-Sampled Time Domain

7.4.3 CLEAN and Compressed Sensing

Throughout this chapter, we have described the criteria for the success or failure of CLEAN intuitively, in terms of several considerations: (1) the random nature of the artifacts produced from random sampling; (2) the low probability that interference between artifacts would make the decomposition of the dirty spectrum degenerate, when approached from strongest component to weakest; and (3) the relative sparseness of the true frequency-domain spectrum translating into a time domain that is not sparse, making it unlikely that the sparse sampling could miss a signal. We also expect intuitively that there are limits on how far sparse sampling may be taken. It is only because of the sparseness of the spectrum that the number of measurements N_K can be less than the size of the spectrum N; determining the true spectrum from sparse data is an underdetermined problem, but the sparseness criterion allows algorithms such as CLEAN or convex l_1-norm minimization to find the correct solution in certain cases. Even with a sparseness criterion and appropriate algorithms, there must be an absolute limit that the number of measurements N_K cannot possibly be lower than the number of components M in the true spectrum, because the number of measurements cannot be lower than the number of nonzero degrees of freedom. Furthermore, we expect that some degree of additional sampling beyond M measurements would be needed, as the algorithm must somehow resolve which positions in the model should be zero and which should be nonzero, and we expect the arrangement of the measurements in the time domain to matter in some way for the ability to find a solution.

Figure 7.6   Eigendecomposition of sparse sampling. The sparse sampling process is equivalent to a linear operator, and this operator can be analyzed in terms of eigenvalues and eigenvectors. We group the eigenvectors into two sets, those with eigenvalues of 1 in set K and those with eigenvalues of 0 in set U. (In panels illustrating the time domain, closed circles indicate measured sampling points, while open circles indicate omitted points.) (a) A frequency-domain spectrum that exactly matches one of the eigenvectors in set K has a time domain consisting of a δ-function at one of the measured sampling points. This time domain is unaltered by sparse sampling, and the resulting FT of the sparse time domain is identical to the original spectrum. (b) A frequency-domain spectrum that exactly matches one of the eigenvectors in set U has a time domain consisting of a δ-function at one of the omitted sampling points. Sparse sampling completely eliminates this signal. (c) A frequency-domain spectrum represented solely by basis vectors in set U would have a corresponding time domain where the signal is present only at omitted sampling points. Sparse sampling completely eliminates this signal. (d) A typical NMR spectrum is sparse in the frequency domain, and therefore fills the entire time domain. Such a spectrum is represented by basis vectors from both sets, and in the time domain has support on both measured and omitted sampling points. Sparse sampling eliminates the omitted time domain points and therefore the basis vectors in set U, but the signals are adequately represented through the measured time domain points and the corresponding basis vectors in set K.

Recent work in the field of information theory allows for more rigorous and quantitative treatment of these questions. The problem of determining the true spectrum from sparsely sampled time domain data in NMR is very similar to problems arising in other fields, including developing more optimal methods of data compression, detecting and correcting errors in signal transmission, and analyzing seismic waves. Problems of this type were traditionally known as *sparse recovery* or *sparse approximation*, but have recently acquired the name *compressed sensing*, and have the following structure. Let $x_0 = \{x_i : i = 1, 2, \ldots, N\}$ be a signal that is sparse, having only M nonzero elements where $M \ll N$. A linear operator $\boldsymbol{\Phi}$ is used to record a set of $N_K < N$ measurements $\mathbf{y} = \{y_i : i = 1, 2, \ldots, N_K\}$, such that:

$$\mathbf{y} = \boldsymbol{\Phi}\mathbf{x}_0 \tag{7.58}$$

where $\boldsymbol{\Phi}$ is a matrix with N_K rows and N columns. The problem is then to determine a vector $\mathbf{x}$ that approximates $\mathbf{x}_0$ from the N_K limited measurements. In the case of NMR, $\mathbf{x}_0$ is the true spectrum $\mathbf{A}$, $\mathbf{x}$ is the model $\mathbf{C}$, and $\boldsymbol{\Phi}$ is an operator relating the measurements to the true spectrum, which can be formulated in several ways. If the set of evolution times recorded in the experiment is $T = \{\mathbf{t}_i : i = 1, 2, \ldots, N_K\}$, we can construct a matrix $\mathbf{F}^{-1}{}_T$ consisting of the N_K rows from the inverse Fourier transform matrix $\mathbf{F}^{-1}$ that correspond to the measured times T. The vector $\mathbf{y}$ is then the set of sparse samples in the time domain, which we will call $\mathbf{d}_T$, and we can write:

$$\mathbf{d}_T = \mathbf{F}_T^{-1}\mathbf{A}. \tag{7.59}$$

Equivalently, we could consider a vector $\mathbf{d}$ of N elements describing the complete time domain—N_K of them measured, and another N_U not measured and instead set to zero—giving:

$$\mathbf{d} = \mathbf{s}_D\mathbf{F}^{-1}\mathbf{A} \tag{7.60}$$

where $\mathbf{s}_D$ is the diagonal matrix representation of the sampling pattern [eqn (7.54)]. This representation is superficially different from the canonical problem in that the operator $\boldsymbol{\Phi} = \mathbf{s}_D\mathbf{F}^{-1}$ is an $N \times N$ matrix with N_K populated rows and N_U rows of zeroes rather than a fully populated $N_K \times N$ matrix. Finally, the problem can be stated in the frequency domain as:

$$\begin{aligned} \mathbf{D} &= \mathbf{F}\mathbf{s}_D\mathbf{F}_T^{-1}\mathbf{A} \\ &= \mathbf{B}\mathbf{A} \end{aligned} \tag{7.61}$$

but now both $\mathbf{A}$ and $\mathbf{D}$ have N elements, and we have to look to the eigenvalues of $\mathbf{B}$ to find that there were only N_K measurements. The goals of compressed sensing theory are to determine whether particular families of measurement operators $\boldsymbol{\Phi}$ (such as all those built from Fourier transform matrices) are compatible with sparse recovery; for those that are compatible, to determine the minimum number of measurements N_K needed to reconstruct a signal $\mathbf{x}_0$ with M nonzero elements out of N total; and to find algorithms that can determine such reconstructions robustly if the requisite criteria are met. We will give here only a brief outline of the main results of compressed sensing theory as they pertain to NMR and CLEAN, as a full discussion is beyond the scope of this chapter. We refer interested readers to the subsequent chapters and references.[15,17,21-25].

The vectors **x** and **y** can be thought of as representations of the same signal but in different bases, with the operator **Φ** carrying out a change in basis from some basis **Ψ**$_x$ used in representation **x** to some basis **Ψ**$_y$ used in representation **y**. The fundamental principle underlying compressed sensing is that it is not possible for a signal to appear sparse in both bases if the bases are *incoherent*, meaning that the basis elements of the two bases are linearly independent.[18-20] For NMR, we take the frequency domain as being in the Dirac basis, in which each element is a δ-function and the signal is a weighted sum of δ-functions at various positions, and the time domain as being in the Fourier basis, in which each element is a complex sinusoid at some frequency and the signal is a weighted sum of complex sinusoids of various frequencies. These two bases happen to be the two bases that are maximally incoherent. One can then impose a constraint on the minimum number of nonzero components across the representations of the signal in the two bases:

$$||\mathbf{x}||_{l_0} + ||\mathbf{y}||_{l_0} \geq 2\sqrt{N}. \tag{7.62}$$

For example, if a spectrum is sparse in the frequency domain, with only M nonzero components, it cannot simultaneously have fewer than $2\sqrt{N} - M$ components in the time domain. This is, in fact, simply a discrete analog of the familiar uncertainty principle:

$$\Delta t \Delta \omega \geq 1 \tag{7.63}$$

stating that a signal that is concentrated in one domain cannot also be concentrated in the other. The constant on the right-hand side of eqn (7.62) is the absolute minimum for all possible signals, but it has been shown that the constant is actually larger for most realistic signals, with the exceptions having unusual structure, as in the case of the comb function.[19-21] Unless the signals in the spectrum are arranged in a very regular pattern, such as a grid, it is safe to state that:

$$||\mathbf{x}||_{l_0} + ||\mathbf{y}||_{l_0} \geq \text{constant} \times \frac{N}{\sqrt{\log N}} \tag{7.64}$$

and it has been shown that the upper bound of the probability of a randomly chosen signal violating this constraint is of the order $N^{-\beta}$, where β is a small positive integer.[19,20] This means that if a spectrum has a small number of nonzero components, its Fourier transform almost certainly has dozens to hundreds of nonzero terms.

The significance of the uncertainty principle for NMR sparse sampling is that it significantly restricts the occasions when degenerate sparse reconstructions could occur.[19,20] If we have two frequency domain models that both fit the data and that are both equally sparse, they must differ at a great number of positions in the time domain. By choosing the number of sampling points and the distribution of sampling points appropriately, we can create conditions in which it is not possible to have two equally sparse models that both fit the data, and an algorithm that is designed to find the sparsest solution will be guaranteed to find a single unique answer. If we can further say, *a priori*, that the true spectrum contains no more than M components, this unique sparse solution is guaranteed to be the correct one. The key to ensuring that these conditions are met for all possible sparse spectra is to carry out the sampling randomly.

Another, more general, way of stating the uncertainty principle is to say that sparse reconstructions will be unique if the mapping from $\mathbf{x}$ to $\mathbf{y}$ carried out by the operator $\mathbf{\Phi}$ is one-to-one for all possible sparse signals $\mathbf{x}$; that is, if every possible sparse signal $\mathbf{x}$ generates a unique $\mathbf{y}$, any sparse $\mathbf{x}$ that we determine from $\mathbf{y}$ must be unique.[22] To assess the mapping behavior of $\mathbf{\Phi}$, Candès and Tao introduced a measure known as the *restricted isometry property*.[14] A given measurement operator $\mathbf{\Phi}$ satisfies the restricted isometry property of order M if the inequality:

$$(1 - \delta_M) \, ||\mathbf{x}||_{l_2}^2 \leq ||\mathbf{\Phi x}||_{l_2}^2 \leq (1 + \delta_M) \, ||\mathbf{x}||_{l_2}^2 \qquad (7.65)$$

is true for all possible signals $\mathbf{x}$ that have no more than M nonzero components, with the constant $\delta_M < 1$ free to vary depending on M. The restricted isometry property requires that the square of the l_2-norm of every $\mathbf{y}$ made from an $\mathbf{x}$ be bounded close to the square of the l_2-norm of the original $\mathbf{x}$. Recall that one way of stating Plancherel's theorem about the energy of a vector $\mathbf{g}$ being the same as the energy of its Fourier transform $\mathbf{G}$ is using squares of l_2-norms:

$$||\mathbf{g}||_{l_2}^2 = ||\mathbf{G}||_{l_2}^2, \qquad (7.66)$$

and this property is sometimes known as *isometry*. Thus the restricted isometry property requires that the energy of a restricted ($N_K \ll N$) set of observations $\mathbf{y}$ be close to that of the sparse signal $\mathbf{x}$, for all such sparse signals $\mathbf{x}$, which is a way of saying that different $\mathbf{x}$ vectors must map to the N_K elements of $\mathbf{y}$ in such a way that changes in $\mathbf{x}$ have a quantitative effect on the generated $\mathbf{y}$. If this is true, and if the constant δ_M is small enough (meaning that the isometry is close enough), the operator $\mathbf{\Phi}$ maps sparse signals in an appropriate way to permit recovery, and in fact the recovery of a signal with no more than M components can be achieved exactly.

The restricted isometry property could also be viewed as a certification that for every subset of M columns taken from the matrix $\mathbf{\Phi}$, those M columns will be linearly independent.[14,22] This property is important because an $\mathbf{x}$ with M nonzero components will be transformed by M columns from $\mathbf{\Phi}$, and we do not know *a priori* which M components in $\mathbf{x}$ will be nonzero and which columns from $\mathbf{\Phi}$ will be needed. If any subset of M columns were singular, there would be a class of sparse signals for which the transformation $\mathbf{\Phi}$ produces degenerate results. There is no question of the columns of $\mathbf{\Phi}$ being singular if $\mathbf{\Phi}$ is the complete Fourier transform matrix of order N, but the answer is not necessarily as clear for an $\mathbf{F}_T^{-1}$ containing only a subset from the N rows. The restricted isometry property assesses whether subsets of columns are singular by measuring quantitatively how different sparse signals are transformed by $\mathbf{\Phi}$.

For sparse sampling in NMR, the relevant inequality is:

$$(1 - \delta_M) \, ||\mathbf{C}||_{l_2}^2 \leq ||\mathbf{F}_T^{-1}\mathbf{C}||_{l_2}^2 \leq (1 + \delta_M) \, ||\mathbf{C}||_{l_2}^2 \qquad (7.67)$$

which can be understood as saying that for every possible spectrum $\mathbf{C}$ with no more than M components, if a substantial part of the energy of $\mathbf{C}$ maps in the time domain to the measured data points K, it will be possible to reconstruct $\mathbf{C}$ uniquely. If, however, there were a class of spectra with M or fewer components whose energy in the time domain were concentrated on the unmeasured points in set U, it would be impossible to achieve a unique reconstruction. As long as our measured points K are also distributed fairly uniformly, and without any regular patterning that would cause us to miss periodic signals, we would expect enough energy to fall on the measured points to allow unique reconstruction. The restricted isometry property quantitatively assesses whether this is true for a particular sampling pattern.

It has been shown that when $\mathbf{\Phi}$ is the Fourier transform over a subset T of possible measurement positions—as in NMR—and when those measurement positions are chosen at random, there is a very high probability that the sampling pattern satisfies the restricted isometry property as long as the number of measurements is sufficient. Unfortunately, despite extensive theoretical work the precise number of samples required is still an open question. Candès *et al.* determined that:

$$N_K \geq \alpha M \log N \qquad (7.68)$$

measurements are enough to provide a reasonably high probability of a unique and correct solution,[19] where α is a constant that depends on the desired probability level, but full satisfaction of the restricted isometry property has only been proven for:

$$N_K \geq \alpha M \log^4 N \qquad (7.69)$$

though it seems likely that $\log N$ is the true dependence.[22–24] To our knowledge no analytical formula for α as a function of desired probability of success has been reported so far, making it difficult to apply eqn (7.68) in practice. However, eqn (7.68) does show that the number of samples needed scales linearly with the complexity of the spectrum, and logarithmically with the size of the spectrum.

We note in passing that compressed sensing theory almost exclusively addresses the case where the randomly chosen samplings are uniformly distributed in the time domain. To our knowledge, there are no proofs at this point of whether sampling patterns with weighted distributions, such as the popular exponentially weighted patterns, satisfy the restricted isometry property, though there are also no particular reasons to believe that they would not.

Given that there is a unique and correct solution to the problem $y = \Phi x$, the next question is whether a particular algorithm such as CLEAN will find it. It is a well-known result of compressed sensing theory that convex l_1-norm minimization (also known as *basis pursuit* in the compressed sensing community) is able to do so whenever the restricted isometry property is satisfied with a small enough constant δ_M.[14] An alternative algorithm known as *matching pursuit* (MP) was also proposed for finding this solution,[25] based on algorithms used for regression in statistics.[26,27] Recall that the solution x is a decomposition of the observations y as a weighted sum of the elements ϕ_j (the column vectors) of the measurement operator Φ. The principle of MP is to find the basis element ϕ_j that most strongly correlates with y:

$$j = \arg \max_k \langle y, \phi_k \rangle, \tag{7.70}$$

decompose y as the contribution from ϕ_j plus a residual r,

$$y = \langle y, \phi_j \rangle \phi_j + r, \tag{7.71}$$

and then repeat this over a number of iterations until the full decomposition has been obtained. Let $J = \{j: 1, 2, \ldots, i-1\}$ be the set of basis indices that have already been fit as of iteration $i-1$. A single iteration i adds one element to the solution vector:

$$x_j \leftarrow \langle y^{(i-1)}, \phi_j \rangle \tag{7.72}$$

where:

$$j = \arg \max_{k, k \notin J} \langle y^{(i-1)}, \phi_k \rangle \tag{7.73}$$

and generates a new residual:

$$y^{(i)} \leftarrow y^{(i-1)} - \langle y^{(i-1)}, \phi_j \rangle \phi_j. \tag{7.74}$$

This is, of course, simply CLEAN with a gain of one ($g = 1$) and with a restriction that no position already visited is to be revisited. In the NMR context, finding the strongest correlation between a single basis element ϕ_j and the observed data y would mean finding the single complex sinusoid $c_j(t)$ that best fits the observed data d, as measured by the inner product:

$$\langle d^{(i-1)}, c_j \rangle. \tag{7.75}$$

That inner product quantitatively measures the contribution of sinusoid $c_j(t)$ at frequency ω_j to the FID, and that value is assigned at that position in the frequency-domain model C:

$$C(\omega_j) \leftarrow \langle d^{(i-1)}, c_j \rangle. \tag{7.76}$$

The contribution of that sinusoid to the observed data:

$$\langle \mathbf{y}^{(i-1)}, \boldsymbol{\phi}_j \rangle \boldsymbol{\phi}_j = \langle \mathbf{d}^{(i-1)}, \mathbf{c}_j \rangle \mathbf{c}_j = C(\omega_j) \mathbf{c}_j \tag{7.77}$$

is then subtracted from the previous residual at the measured points K to produce an updated one missing that sinusoidal component:

$$d_{k,k \in K}^{(i)} \leftarrow d_{k, k \in K}^{(i-1)} - \hat{c}_{k,k \in K}, \quad \hat{\mathbf{c}} = C(\omega_j) \mathbf{c}_j. \tag{7.78}$$

These steps are equivalent to those of the frequency-domain formulation of CLEAN given at the start of the chapter. Fitting a sinusoid in the time domain based on maximizing the inner product is equivalent to positioning a δ-function in the frequency domain at the position in the spectrum with strongest intensity ω_{max}. Subtracting a weighted sinusoid in the time domain at the measured points K is equivalent to subtracting a weighted point response $\mathbf{S}$ centered at ω_{max} in the frequency domain. The only differences are that CLEAN canonically uses a gain $g < 1$ and that CLEAN will revisit the same position multiple times as needed. MP's use of $g - 1$ makes it a so-called "greedy" algorithm,[28] building a solution $\mathbf{x}$ as quickly as possible by fully fitting and subtracting the maximal component in each iteration.

Shortly after MP was introduced in the information theory community, it was found to have a major weakness.[29] Because the fitting of a basis element $\boldsymbol{\phi}_j$ to $\mathbf{y}$ can only be carried out over the measured data points, the inner product between $\boldsymbol{\phi}_j$ and those points at any given iteration is frequently an inaccurate estimate of the correct weight for $\boldsymbol{\phi}_j$ in the true solution $\mathbf{x}_0$. Once more components have been subtracted out, it is frequently found that the initial estimate for the weighting of some early component $\boldsymbol{\phi}_j$ was an over- or underestimate, but because of the stipulation that no $\boldsymbol{\phi}_j$ is to be revisited, those errors remain in the solution and can lead to compounding problems for estimations for other basis elements. This is, of course, simply the time domain equivalent to the frequency domain artifactural interference problem described earlier in this chapter, whereby artifactual contributions from a component at one position add or subtract from the intensity measured at some other position. CLEAN solves this by revisiting positions as needed on subsequent iterations. An alternative approach was proposed in the information theory community,[30–33] whereby the weightings at all previously chosen positions J are readjusted after each iteration through a least-squares fit to the original data $\mathbf{y}$:

$$x_{j,j \in J}^{(i)} \leftarrow \underset{\hat{x}_{j,j \in J^{(i)}}}{\arg \min} \| \mathbf{y} - \Phi \hat{\mathbf{x}} \|_{l_2} \tag{7.79}$$

with the updated solution values used to prepare a new residual from the original data:

$$\mathbf{y}^{(i)} \leftarrow \mathbf{y} - \Phi \mathbf{x}^{(i)}. \tag{7.80}$$

This revision to MP was named *orthogonal matching pursuit* (OMP) as it keeps the residual $\mathbf{y}^{(i)}$ after iteration i orthogonal to the solution vector $\mathbf{x}^{(i)}$ developed up to that point.

Empirical testing has shown that OMP works very well in many cases.[22,34–38] It is much faster than l_1-norm minimization, finding the solution in only M iterations, and it is also in some cases more numerically stable than l_1-norm minimization. Its only remaining weakness is that it can go astray when the component ϕ_j that seems to fit the residual $\mathbf{y}^{(i)}$ best is not, in fact, a component of the true solution $\mathbf{x}_0$—as could occur if the number of measurements are not sufficient to distinguish basis elements, or if the measurements are positioned in such a way that the basis elements become degenerate on the set of measurements (as could easily happen with sinusoids if the measurements were spaced on a regular grid). This is the same case where CLEAN is known to fail: the case of artifacts interfering to produce a false signal in the frequency domain, which is mistakenly incorporated into the model. Intuitively, we might expect such cases to be problematic for l_1-norm minimization as well, and that perhaps there simply isn't enough information present in the data to solve such cases. Theoretical analysis of OMP has been very difficult, but several proofs have now been presented which come close to answering these questions.[36,39,40] If the measurement operator $\mathbf{\Phi}$ is a subset of the Fourier transform operator—as it is in NMR—and if the measurement positions are chosen randomly, it has been shown[36] that OMP will find the correct solution with overwhelming probability as long as the measurements are in the order of:

$$N_K \geq \alpha M \log N \tag{7.81}$$

which is the same constraint proposed by Candès *et al.* for l_1-norm minimization.[19] It was further proven that one variant of OMP known as regularized OMP will find the correct solution as long as the sampling pattern satisfies the restricted isometry property.[22,35] Though the proofs reported to date do not provide quite the same level of guarantees as those for l_1-norm minimization, these results suggest fairly strongly that OMP should work in the same cases.

Surprisingly, CLEAN's methods of solving the nonorthogonality problem—using a gain that is less than one, and revisiting the same positions in multiple iterations—do not seem to been considered in the information theory community, and we are unaware of any theoretical analysis of an MP variant that works like CLEAN. As such, the proof that OMP will find the correct answer with very high probability if the sampling is random and meets eqn (7.81) is not necessarily applicable to CLEAN. However, it is reasonable to speculate that the methods are equivalent and that the same sampling requirements would apply.

7.5   Implementations of CLEAN in NMR

7.5.1   Early Uses of CLEAN in NMR

The first applications of CLEAN to NMR were some time ago, and they had a very different purpose from the sparse recovery problems discussed throughout this chapter. A number of groups were exploring post-processing methods to enhance the quality of conventionally sampled spectra, and CLEAN—as a general deconvolution method—was identified by the Freeman group in Oxford as a possible approach to address several convolution phenomena that were limiting spectral quality.

They began with the problem of phase-twist lineshapes, which arise when absorptive and dispersive components become overlaid, as in the classic *J*-resolved experiments.[41] These are not instrumental responses or sampling artifacts, but rather the natural lineshape components of the resonance, and the problem could be described as:

$$\mathbf{D} = \mathbf{L}_A * \mathbf{A} + \mathbf{L}_D * \mathbf{A} = (\mathbf{L}_A + \mathbf{L}_D) * \mathbf{A} \tag{7.82}$$

where $\mathbf{L}_A$ and $\mathbf{L}_D$ represent the natural absorptive and dispersive lineshapes, respectively. The goal is to decompose $\mathbf{D}$ into a set of δ-functions $\mathbf{C}_I$, and then build a final absorptive mode spectrum:

$$\mathbf{C} = \left(\sum_i \mathbf{L}_A * \mathbf{C}_i \right) + \mathbf{R}. \tag{7.83}$$

A significant difference with the application to sparse sampling is that the lineshapes, being properties of the resonances and not of the NMR procedure or apparatus, are not guaranteed to be the same for all resonances, and must either be estimated from the data or derived from *a priori* assumptions. The decomposition and reconstruction are therefore unlikely to be perfect. In addition, the presence of artifacts that do not obey the convolution relationship in eqn (7.82), such as T_1 ridges, can be problematic for the algorithm. Nonetheless, they found that the method was effective at removing dispersive lineshape components.

Keeler presented a second use for CLEAN, as a way of removing truncation artifacts.[42] Whenever the time domain sampling of an NMR signal ends before the signal has naturally decayed to zero, it is equivalent to multiplying the signal by a rectangular cutoff function, traditionally denoted as Π. This multiplication in the time domain:

$$\mathbf{d} = \Pi \mathbf{a}, \tag{7.84}$$

leads to convolution with the sinc function in the frequency domain:

$$\mathbf{D} = \mathbf{sinc} * \mathbf{A} \tag{7.85}$$

producing "wiggles" stretching out from each signal. Truncation is traditionally prevented by apodization, but apodization discards some of the known high-resolution information about the spectrum, and one could imagine instead using CLEAN to remove the truncation artifacts. In this case the convolved lineshape *is* known exactly, and it is the same for all signals in the spectrum. For the reconstruction, an artificial Lorentzian lineshape was used, with the width chosen to match the expected natural linewidth. Keeler applied the method to the coupling dimension of a *J*-resolved experiment recorded with only five increments, which suffered from severe truncation artifacts, with excellent results.

The third application demonstrated by the Oxford group was enhancing spectral resolution to correct for instrumental line broadening.[43] Obviously CLEAN cannot divine (correct) fine structure out of nothing: the FID must be observed out to a long enough evolution time to capture high-resolution information, and the decay function must not be strong enough to annihilate this information entirely from the measurements. But provided the information is present in some form in the data, Davies *et al.* showed that CLEAN can successfully recover it (Figure 7.7).[43] Interestingly, in this application the gain had to be extremely low, on the order of $g = 0.02$–0.05; otherwise, the method failed.

All three of these applications of CLEAN differ considerably from the use in sparse sampling. Because the measurement operators (Φ) for these cases have a different structure from the operators that apply in sparse sampling, the theoretical analysis above does not apply. These results suggest that CLEAN may be effective as a general deconvolution tool, beyond the areas of random sparse sampling that have been considered by compressed sensing theory.

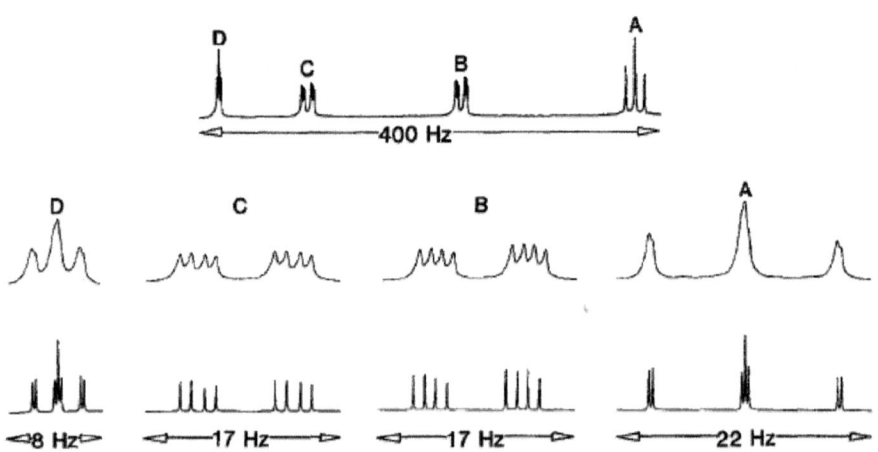

Figure 7.7 Resolution enhancement *via* CLEAN. One of the earliest applications of CLEAN to NMR was for resolution enhancement. Here, a six-fold improvement in resolution is obtained from applying CLEAN to the 400 MHz proton spectrum of 3-bromonitrobenzene. The full spectrum before application of CLEAN is shown at the top, and the middle row shows each signal multiplet on an expanded scale. The bottom row shows the same multiplets after resolution enhancement by CLEAN. (Reprinted from *J. Magn. Reson.*, **76**, S. J. Davies, C. Bauer, P. J. Hore and R. Freeman, Resolution enhancement by nonlinear data processing. "HOGWASH" and the maximum entropy method, 476–493, Copyright 1988 Academic Press, Inc., with permission from Elsevier.)

7.5.2 Projection–reconstruction NMR

In 2005, Freeman's group in Oxford again made use of CLEAN, this time in the context of reconstructing NMR spectra from projections.[44] The measurement of projections from radial samples is a form of sparse sampling, but it differs considerably from the cases analyzed in this chapter in that the sampling pattern has regular structure rather than being randomized. The formulation of CLEAN in this study was somewhat different from the canonical CLEAN in that it was integrated into repeated cycles of tomographic reconstruction. An HNCO spectrum was successfully processed by the method.

7.5.3 CLEAN and Randomized Sparse Nonuniform Sampling

Though Barna *et al.* carried out a few trials of CLEAN on some of the very first datasets collected with randomized sparse sampling in the late 1980s,[45] the method was not routinely applied to this type of data until 2007 and 2008, when two groups reintroduced it.

The present authors proposed the use of the CLEAN algorithm as described in the sections above, where the spectrum is modeled as a set of δ-functions, where all operations are carried out in the frequency domain, and where a gain $g < 1$ is used to relax away interference between artifact patterns.[46] This is the form of CLEAN used by the radioastronomy community, and it is the form of CLEAN to which the theoretical analyses above apply. The decomposition into δ-functions allows the algorithm to model *any* and *all* signals encountered in NMR spectra, including overlapping signals and signals with different lineshapes, without distortions to peak shapes, intensities, or volumes. CLEAN proved highly effective on a 4-D HCCH-TOCSY experiment,[46] on 4-D sequential assignment experiments,[47] and on 4-D diagonal-suppressed NOESY experiments[48,49] (an example is shown in Figure 7.8). NOESY results on spectra with full-strength diagonals initially pointed to a significant weakness, however, in terms of the dynamic range that could be achieved.[50] Consider a strong signal $\mathbf{A}_s$ with intensity I_s at position ω. If the gain $g = 1$, this signal would be processed completely in a single iteration. However, we have already seen that there are many reasons the gain should be set to less than one, in which case it will take some number of iterations to process this peak. After n iterations applied at position ω, the residual intensity of this signal will be:

$$I_s^{(n)} = I_s(1 - g)^n \qquad (7.86)$$

and the entire spectrum will still be affected by artifacts owing to $\mathbf{A}_s$, scaled to a fraction $I_s^{(n)}/I_s$ of their original intensity. If, however, $I_s^{(n)}$ and all other signals are now at a level on a par with the apparent noise level (the thermal noise plus the residual artifacts from all other peaks), such that CLEAN can no longer identify which positions are signals and which are noise, the processing of $\mathbf{A}_s$ must stop and the remaining artifacts from $\mathbf{A}_s$ will not be removed. This has the effect of limiting the dynamic range of the spectrum,

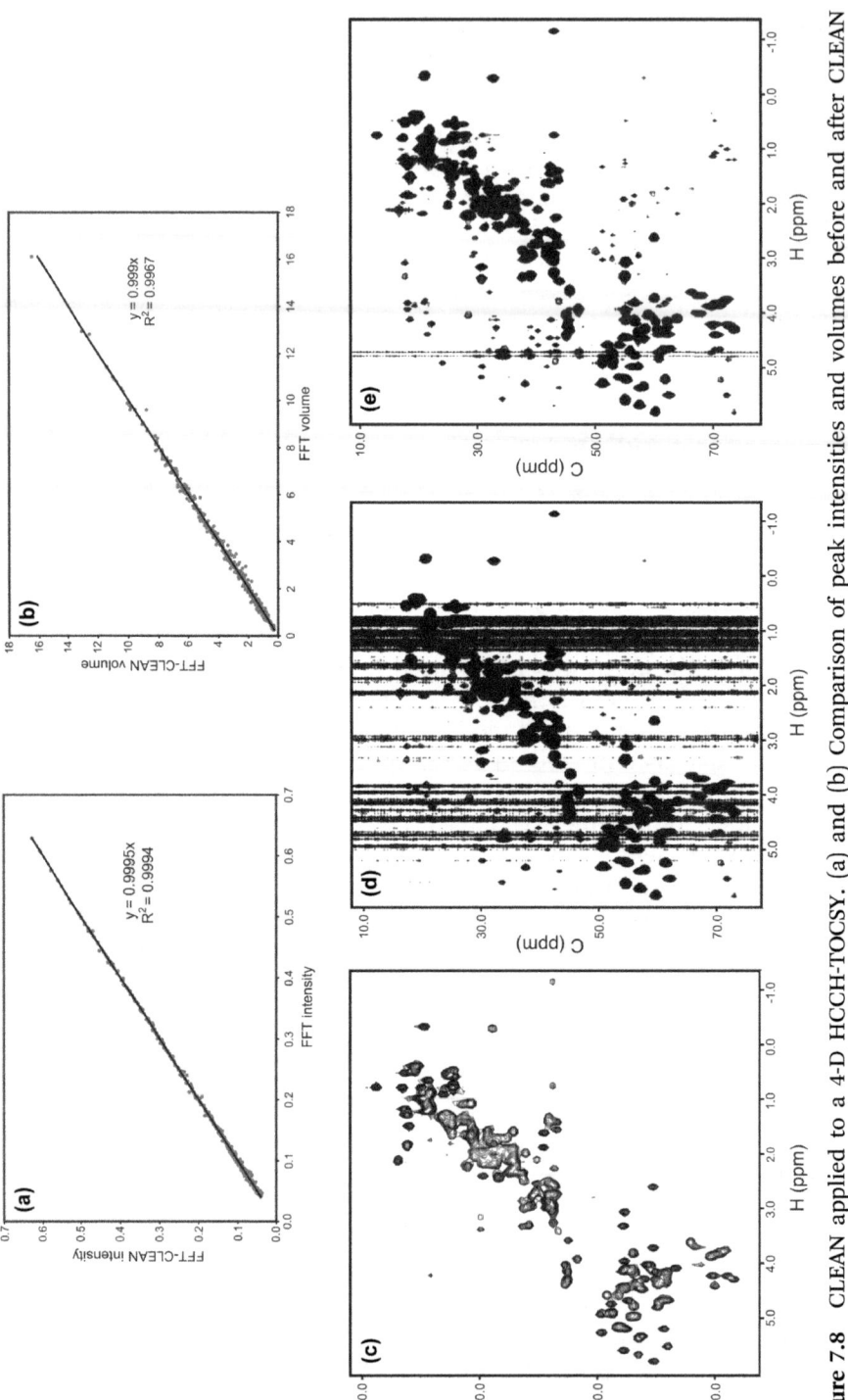

Figure 7.8   CLEAN applied to a 4-D HCCH-TOCSY. (a) and (b) Comparison of peak intensities and volumes before and after CLEAN processing. (c) Control ^1H–^{13}C HSQC spectrum. (d) 2D projection of the sparsely sampled 4D HCCH-TOCSY experiment processed with the FFT, with artifacts present. (e) Removal of aliasing artifacts with CLEAN. (Panels (c), (d), and (e) adapted from *J. Biomol. NMR*, High resolution 4-D spectroscopy with sparse concentric shell sampling and FFT-CLEAN, **42**, 2008, 225–239, B. E. Coggins and P. Zhou, (Copyright 2008 Springer Science + Business Media B.V.) with permission of Springer.)

and it was found empirically that the dynamic range achieved is in some cases insufficient for experiments of interest.

At approximately the same time, the Kozminski group in Warsaw developed a modified version of CLEAN that they called the *signal separation algorithm* (SSA).[51-53] This algorithm decomposes the spectrum into a set of peaks rather than a set of δ-functions. Each peak in the artifact-corrupted spectrum **D** is identified from the noise by statistical criteria and listed as a set $P^{(1)} = \{P_j : j = 1, 2, \ldots, n_1\}$. An individual peak P_j and its immediate local region of the spectrum are extracted as a matrix $\mathbf{P}_j$ and processed by inverse-DFT to give a time domain representation $\mathbf{p}_j$ of the entire NMR resonance, which is subtracted from the time domain FID to give an updated FID:

$$\mathbf{d}^{(1,j)} \leftarrow \mathbf{d}^{(1,j-1)} - \mathbf{p}_j. \tag{7.87}$$

Alternatively, a fitting procedure can be used to fit the peak with Lorentzian lineshapes, and a $\mathbf{p}_j$ can be constructed from the fitted parameters. After processing all n_1 identified peaks, the residual FID $\mathbf{d}^{(1,n)}$ is Fourier transformed to give a new spectrum $\mathbf{D}^{(2)}$ with reduced artifacts, and the process is repeated. Thus the algorithm can be considered to operate on batches of peaks. Clearly the biggest challenges of this approach have to do with determining the peak shapes, whether by the extraction and inverse-DFT processing of each peak region—which requires isolating individual peaks and defining their boundaries clearly—or by fitting lineshapes with analytical functions. The authors report difficulties when overlapping signals are encountered, as would be expected. The method uses a gain of one, and the authors also report some difficulties from interfering artifact patterns. Nonetheless, the method was demonstrated to work well on 3D ^{15}N- and ^{13}C-NOESY, 4-D HCCH-TOCSY, and 4-D ^{13}C/^{15}N-NOESY (Figure 7.9). Because a gain of one is used, the problem of residual artifacts is avoided, and a better dynamic range is achieved than for the classic CLEAN.

Most recently, a modified version of CLEAN was introduced under the name of SCRUB, which solves many of the problems encountered by both classic CLEAN and by the SSA.[50] As in classic CLEAN, by using a decomposition into δ-functions, the difficulty of fitting or isolating peaks is avoided, and by using a gain $g < 1$, the problem of artifact interference is resolved. SCRUB differs from CLEAN, however, in that it completely eliminates the artifacts from all identified signals. As in the SSA, this is accomplished by working in batches and taking all components from a given batch to zero before moving on to the next batch. Importantly, because these positions have been identified as signals rather than noise, they can be processed all the way to zero rather than merely to the noise level, leaving no residual artifacts. The algorithm can be summarized as follows:

SCRUB
1. **Initialize:** Set $\mathbf{D}_{\text{residual}} \leftarrow \mathbf{D}$ and $\mathbf{C} \leftarrow 0$.
2. **Decomposition:** If any positions in $\mathbf{D}_{\text{residual}}$ can be identified as clearly being signals rather than noise, process a batch i:
   a. Identify all n_i positions in $\mathbf{D}_{\text{residual}}$ that are clearly above the noise level and designate these positions as set $J_i = \{j_i : 1, 2, \ldots, n_i\}$.

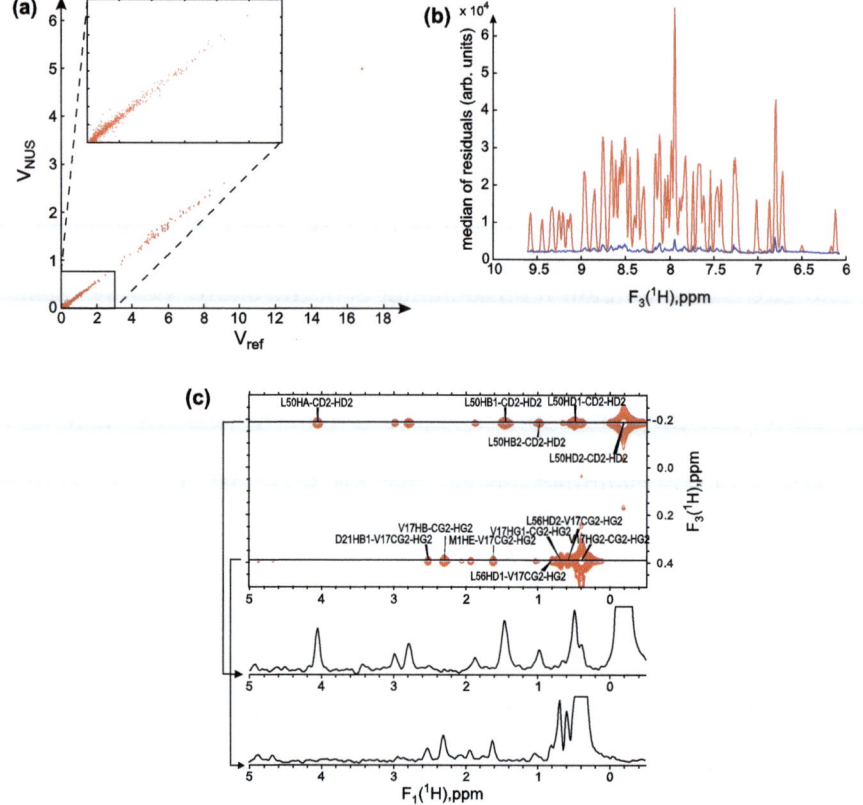

Figure 7.9   SSA applied to 3-D ^{15}N- and ^{13}C-NOESY on ubiquitin. (a) Correlation plot
of peak volumes between the SSA-processed ^{15}N-NOESY spectrum and a
reference spectrum. (b) Pre-SSA (upper curve) and post-SSA (lower curve)
apparent noise (thermal noise + artifacts) in the ^{15}N-NOESY as a function of direct dimension position, showing suppression of artifacts. (c)
Contour plot and cross-sections from an F_1F_3 plane of the ^{13}C NOESY.
(Figure adapted from *J. Biomol. NMR*, Iterative algorithm of discrete
Fourier transform for processing randomly sampled NMR data sets, **47**,
2010, 65–77, J. Stanek and W. Kozminski, (Copyright 2010 Springer
Science + Business Media B.V.) with permission of Springer.)

b. Carry out CLEAN on the positions in set J_i by repeating the following
   steps until the intensities of the points in $\mathbf{D}_{\text{residual}}$ corresponding to
   J_i are all less than or equal to a threshold τ close to zero (and below
   the noise level):
   i. Identify the position ω_{max} in $\mathbf{D}_{\text{residual}}$ of strongest intensity,
      where ω_{max} must correspond to one of the positions in J_i.
   ii. Add   a   component   to   **C**: $\mathbf{C} \leftarrow \mathbf{C} + \mathbf{C}_{\text{new}}$,   where
       $\mathbf{C}_{\text{new}}(\omega) = gI_{\text{max}}\delta(\omega - \omega_{\text{max}})$.
   iii. Update the residual by subtracting the artifacts corresponding
        to $\mathbf{C}_{\text{new}}$: $\mathbf{D}_{\text{residual}} \leftarrow \mathbf{D}_{\text{residual}} - \mathbf{C}_{\text{new}} * \mathbf{S}$.
3. **Reconstruction:** Generate the output spectrum: $\mathbf{C} \leftarrow \mathbf{C} * \mathbf{H} + \mathbf{D}_{\text{residual}}$.

SCRUB has been shown to achieve a much better dynamic range than CLEAN both in simulations and on experimental data while preserving the quantitative accuracy of peak intensities and volumes, and it successfully reduced the artifacts in 4-D NOESY experiments (^{13}C/^{13}C, ^{13}C/^{15}N, ^{15}N/^{13}C, ^{15}N/^{15}N) to the thermal noise level in trials on two proteins between 20 and 30 kDa in size[50] (Figure 7.10). The quality of this data was sufficient to permit automatic assignment and structure calculation using CYANA. More recently, it was employed in 4-D intermolecular NOE experiments.[54] It is interesting to note that the batchwise approach together with the decomposition into δ-functions makes SCRUB very similar to some of the OMP variants for which theoretical guarantees are available.[22,35,37]

7.6 Using CLEAN in Biomolecular NMR: Examples of Applications

The CLEAN and SCRUB algorithms for high-resolution reconstruction of sparsely-sampled NMR data have been implemented in a standalone software package and can be called from standard processing programs such as NMRPipe.[50,55] Data processing is highly efficient, typically running from a couple of minutes to several hours for reconstruction of 3-D and 4-D spectra on a standard personal computer. These traits make CLEAN/SCRUB reconstruction a powerful tool for routine biomolecular NMR applications. We conclude by offering several examples, drawn from our own work, of CLEAN applied in biomolecular NMR. In all of these examples, we utilized sparse nonuniform sampling to collect 4-D spectra at very high resolution, substantially reducing assignment ambiguity over conventional high-resolution 3-D or low-resolution 4-D experiments. High-quality spectra were obtained by use of CLEAN and/or SCRUB for artifact suppression.

Triple-resonance backbone experiments are often the first step toward resonance assignment and dynamic and/or structural analysis of proteins. These experiments are particularly suitable for sparse NUS due to the sensitivity and sparsity of the signals, which makes high-quality spectral reconstruction feasible even with very few measurement points. Figure 7.11 shows 4-D HNCACB, HN(CO)CACB, HA(CA)CONH and HA(CA)CO(CA)NH sequential assignment experiments recorded on a 7.8 kDa intrinsically disordered protein, SKIPN, at 1 mM, and processed using CLEAN.[47] Only 201 sampling points were used for each experiment, adapted to a 50×50×50 grid for a sampling density of 0.16%. Each 4-D experiment required 2.6 hours of instrument time.

4-D TOCSY experiments such as HCccoNH/HccNH and HCCH-TOCSY have also been demonstrated, initially for projection–reconstruction and more recently for CLEAN processing.[46,56] Figure 7.12 shows a 4-D HCCH-TOCSY experiment on 1 mM GB1 (6.2 kDa), sampled at a density of 1.2% and processed with CLEAN, with comparisons before and after CLEAN processing to show the extent of artifact reduction.[46]

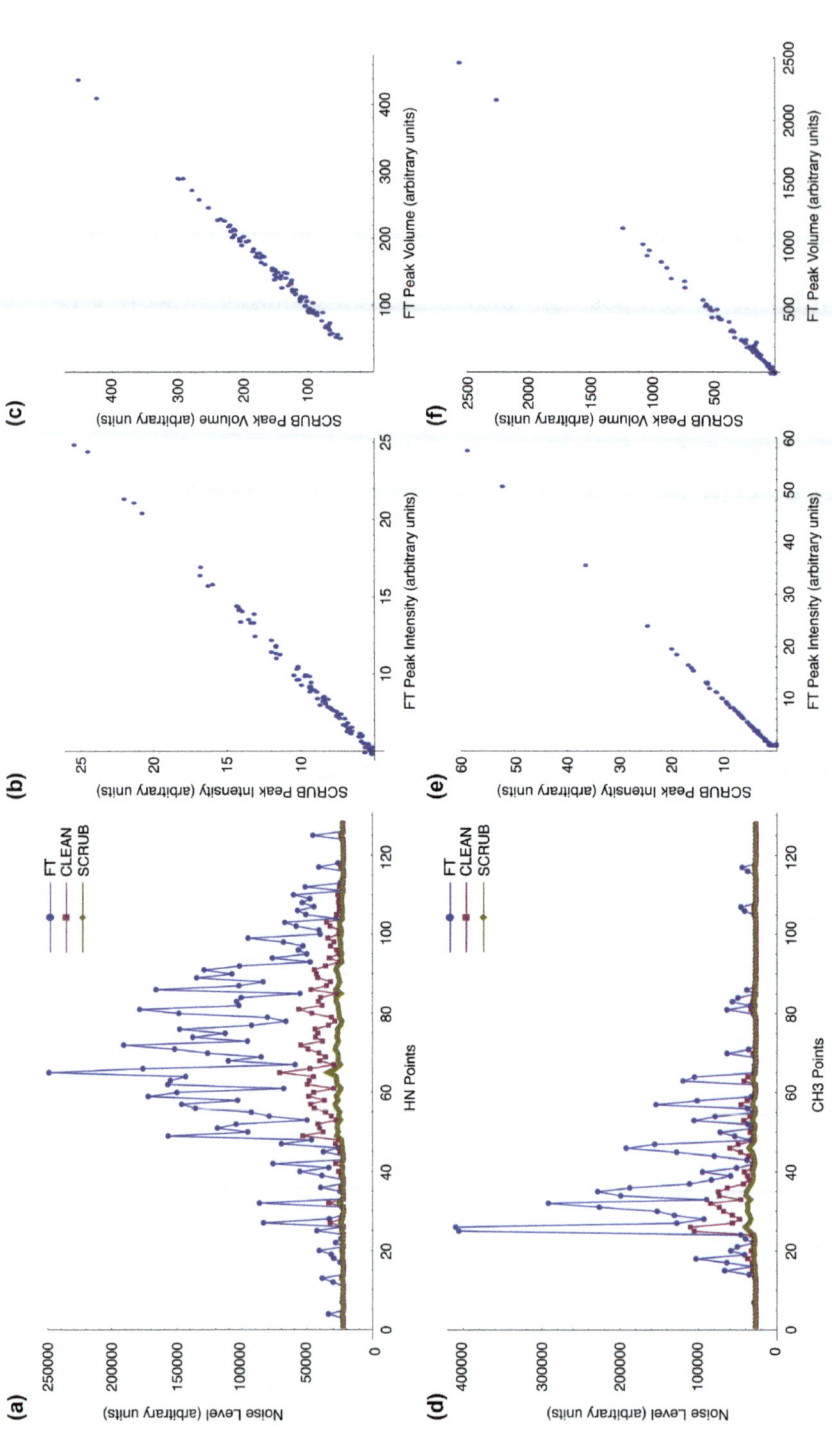

Figure 7.10 SCRUB applied to 4-D NOESY. (a) and (d) Apparent noise levels (thermal noise + artifacts) for the amide and methyl NOESY spectra processed using the FFT alone, FFT + CLEAN, or FFT + SCRUB, showing artifact suppression by SCRUB. (b), (c), (e), (f) Correlation plots comparing peak intensities and volumes before and after SCRUB processing. (Reprinted with permission from B. E. Coggins, J. W. Werner-Allen, A. Yan and P. Zhou, Rapid Protein Global Fold Determination Using Ultrasparse Sampling, High-Dynamic Range Artifact Suppression, and Time-Shared NOESY, *J. Am. Chem. Soc.*, 2012, **134**, 18619–18630. Copyright 2012 American Chemical Society.)

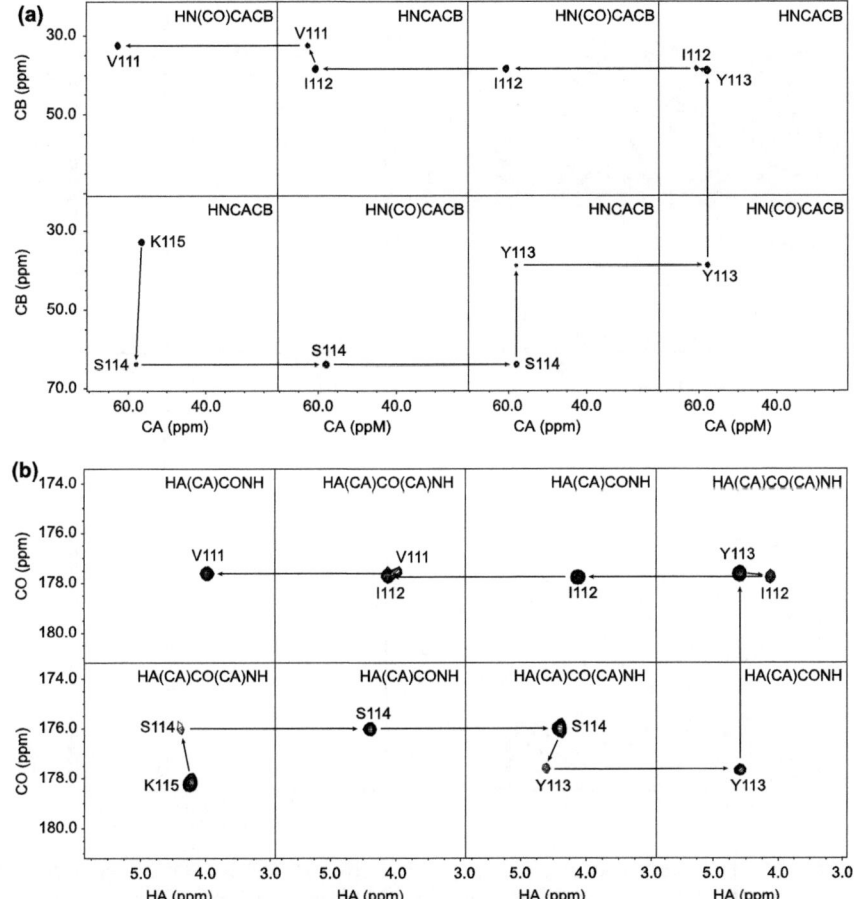

Figure 7.11 Sequential connectivity with 4D triple-resonance experiments. (a) Connectivity of HNCACB and HN(CO)CACB. (b) Connectivity with HA(CA)-CONH and HA(CA)CO(CA)NH. The experiment was collected on the intrinsically disordered protein SKIPN.
(Reprinted from *J. Magn. Reson.*, **209**, J. Wen, J. Wu and P. Zhou, Sparsely sampled high resolution 4-D experiments for efficient backbone resonance assignment of disordered proteins, 94–100, Copyright (2010) with permission from Elsevier.)

If strong diagonal peaks are suppressed in the pulse sequence, it is possible to use CLEAN on NOESY experiments. Figure 7.13 illustrates a diagonal-suppressed 4-D TROSY-NOESY-TROSY experiment on the 23 kDa phosphatase Ssu72.[48] Data were collected with a 1 mM ^{2}H/^{13}C/^{15}N-labeled sample of the C13S mutant at a sampling density of 1.2%. The total measurement time was 48 hours. The advantages of 4-D spectroscopy over 3-D are readily apparent in this example, where the high-resolution 4-D experiment clearly resolves a number of crosspeaks that overlap in the 3-D control experiments.

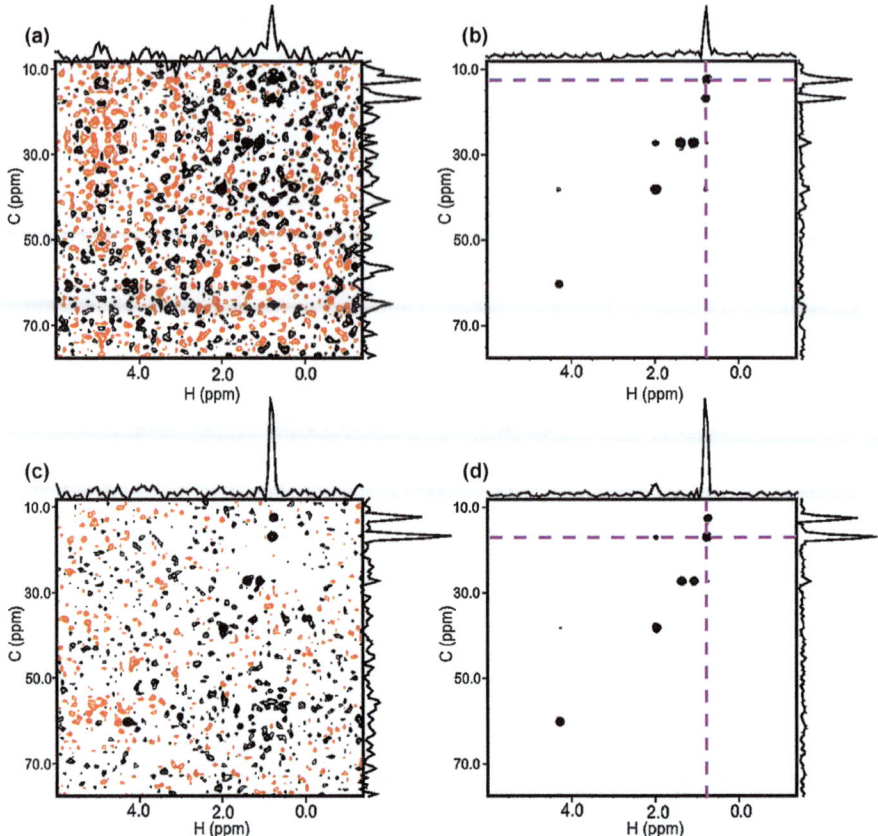

Figure 7.12   HC planes from the reconstructed 4-D HCCH-TOCSY spectrum of
protein G B1 domain. (a) The 2D plane centered at Ile6 CD1/QD1 at
12.6 ppm/0.73 ppm in [F3/F4] processed with the FFT. (b) The same
plane as (a), processed with CLEAN. (c) The 2D plane centered at Ile6
CG2/QG2 at 16.9 ppm/0.78 ppm in [F3/F4] processed with the FFT.
(d) The same plane as (c), processed with CLEAN.
(Adapted from *J. Biomol. NMR*, High resolution 4-D spectroscopy with
sparse concentric shell sampling and FFT-CLEAN, **42**, 2008, 225–239,
B. E. Coggins and P. Zhou, (Copyright 2008 Springer Science + Business
Media B.V.) with permission of Springer.)

Diagonal suppression is not necessary with SCRUB, as demonstrated
with the 4-D time-shared NOESY experiment[50] discussed briefly above, and
illustrated in Figure 7.14. Data were collected for the 23 kDa Ssu72 and the
29 kDa HCA II proteins, in both cases using $^{13}C/^{15}N$-labeled samples
protonated at ILV methyl positions and otherwise deuterated. A single
72-hour time-shared NOESY experiment at 1.2% sampling density for each
protein provided sufficient structural restraints for global fold determination, at a sufficient quality level to permit automatic assignment *via*
CYANA. It should be noted that CYANA failed to converge properly when

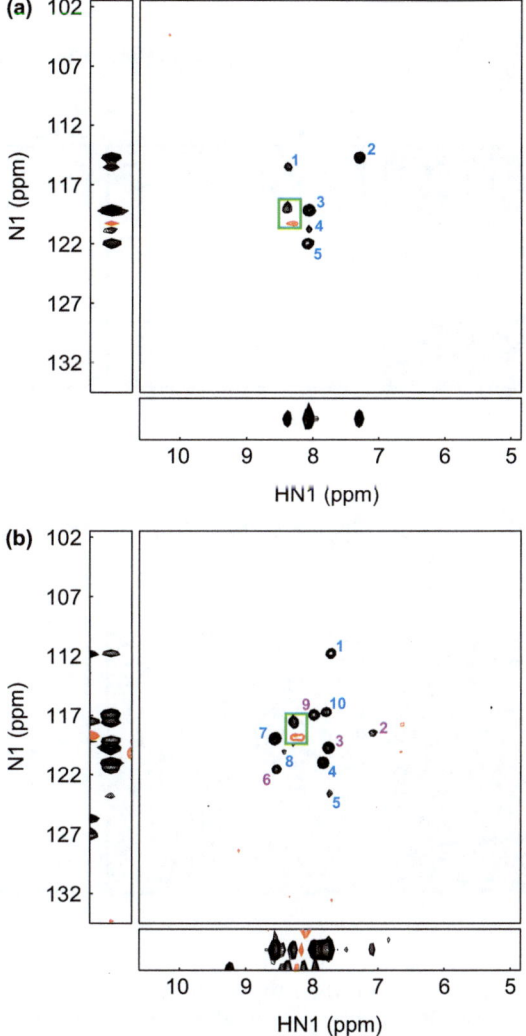

Figure 7.13 Representative planes from the 4-D diagonal-suppressed TROSY-NOESY-TROSY (ds-TNT) dataset. Corresponding strips are from 3-D ds-TNT control spectra collected with conventional sampling. Residual diagonal signals are boxed in green. Panel (a) shows crosspeaks from residue I176 of C13S Ssu72 to D173 (1), N174 (2), D175 (3), D177 (5) and E178 (4). Panel (b) contains crosspeaks from two overlapped residues, L72 and N92. Blue numbers denote crosspeaks from L72 to Y69 (8), R70 (10), D71 (4), E73 (7), S74 (1) and K75 (5) and purple numbers correspond to crosspeaks from N92 to D90 (6), R91 (3), R93 (9) and R94 (2). (Adapted from *J. Magn. Reson.*, **204**, J. W. Werner-Allen, B. E. Coggins and P. Zhou, Fast acquisition of high resolution 4-D amide–amide NOESY with diagonal suppression, sparse sampling and FFT-CLEAN, 173–178, Copyright (2010) with permission from Elsevier.)

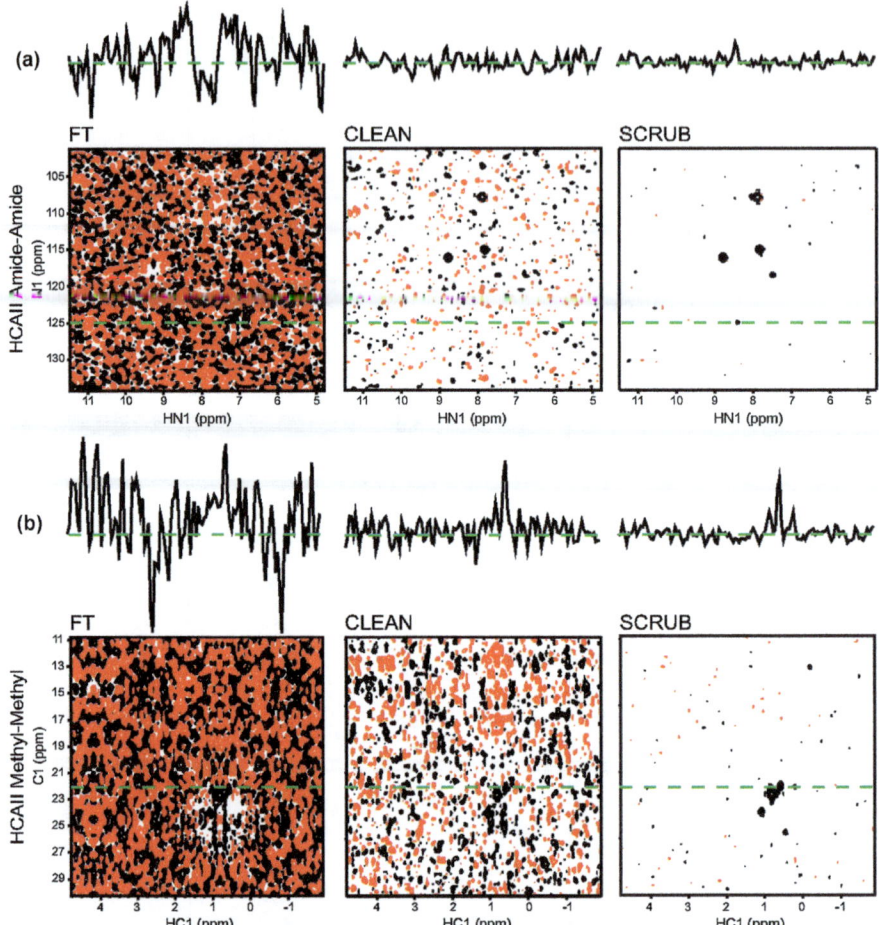

Figure 7.14   Artifact removal with SCRUB in 4-D TS NOESY spectra. Representative F1/F2 planes from the amide–amide and methyl–methyl NOESY spectra of HCAII are shown after processing with either the FFT alone (left panel), CLEAN (middle panel), or SCRUB (right panel). Panels are plotted at identical contour levels, and 1-D slices along the green dashed lines are shown above each panel to highlight long-range crosspeaks that are lost to aliasing noise in the FFT- and CLEAN-processed spectra. Planes in parts (a) and (b) are taken from F1/F2/F3 cubes #75 and #31, respectively, which have 5.52-fold and 7.73-fold reductions in artifact noise with SCRUB processing, respectively.
(Reprinted with permission from B. E. Coggins, J. W. Werner-Allen, A. Yan and P. Zhou, Rapid Protein Global Fold Determination Using Ultrasparse Sampling, High-Dynamic Range Artifact Suppression, and Time-Shared NOESY, *J. Am. Chem. Soc.*, 2012, **134**, 18619–18630. Copyright 2012 American Chemical Society.)

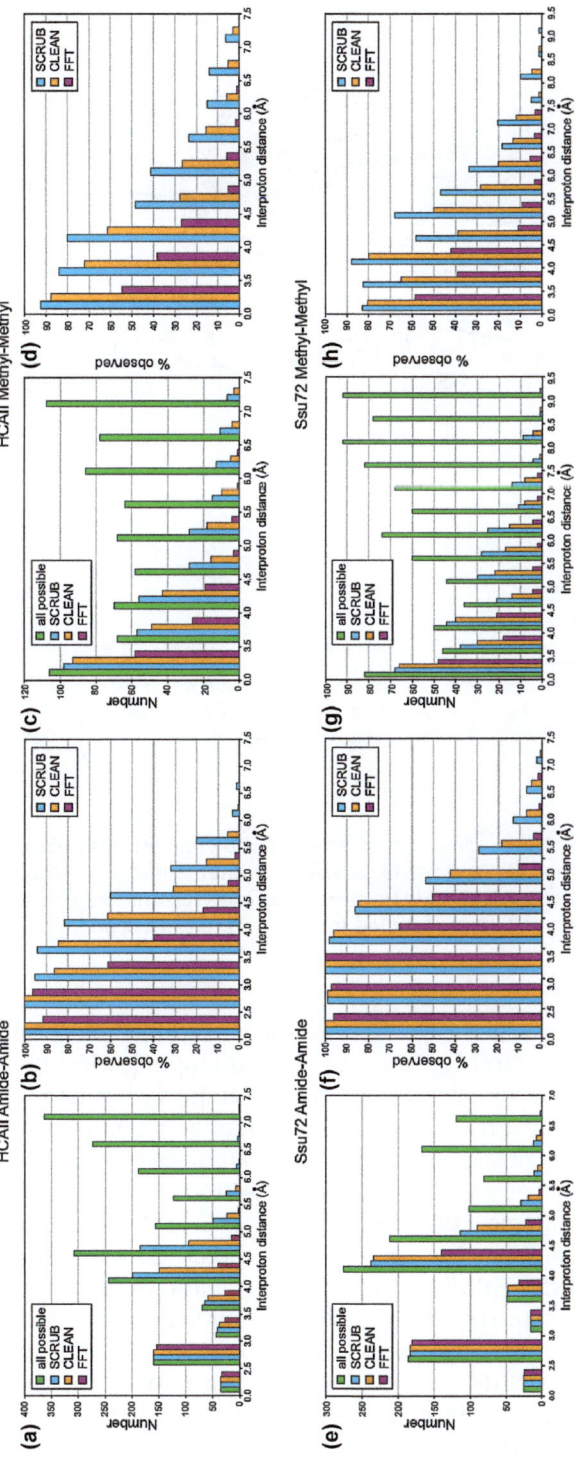

Figure 7.15 Removal of aliasing artifacts promotes peak identification in the 4-D TS NOESY spectra for HCAII (top) and Ssu72 (bottom). NOE crosspeaks with intensities over four times the noise level in spectra processed with either FFT, CLEAN, or SCRUB were matched to their corresponding interproton distances, and unique distances are plotted as histograms in parts (a), (c), (e), and (g). Each interproton distance was counted as observed if either corresponding NOE ($H_i \rightarrow H_j$ or $H_j \rightarrow H_i$) surpassed the noise threshold. "All possible" refers to the total number of interproton distances calculated from a reference crystal structure (PDB code 2ILI for HCAII and PDB code 3P9Y for Ssu72). The histograms in parts (b), (d), (f), and (h) show the percent of observed distances relative to all expected. Short interproton distances that were not identified in the SCRUB-processed methyl – methyl spectra are the result of signal degeneracy. The small number of very long distances (>8.0 Å) observed in the methyl–methyl spectrum of Ssu72 likely result from incorrect rotamer assignments in the reference crystal structure due to the insufficient resolution of the model (2.1 Å) for precisely defining ILV side chain orientations. (Reprinted with permission from B. E. Coggins, J. W. Werner-Allen, A. Yan and P. Zhou, Rapid Protein Global Fold Determination Using Ultrasparse Sampling, High-Dynamic Range Artifact Suppression, and Time-Shared NOESY, *J. Am. Chem. Soc.*, 2012, **134**, 18619–18630. Copyright 2012 American Chemical Society.)

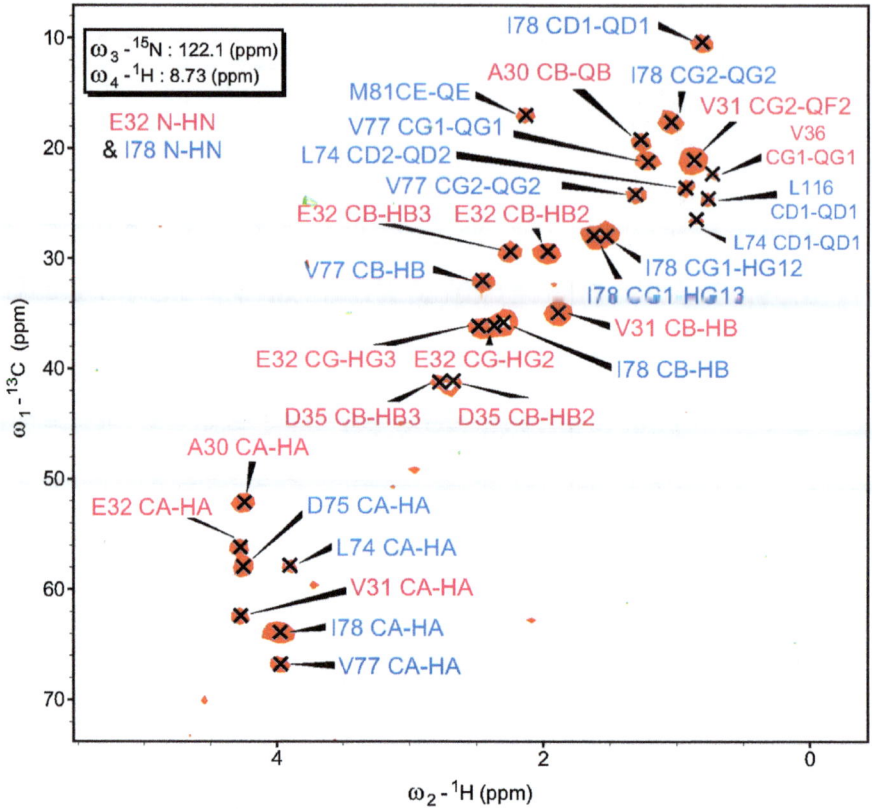

Figure 7.16 A 4-D ^{13}C HMQC-NOESY-^{15}N HSQC spectrum, showing the H-C plane centered at 122.1 ppm/8.73 ppm in N and H dimensions. The plane contains signals from NOE signals to the overlapping resonances of E32 and I78. NOE signals to E32 are labeled in pink and to I78 are labeled in blue.

attempting to automatically assign 3-D conventionally sampled control NOESY spectra due to excessive ambiguous assignments, highlighting the value of an extra dimension of signal separation. The improved dynamic range from SCRUB's additional artifact suppression made it possible to obtain considerably more long-range structural restraints (Figure 7.15). Another example of the use of SCRUB with 4-D NOESY is found in Figure 7.16, which shows a single H/C plane from a ^{13}C HMQC-NOESY carried out at 0.8% sampling on the 16.2 kDa complex between the Rev1 C-terminal domain and the Pol k/Rev1-interacting region, in which an extraordinary number of crosspeaks are resolved.[57] Finally, sparse NUS data collection and SCRUB processing have recently been employed successfully in the intermolecular difference NOE (omit) experiment for the heterodimeric human UBM1–ubiquitin complex and for the homotrimeric foldon complex shown in Figures 7.17 and 7.18, respectively.[54] It is

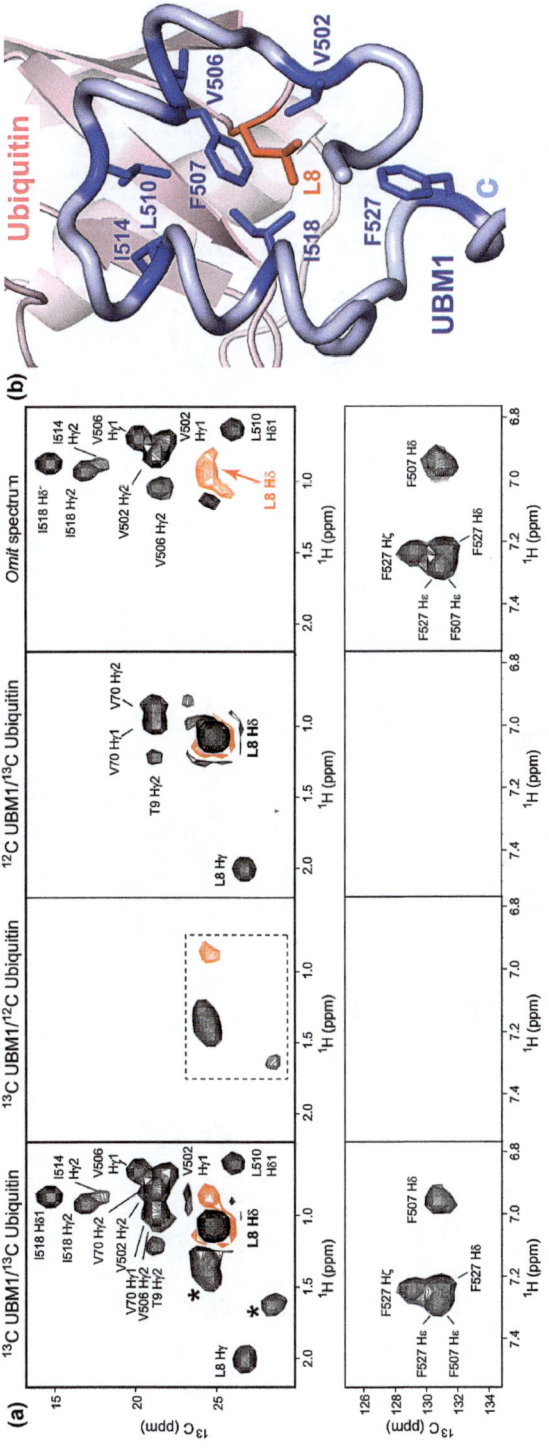

Figure 7.17 Omit spectrum of the human Pol ι UBM1–ubiquitin complex showing intermolecular NOEs. 4-D 13C HMQC-NOESY-HSQC spectra are collected for the gb1UBM1–ubiquitin complex with both components or with individual components ¹³C-labeled. Reconstruction of the difference time domain signals of the uniformly labeled protein complex from component-labeled samples generates an omit spectrum containing only intermolecular NOEs. Slight over-subtraction of time domain data from individual components generates negative diagonal signals (red) in the omit spectrum and ensures all of the positive cross-peaks originate from intermolecular NOEs. Panel (a) (upper, aliphatic regions; lower, aromatic regions) shows sections of F1–F2 slices of the corresponding 4-D spectra centered at 24.53 ppm in F3 and 1.05 ppm in F4, displaying NOEs to the ubiquitin L8 methyl groups. Boxed peaks or peaks labeled with asterisks are off-plane signals. (b) Interface of the human Pol ι UBM1–ubiquitin complex, showing an interaction network centered at L8 of ubiquitin. (Adapted from *J. Biomol. NMR*, Sparsely-sampled, high-resolution 4-D omit spectra for detection and assignment of intermolecular NOEs of protein complexes, **59**, 2014, 51–56, S. Wang and P. Zhou, (Copyright 2014 Springer Science + Business Media Dordrecht) with permission of Springer.)

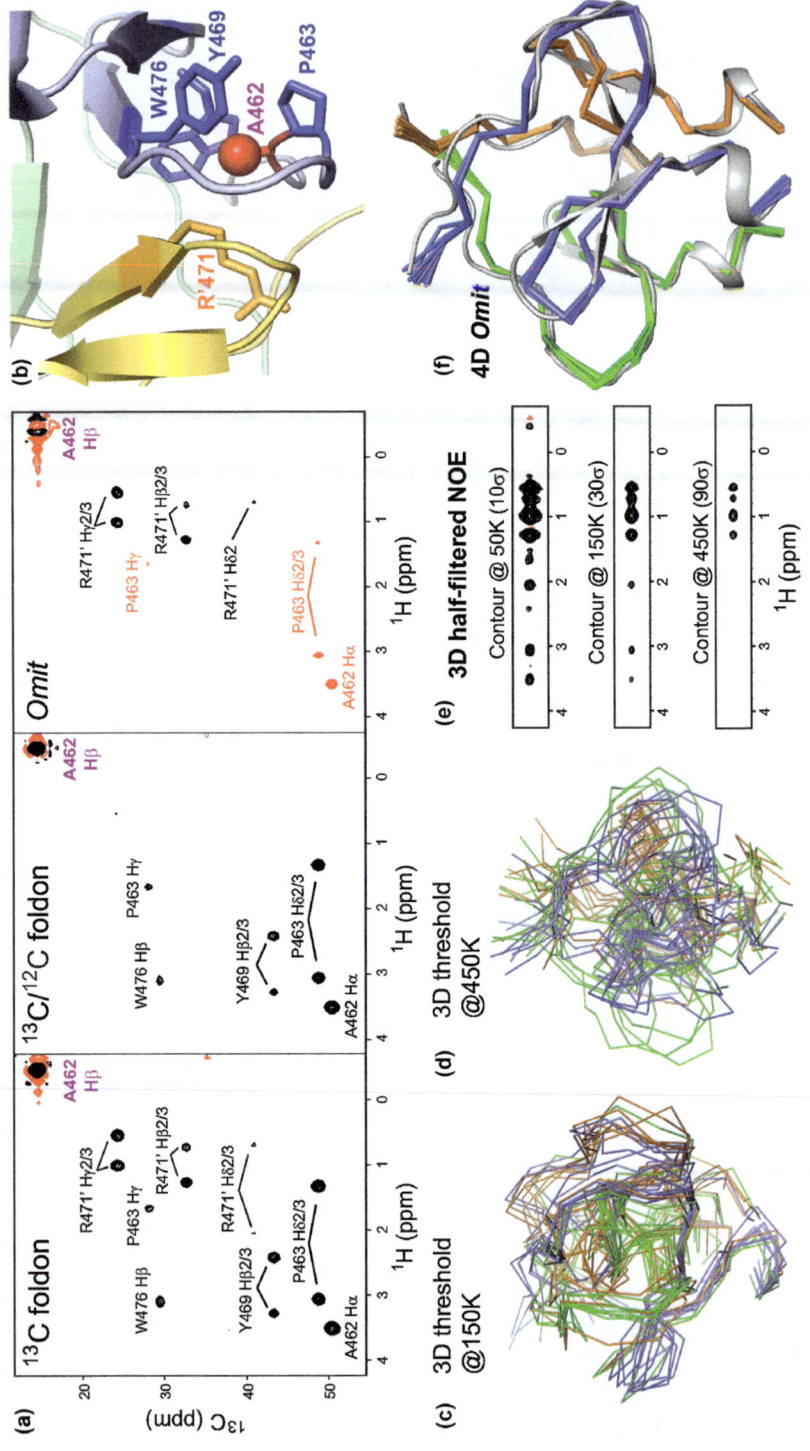

interesting to note that collection of high-resolution 4-D intermolecular NOEs largely eliminated assignment ambiguity and enabled automated structural analysis for the foldon trimer, whereas similar analysis based on 3-D filtered NOEs failed to generate converged structures.

7.7 Conclusions

Processing sparsely sampled, nonuniform NMR data directly with the FFT is an attractive option owing to the FFT's simplicity, speed, and linearity, but the resulting sampling artifacts substantially reduce the spectrum's dynamic range. The CLEAN method provides a solution by reducing or eliminating these artifacts. Developed in the field of radioastronomy, where its successes over the last 40 years have led to its establishment as standard operating procedure, CLEAN has recently been reintroduced in NMR with enhanced versions that achieve excellent performance on challenging 3-D and 4-D experiments. CLEAN has been shown to conserve spectral energy and therefore replicates peak intensities, volumes, and lineshapes correctly, making it ideal for quantitative applications. Recent theoretical work in the field of compressed sensing has proven that methods very similar to CLEAN succeed under the same conditions as l_1-norm minimization, providing an analytical confirmation to the decades of empirical evidence of this method's reliability. The combination of the FFT and CLEAN makes for a fast, straightforward, and robust means to process nonuniformly sampled NMR data.

Figure 7.18 Omit spectrum of the foldon trimer. (a) F1–F2 slices of the corresponding 4-D spectra centered at 14.12 ppm in F3 and -0.49 ppm in F4, displaying NOEs to the A462 methyl group. 4-D ^{13}C HMQC-NOESY-HSQC spectra are collected for foldon with uniformly (left) or 25% (middle) ^{13}C-labeled samples. Subtraction of these two spectra generates an omit spectrum (right) containing positive intermolecular NOE and negative (red) diagonal signals and intramolecular NOEs. (b) Interface of two subunits of the foldon trimer, showing an intersubunit interaction between R471 and A462 and intrasubunit interactions between W476, Y469, P463 and A462. Panels (c) and (d) show the results of structure calculation using the 3-D half-filtered NOE peak list with intensity thresholds of 150K and 450K, respectively. (e) A slice of the 3-D half-filtered NOE spectrum centered at the A462 methyl group, plotted at contour levels of 50K, 150K and 450K, respectively. (f) The structural ensemble of trimeric foldon (subunits colored in blue, green and orange) calculated from the CYANA-assigned 4-D intermolecular NOE peak list from the omit spectrum superimposed with the crystal structure (grey; PDB ID: 1OX3).
(Adapted from *J. Biomol. NMR*, Sparsely-sampled, high-resolution 4-D omit spectra for detection and assignment of intermolecular NOEs of protein complexes, **59**, 2014, 51–56, S. Wang and P. Zhou, (Copyright 2014 Springer Science + Business Media Dordrecht) with permission of Springer.)

Acknowledgements

This work was partially supported by NIH grant AI-055588 and a bridge fund from Duke University Medical Center, both to P.Z.

References

1. R. S. Dixon and J. D. Kraus, *Astron. J.*, 1968, **73**, 381–407.
2. J. A. Högbom, *Astron. Astrophys., Suppl. Ser.*, 1974, **15**, 417.
3. J. A. Högbom, in *Radio Astronomy at the Fringe*, ed. J. A. Zensus, M. H. Cohen and E. Ros, ASP Conference Series, Astronomical Society of the Pacific, San Francisco, 2003, vol. 300, pp. 17–20.
4. W. N. Christiansen and J. A. Högbom, *Radiotelescopes*, Cambridge University Press, Cambridge, 1969.
5. D. H. Rogstad and G. S. Shostak, *Astron. Astrophys.*, 1971, **13**, 99–107.
6. T. J. Cornwell, *Astron. Astrophys.*, 2009, **500**, 65–66.
7. U. J. Schwarz, *Astron. Astrophys.*, 1978, **65**, 345–356.
8. J. C. J. Barna and E. D. Laue, *J. Magn. Reson.*, 1987, **75**, 384–389.
9. J. C. J. Barna, E. D. Laue, M. R. Mayger, J. Skilling and S. J. P. Worrall, *J. Magn. Reson.*, 1987, **73**, 69–77.
10. A. S. Stern, D. L. Donoho and J. C. Hoch, *J. Magn. Reson.*, 2007, **188**, 295–300.
11. K. Kazimierczuk and V. Y. Orekhov, *J. Magn. Reson.*, 2012, **223**, 1–10.
12. J. C. Hoch and A. S. Stern, *Methods Enzymol.*, 2001, **338**, 159–178.
13. K. A. Marsh and J. M. Richardson, *Astron. Astrophys.*, 1987, **182**, 174–178.
14. E. J. Candès and T. Tao, *IEEE Trans. Inf. Theory*, 2005, **51**, 4203–4215.
15. R. N. Bracewell, *The Fourier Transform and Its Applications*, McGraw-Hill, Boston, 2000.
16. R. M. Gray, *Toeplitz and Circulant Matrices: A Review*, Now Publishers, Norwell, MA, 2006.
17. G. E. Forsythe and W. R. Wasow, *Finite-Difference Methods for Partial Differential Equations*, John Wiley and Sons, New York, 1960.
18. M. Elad and A. M. Bruckstein, *IEEE Trans. Inf.*, 2002, **48**, 2558–2567.
19. E. J. Candès, J. Romberg and T. Tao, *IEEE Transactions Inf. Theory*, 2006, **52**, 489–509.
20. E. J. Candès and J. Romberg, 2008, arXiv:math/0411273v1.
21. J. A. Tropp, 2010, arXiv:1003.0415v1.
22. D. Needell and R. Vershynin, *Found. Comput. Math.*, 2009, **9**, 317–334.
23. M. Rudelson and R. Vershynin, *Commun. Pure Appl. Math.*, 2008, **61**, 1025–1045.
24. D. L. Donoho and J. Tanner, *Philos. Trans. R. Soc., A*, 2009, **367**, 4273–4293.
25. S. G. Mallat and Z. Zhang, *IEEE Trans. Signal Process.*, 1993, **41**, 3397–3415.
26. J. H. Friedman and W. Wuetzle, *J. Am. Stat. Assoc.*, 1981, **76**, 817–823.
27. P. J. Huber, *Ann. Stat.*, 1985, **13**, 435–475.

28. V. N. Temlyakov, *J. Approximation Theory*, 1999, **98**, 117–145.
29. S. S. Chen, D. L. Donoho and M. A. Saunders, *SIAM J. Sci. Comput.*, 1998, **20**, 33–61.
30. S. Chen, S. A. Billings and W. Luo, *Int. J. Control*, 1989, **50**, 1873–1896.
31. Y. C. Pati, R. Rezaiifar and P. S. Krishnaprasad, in *Proceedings of the 27th Asilomar Conference on Signals, Systems and Computers*, ed. A. Singh, IEEE Comput. Soc. Press, Los Alamitos, CA, 1993.
32. G. Davis, S. G. Mallat and Z. Zhang, *Opt. Eng.*, 1994, **33**, 2183–2191.
33. S. Qian and D. Chen, *Signal Process.*, 1994, **36**, 329–355.
34. J. A. Tropp and A. C. Gilbert, *IEEE Trans. Inf. Theory*, 2007, **53**, 4655–4666.
35. D. Needell and R. Vershynin, *arXiv*, 2007, **0712**, 1360v1.
36. H. Rauhut, *IEEE Trans. Inf. Theory*, 2008, **54**, 5661–5670.
37. D. Needell and J. A. Tropp, *Appl. Comput. Harmonic Anal.*, 2009, **26**, 301–321.
38. D. L. Donoho, Y. Tsaig, I. Drori and J.-L. Starck, *IEEE Trans. Inf. Theory*, 2012, **58**, 1094–1121.
39. J. A. Tropp, *IEEE Trans. Inf. Theory*, 2004, **50**, 2231–2242.
40. R. Gribonval and P. Vandergheynst, *IEEE Trans. Inf. Theory*, 2006, **52**, 255–261.
41. A. J. Shaka, J. Keeler and R. Freeman, *J. Magn. Reson.*, 1984, **56**, 294–313.
42. J. Keeler, *J. Magn. Reson.*, 1984, **56**, 463–470.
43. S. J. Davies, C. Bauer, P. J. Hore and R. Freeman, *J. Magn. Reson.*, 1988, **76**, 476–493.
44. E. Kupče and R. Freeman, *J. Magn. Reson.*, 2005, **173**, 317–321.
45. J. C. J. Barna, S. M. Tan and E. D. Laue, *J. Magn. Reson.*, 1988, **78**, 327–332.
46. B. E. Coggins and P. Zhou, *J. Biomol. NMR*, 2008, **42**, 225–239.
47. J. Wen, J. Wu and P. Zhou, *J. Magn. Reson.*, 2011, **209**, 94–100.
48. J. W. Werner-Allen, B. E. Coggins and P. Zhou, *J. Magn. Reson.*, 2010, **204**, 173–178.
49. J. Wen, P. Zhou and J. Wu, *J. Magn. Reson.*, 2012, **218**, 128–132.
50. B. E. Coggins, J. W. Werner-Allen, A. Yan and P. Zhou, *J. Am. Chem. Soc.*, 2012, **134**, 18619–18630.
51. K. Kazimierczuk, A. Zawadzka, W. Kozminski and I. Zhukov, *J. Magn. Reson.*, 2007, **188**, 344.
52. J. Stanek and W. Kozminski, *J. Biomol. NMR*, 2010, **47**, 65–77.
53. J. Stanek, R. Augustyniak and W. Kozminski, *J. Magn. Reson.*, 2011, **214**, 91–102.
54. S. Wang and P. Zhou, *J. Biomol. NMR*, 2014, **59**, 51–56.
55. B. E. Coggins, SCRUB (Version 1.0), Durham, NC, 2013.
56. L. Jiang, B. E. Coggins and P. Zhou, *J. Magn. Reson.*, 2005, **175**, 170–176.
57. J. Wojtaszek, J. Liu, S. D'Souza, S. Wang, Y. Xue, G. C. Walker and P. Zhou, *J. Biol. Chem.*, 2012, **287**, 26400–26408.

CHAPTER 8

Covariance NMR

KIRILL BLINOV,[a] GARY MARTIN,[b] DAVID A. SNYDER*[c] AND
ANTONY J. WILLIAMS[d]

[a] Molecule Apps, LLC, 2711 Centerville Road, Suite 400, Wilmington,
DE 19808, USA; [b] Merck Research Laboratories, Process & Analytical
Chemistry, NMR Structure Elucidation, 126 E. Lincoln Ave., Rahway,
NJ 07065, USA; [c] Department of Chemistry, William Paterson University,
300 Pompton Road, Wayne, NJ 07470, USA; [d] ChemConnector,
513 Chestnut Grove Court, Wake Forest, NC 27587, USA
*Email: snyderd@wpunj.edu

8.1 Introduction

NMR spectroscopy is one of the most powerful sources of data for elucidating molecular connectivity, conformation and dynamics.[1] However, even
NMR spectroscopy is relatively insensitive and obtaining high signal-to-noise
multidimensional spectra with adequate spectral resolution often requires
impractically long measurement times.[2] Spectroscopists have developed a
variety of techniques to optimally utilize the data recordable within a given
acquisition time by either avoiding or supplementing the Fourier transform
typically used in processing NMR spectra. Covariance NMR comprises a
family of techniques that use the properties of matrix algebra to transform
NMR spectra. Covariance NMR methods use the symmetry present within
NMR spectra,[3,4] or between pairs of NMR experiments,[2,5,6] in order to
maximally leverage the information content present in NMR spectra.

Key advantages of covariance NMR include its ease of implementation and
its applicability to many classes of NMR experiments and multiple data acquisition schemes, ranging from traditional Fourier Transform spectra to

New Developments in NMR No. 11
Fast NMR Data Acquisition: Beyond the Fourier Transform
Edited by Mehdi Mobli and Jeffrey C. Hoch
© The Royal Society of Chemistry 2017
Published by the Royal Society of Chemistry, www.rsc.org

GFT-NMR datasets.[7] The full resolution enhancement powers of covariance NMR and sensitivity advantages of covariance reconstruction are obtained even when applying covariance techniques to standard, uniformly sampled Fourier transform NMR data. Thus, spectra recorded for covariance NMR can also be analyzed using standard processing techniques, and optimal use of covariance NMR does not require implementation of special sampling or data processing schemes. On all but the largest datasets (*e.g.* 4D NMR), covariance processing takes minutes or even seconds using readily available software, such as the Covariance Toolbox for MATLAB,[8] ACD/Spectrus v14.0,[9] or Mestrelab MNOVA.[10]

This chapter summarizes the theoretical underpinnings of covariance NMR, with special attention to issues related to rapid collection of data used to produce high signal-to-noise and high-resolution NMR spectra. It also describes several variations on the basic covariance NMR technique, such as direct, indirect, unsymmetrical, and generalized indirect covariance. The relationship between covariance NMR and other techniques that facilitate the accelerated collection and rapid, yet robust, analysis of NMR data are also discussed, with the goal of presenting both informational tidbits as well as the big picture of how covariance NMR can facilitate structure elucidation.

8.2 Direct Covariance NMR

Consider an $N_1 \times N_2$ NMR dataset T processed (including Fourier transformation and phasing) along the (second) direct dimension but not along the (first) indirect dimension and that same dataset S following the (unitary) Fourier transform along the indirect dimension. According to Parseval's theorem[11] (where * denotes the conjugate transpose):

$$T * T = S * S \qquad (8.1)$$

For spectra without any zero-frequency signals (*e.g.* spectra following deconvolution of the carrier frequency and time-domain polynomial baseline correction to remove a centered solvent signal) $T * T / N_1$ is a covariance matrix of the columns of T. Suppose data acquisition proceeded through the collection of $N_1 = N_2$ data points along the indirect dimension: for spectra such as homonuclear NOESY, TOCSY and COSY spectra with appropriate phasing and regularization, if needed,[12] the real part of matrix S is, ignoring noise and artifacts, a positive definite, symmetric matrix S_R and hence:

$$\mathrm{real}\left[\sqrt{(T * T)}\right] = S_R \qquad (8.2)$$

In other words, the matrix square root of the *covariance* matrix (without the normalization factor of $N_1 = N_2$) representing (in the mixed time-frequency domain) a fully "symmetric" NMR spectrum is equal to the Fourier transformed spectrum. This observation lies at the root of the family of techniques collectively called *Covariance NMR*.

The immediate significance of this observation is the application of this covariance processing technique to otherwise symmetric spectra sampled at a random set of increments along the indirect dimension. Consider a mixed-time frequency domain spectrum T_1 obtained in the same manner as the $N_2 \times N_2$ spectrum T but acquired with only a random subset of N_1 out of N_2 increments along the indirect dimension. The covariance matrix for T_1 is an efficient and unbiased estimator of T in the context of the extrinsic geometry of the cone of $N_2 \times N_2$ covariance matrices within $\mathbf{R}^{N_2 \times N_2}$.[13] Thus (up to a scalar multiplier), $\mathrm{real}\left[\sqrt{(T_1 * T_1)}\right]$ is an $N_2 \times N_2$ matrix estimating the spectrum S_R. That is, covariance processing allows for recovery of full resolution data with reduced acquisition time.

In practice, it is often more convenient to sample the first (uniformly spaced) N_1 time increments along the indirect dimension rather than selected N_1 increments at random. This allows for Fast Fourier transformation (along with phasing and baseline correction) of the acquired dataset along both the direct and indirect dimensions. Additionally, owing to relaxation, the initial increments have higher signal to noise value than later increments; uniform sampling ensures inclusion of these higher signal to noise increments. Subjecting this real valued spectrum S_P to the transformation [analogous to the transformation on the right hand side of eqn (8.1) followed by a matrix square root]:

$$C = \sqrt{S_P{}^T S_P} \tag{8.3}$$

results in an $N_2 \times N_2$ matrix. This "frequency domain covariance matrix" typically is a better reconstruction of S_R than is the matrix obtained from time-domain covariance processing from a spectrum obtained with randomly sampled t_1 increments (Figure 8.1). A further advantage of applying covariance NMR to frequency domain spectra is the ease of removing baseline artifacts, which could lead to spurious covariance signals, in the frequency domain relative to removing them in the time-domain.

Substantial resolution gains are possible with direct covariance NMR. For example, direct covariance accurately recovered a full resolution 2048×2048 point TOCSY spectrum of the cyclic decapeptide, antamanide with as few as 48 t_1 increments. In this case, *covariance NMR can reduce the time required* to obtain an NMR spectrum at a maximal resolution *by a factor of 40.*[14] Additionally, sensitivity gains are also possible as covariance NMR averages out asymmetric noise in producing a symmetric spectrum. Furthermore, direct covariance NMR reconstructs high resolution spectra without requiring acquisition of increments with high values of t_1 and hence avoids the use of scans with low sensitivity owing to relaxation during t_1. Direct covariance avoids the use of low sensitivity (high value of t_1) scans by endowing the indirect dimension with the resolution and sweep-width of the direct dimension. For example, in a correlation spectrum of microcrystalline catabolite repressor phosphoprotein (Crh) obtained using PARIS recoupling,

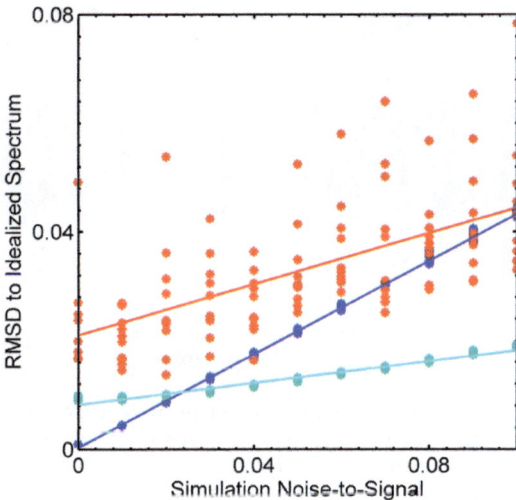

Figure 8.1 Quality of covariance reconstruction of a simulated TOCSY in the presence of noise. The error propagation results depicted in this figured started with simulation of a TOCSY-type spectrum of a mixture of NADH and NAD with random peak heights and line-widths of 5.2 Hz (0.013 ppm on a 400 MHz spectrometer). Four simulated spectra were calculated for each of ten calculations: (1) a noise-free, *"idealized"* 1024×1024 frequency domain spectrum, (2) a mixed time/frequency domain 1024×1024 spectrum with Gaussian distributed white noise, (3) a spectrum obtained by random sampling along t_1 in the mixed time/frequency domain spectrum and (4) a 64×1024 "low resolution" spectrum with Gaussian distributed white noise. In spectrum 4, a line-width of 32 Hz was used in the indirect dimension, which is a typical line-width for such a spectrum processed with typical apodization prior to Fourier transform along each dimension. Spectra 2–4 were subjected to direct covariance and each was compared with spectrum 1. Each blue point plots the RMSD between spectra 1 and 2 as a function of the "noise-to-signal" (the reciprocal of signal-to-noise, where minimum peak height in the Fourier transformed version of spectrum 2/median absolute value of noise defines the signal-to-noise) of spectrum 2. Each red point similarly plots the RMSD between spectra 1 and 3 as a function of the noise-to-signal in the corresponding spectrum 2 for that simulation, and each cyan point plots the RMSD between spectra 1 and 4 as a function of the noise-to-signal of spectrum 4. Best fit lines for each comparison are shown to guide the eye. Note that as the signal-to-noise decreases below about 40 (noise-to-signal > 0.025), frequency domain covariance with a lower resolution spectrum (cyan) actually has the lowest RMSD. This is because covariance accurately recovers information along the indirect dimension while only using the first 64 t_1 increments, which provide the highest signal relative to baseline noise.

direct covariance NMR not only provided significant and useful resolution enhancements but also resulted in increased sensitivity (Figure 8.2).

While the direct calculation of the frequency domain covariance matrix according to eqn (8.3) is computation and memory intensive, use of the

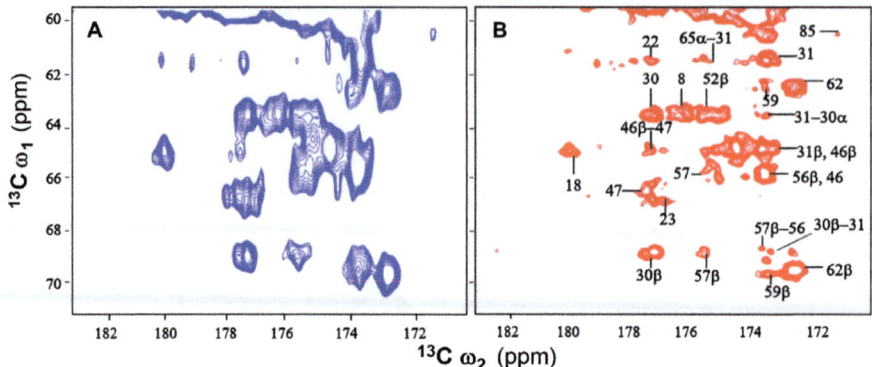

Figure 8.2   Solid state NMR correlation spectrum with PARIS recoupling (A) before and (B) after covariance. (A) Ca–C' region of a Fourier Transform NMR correlation spectrum of microcrystalline Crh protein obtaining using PARIS recoupling and 256 t_1 increments. (B) The same region of the spectrum following covariance. Numbers indicate the intra- and inter-residue correlations from which the numbered peaks arise. Note that covariance not only increases resolution but also results in increased sensitivity leading to the detection of additional correlation peaks such as the one circled in green.
Adapted from Ref. 76 with permission from The Royal Society of Chemistry.

singular value decomposition provides substantial savings in computational time and resources. Let $U*D*V^T = S_P{}^T$ be the SVD (Singular Value Decomposition) transpose of the $N_1 \times N_2$ frequency domain spectrum S_P. Then:

$$C = U*D*U^T \qquad (8.4)$$

follows from the orthogonality of U and V. In a 2004 study of time domain covariance processing introducing the SVD method, the SVD implementation sped up covariance calculations typically by a factor of ten or more in cases where N_1 was less than 1/10th N_2.[15] On a modern computer (2.2 GHz processor speed, 4.00 GB memory), SVD-based direct covariance processing of 2D spectra ($N_1 = 512$, $N_2 = 2048$) takes less than a second in MATLAB R11b. SVD of a typically sized, $8 \times 32 \times 16 \times 1024$ 4D NMR spectrum (*vide infra*) takes approximately 1 second on the same platform, although reconstruction of the covariance spectrum in such a case can be memory intensive and time-consuming.[8]

While matrices naturally represent 2D NMR spectra, direct covariance also applies to 4D NMR experiments, where it can provide substantial increases in resolution. Consider an $N_1 \times N_2 \times N_3 \times N_4$ spectrum where the first two dimensions are "donor" dimensions and the third and fourth (direct) dimension are "acceptor" dimensions. For example, consider a 4D ^{13}C–^{1}H-HSQC-NOESY-^{13}C–^{1}H-HSQC experiment, in which the projections onto the donor and acceptor dimensions are both HSQC spectra and cross-peaks are between two proton/carbon chemical shift pairs such that the donor carbon is bonded to the donor proton, the acceptor proton is bonded to the acceptor carbon and the donor and acceptor protons experience an NOE. Reshaping this spectrum into an $(N_1 * N_2) \times (N_3 * N_4)$ matrix representing correlations between carbon/proton pairs allows for the application of covariance NMR. Covariance transformation yields an $(N_3 * N_4) \times (N_3 * N_4)$ matrix and hence an $N_3 \times N_4 \times N_3 \times N_4$ spectrum.

Direct covariance NMR endows the donor dimensions of a 4D NMR spectrum with the same resolution as the acceptor dimensions, the latter of which include the high-resolution directly detected dimension.[16] Application of direct covariance to a typically sized, $8 \times 32 \times 16 \times 1024$ 4D NMR spectrum, that takes 130 hours (almost five and a half days) to acquire, results in a 64-fold gain in digital resolution and produces a $16 \times 1024 \times 16 \times 1024$ spectrum (Figure 8.3). Experimentally acquiring such a spectrum would take almost a year! Note that reconstruction of the full resolution covariance NMR 4D spectrum, however, requires care due to the large amount of memory that would be required to store such a spectrum.[8]

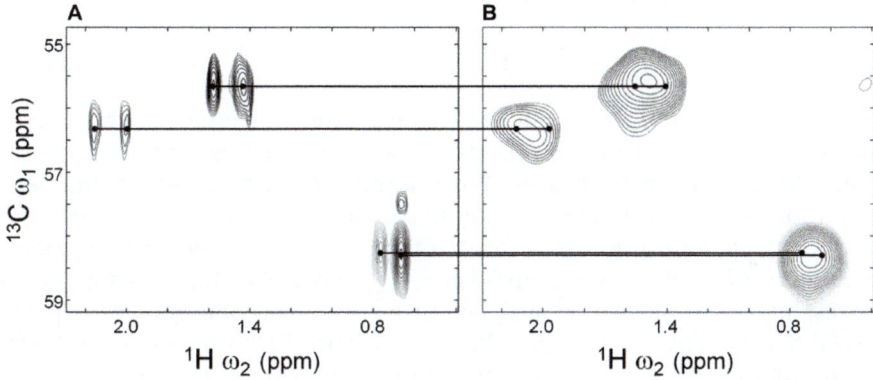

Figure 8.3 Comparison between shared-evolution 4D ^{13}C/^{15}N-edited NOESY donor planes before (B) and after (A) covariance. Portion of the Q70 N–H plane corresponding to a ^{15}N chemical shift of 125.6 ppm and ^{1}H chemical of shift 7.80 ppm of a shared-evolution 4D ^{13}C/^{15}N-edited NOESY of the 24 kDa protein DdCAD-1. In particular, covariance NMR distinguishes between the signal owing to the long-range I55HD1-Q70NH NOE (left peak in the bottom right pair of peaks in panel A) and the signal due to the long-range I84HD1-Q70NH interaction (right peak of the pair). The underlying shared-evolution Fourier transform spectrum was recorded at $T = 303$ K with 22 complex points in each hetero-atom dimension and 32 complex points in the indirect (donor) proton dimension, which took approximately 149 hours (6.2 days) to record on a 800 MHz spectrometer equipped with a standard TXI probe. Recording a Fourier transform spectrum matching the resolution available *via* covariance would require over 3 months of measurement time.

Adapted with permission from D. A. Snyder, Y. Xu, D. Yang and R. Brüschweiler, Figure S1 of Resolution-Enhanced 4D 15N/13C NOESY Protein NMR Spectroscopy by Application of the Covariance Transform, *J. Am. Chem. Soc.*, 2007, **129**, 14126, Copyright 2007 American Chemical Society.

Direct covariance can even start from non-symmetric input spectra such as shared-evolution 4D $^{13}C/^{15}N$-edited NOESY spectra. The resulting direct covariance dataset *is* symmetric and establishes through-space correlations not only between $^1H,^{13}C$ pairs and $^1H,^{15}N$ pairs but also amongst both the $^1H,^{13}C$ pairs and the $^1H,^{15}N$ pairs.[16] As with 4D covariance in general, covariance provides substantial resolution gains facilitating the identification and assignment of conformationally restricted long-range NOESY constraints important in protein structure determination. Covariance NMR facilitates the use of week-long experiments to resolve the same correlations that would take months of data collection to resolve in a purely Fourier transformed spectrum.

8.3   Indirect Covariance NMR

8.3.1   Principle

Covariance does not have to be applied only to the direct dimension since eqn (8.3) can be modified in the following manner:

$$C = \sqrt{S_P S_P^T} \qquad (8.5)$$

This calculation usually does not make sense for homonuclear spectra because the resultant matrix (spectrum) will have the resolution of the indirect dimension on both axes. For example, if an NMR spectrum has a resolution of 1024 $(F_2) \times 256$ (F_1) points then a direct covariance calculation produces a spectrum of resolution 1024×1024 while an indirect covariance calculation gives us a spectrum with resolution 256×256. This manner of processing the data is obviously of little value for homonuclear spectra (COSY, TOCSY, *etc.*). However, indirect covariance (IC) processing applied to heteronuclear spectra can produce a much improved data representation. Application of IC to a HSQC-TOCSY spectrum, which contains both short- and long-range C–H correlations, produces a spectrum that contains the corresponding carbon–carbon correlations (Figure 8.4).[3] The only useful way in which these data can be obtained directly is by the acquisition of an inverted $^1J_{CC}$ 1,n-ADEQUATE spectrum.[17]

It should be noted that, as with direct covariance processing, indirect covariance calculations do not produce any new information. The covariance calculation only converts the data to a more "convenient" representation for interpretation.

8.3.2   Unsymmetrical Indirect Covariance (UIC) and Generalized Indirect Covariance (GIC) NMR

If we replace eqn (8.5) with the following equation:

$$C = S1_P S2_P^T \qquad (8.6)$$

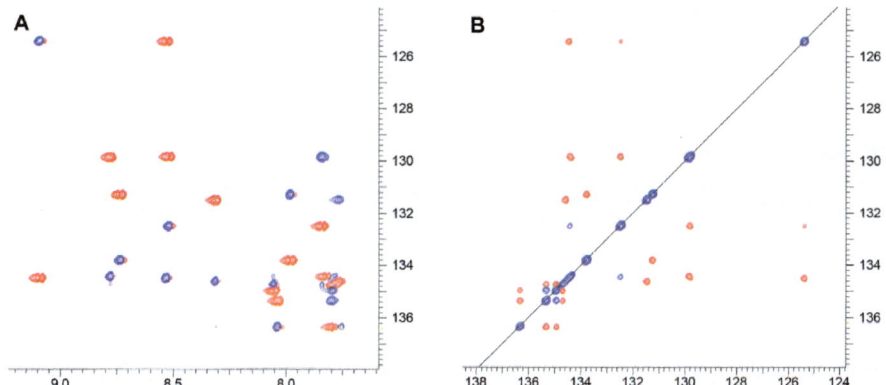

Figure 8.4 HSQC-TOCSY and indirect covariance HSQC-TOCSY spectra. (A) Inverted direct response (IDR) HSQC-TOCSY spectrum. Positive signals (red) correspond to long range correlations while negative signals (blue) correspond to direct CH correlations. (B) Corresponding indirect covariance spectrum. Non-diagonal peaks correspond to C–C bonds. Positive (red) non-diagonal peaks correspond to the real correlations. Negative (blue) non-diagonal peaks correspond to the artifacts.

where $S1_P$ and $S2_P$ are matrices from different spectra then this simple matrix multiplication, historically called "Unsymmetrical Indirect Covariance" (UIC)[2,6] to emphasize the relationship to indirect covariance, provides us with new ways to review the data encoded into the two separate 2D data matrices.

According to matrix multiplication rules $S1$ and $S2$ should have equal size in the row dimension that corresponds to the F_2 (^{1}H in ^{1}H-detected spectra) dimension in 2D NMR spectra. The resultant matrix corresponds to a new spectrum that in many cases is generally inaccessible by conventional NMR detection methods. Interpretation of the result of this calculation depends on the spectra used in the calculation. The matrix obtained as a result of UIC is non-square in general, and the matrix square root operation cannot be applied. To overcome this restriction, a method called "Generalized Indirect Covariance" has been proposed.[5] The Generalized Indirect Covariance (GIC) method contains four main stages. First, all matrices are combined into one "vertical" matrix. The resultant matrix contains vertically stacked original matrices multiplied by their transposed matrices (*i.e.* conventional Indirect Covariance is applied). The result of matrix multiplication is a square matrix and the matrix square root, or more generally, the matrix power operation can be applied to this matrix in the third stage. Finally, the result of multiplication is partitioned. Each part corresponds to some IC spectrum. The simplest case of two matrices is described in Figure 8.5. The final matrix contains all four possible combinations of IC which can be applied to two initial matrices. A schematic representation of conventional unsymmetrical indirect covariance is added for comparison.

(a) Unsymmetrical Indirect Covariance

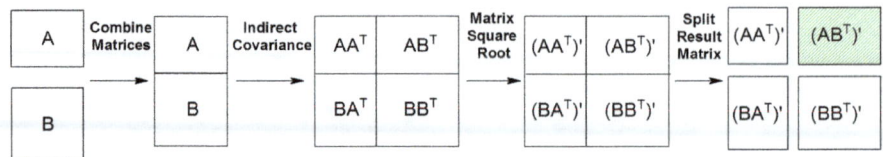

(b) Generalized Indirect Covariance

Figure 8.5 Comparison between unsymmetrical indirect covariance (UIC) and generalized indirect covariance (GIC). (a) Unsymmetrical indirect covariance. It is equal to conventional matrix multiplication, and the matrix square root procedure isn't applicable to result matrix in all cases. (b) Generalized indirect covariance. GIC consists of three stages. The first stage combines initial matrices into a special "stacked" matrix. The second stage applies matrix multiplication (or indirect covariance) to the stacked matrix, followed by the matrix square root calculation. The third stage splits the result matrix into several submatrices. Each submatrix corresponds to some type of indirect covariance spectrum, with the highlighted submatrix corresponding to the desired unsymmetrical spectrum.

The main advantage of GIC relative to conventional UIC is the ability to apply a matrix power operation during the calculation. The matrix power operation is useful for artifact removal, as described later in the section regarding artifacts. Additionally, the method produces all possible IC spectra calculated at once, which may be convenient in some cases. The main disadvantage of the method is that it requires intensive computation that increases significantly with matrix size.

8.3.3 Signal/Noise Ratio in Covariance Spectra

The goal of unsymmetrical and generalized indirect covariance NMR is to reconstruct NMR spectra whose experimental acquisition is impractical or even impossible owing to low sensitivity. An example would be $^{13}C-^{15}N$ heteronuclear correlation at natural abundance. In particular, by combining ^{1}H-detected spectra, the inputs to covariance processing can have up to approximately 700-fold higher signal-to-noise than do spectra requiring the correlation of insensitive nuclei at natural abundance (*vide infra*). Moreover, spectra reconstructed using unsymmetrical covariance processing, which involves summing squares, typically have *higher nominal signal-to-noise* values than do the spectra input to covariance processing.[18]

A detailed statistical analysis of unsymmetrical covariance spectra, which also applies to off-diagonal peaks in direct and indirect covariance spectra prior to calculation of the matrix square-root, shows that: (A) additive, Gaussian noise in input spectra propagates to additive, Gaussian noise in

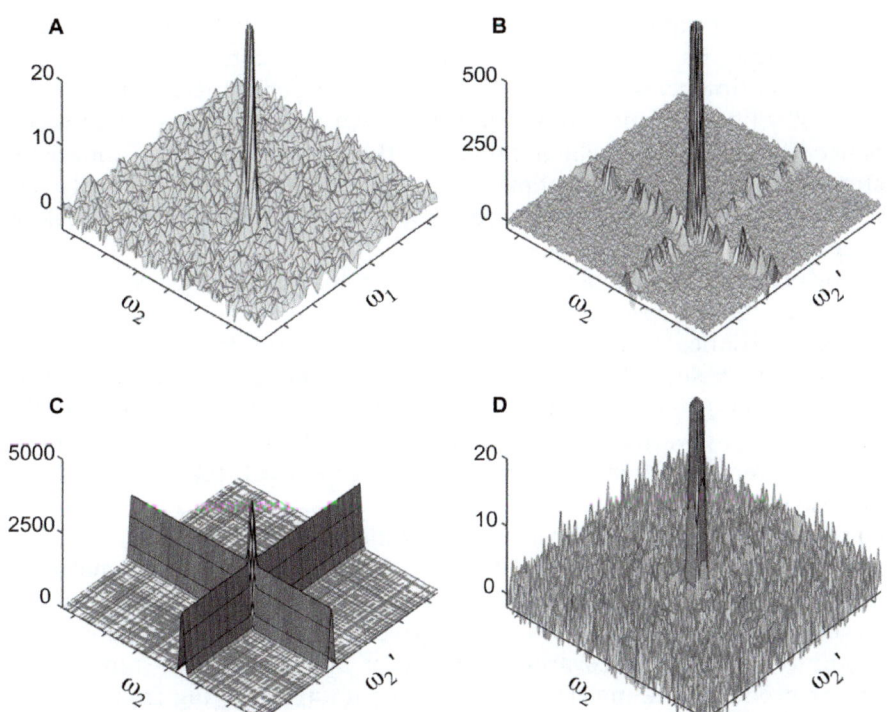

Figure 8.6 Propagation of noise through unsymmetrical indirect covariance.
(A) Simulated "noisy" input spectrum **A** with S/N = 45. (B) Covariance
spectrum $\mathbf{A^TB}$, where **B** has the same signal peak as well as noise level as
A. Note the noise ridges meeting at the covariance peak. (C) The expected
variance calculated for each point in $\mathbf{A^TB}$ using the error propagation
formula described in Snyder *et al.* [Z-Matrix Paper]. (D) The Z-matrix
calculated as described in Snyder *et al.* [Z-Matrix Paper]. Note that the
noise ridges disappear upon transformation to the Z-matrix.

Reproduced with permission from Snyder *et al.*, Z-matrix formalism for
quantitative noise assessment of covariance nuclear magnetic resonance
spectra, *J. Chem. Phys.*, **129**, 104501. Copyright 2008, AIP Publishing LLC.

covariance spectra;[7] (B) the noise floor of unsymmetrical covariance spectra
is uneven and possesses ridges similar to those found in maximum entropy
processing of data acquired using certain sampling and processing
schemes;[19] and (C) the overall sensitivity of unsymmetrical covariance
spectra is on the same order as the sensitivity of the spectra subjected to
covariance processing (Figure 8.6).[7]

The sensitivity gains realized *via* covariance reconstruction can be substantial. For example $^{13}C-^{13}C$ spectra reconstructed from proton detected
spectra maintain the $(\gamma_H/\gamma_C) \approx$ four-fold increase of proton detection sensitivity compared with carbon detection. Moreover, an experiment correlating
two ^{13}C atoms acquired on a sample with natural (1.1%) ^{13}C abundance
has only 0.012% of the signal of the same experiment performed on an

isotopically pure sample. On the other hand, a heteronuclear experiment on such a sample has 1.1%, 90 times the factor associated with the homonuclear experiment, of the signal of the same experiment acquired on an isotopically pure sample. Combining these two factors suggests that a covariance reconstruction, which preserves the sensitivity of its underlying heteronuclear experiments,[7] allows for as much as a 360-fold sensitivity gain over direct detection of homonuclear experiments. Of course, this "back-of-the-envelope" calculation does not consider the larger proton line-widths and the presence of proton–proton *J*-couplings that decrease peak intensities in proton-detected spectra and thus reduce the expected sensitivity advantage of covariance NMR. Nevertheless, even without the use of recently developed pure-shift HSQC experiments[20–27] that avoid the impact of proton–proton *J*-couplings on the sensitivity of proton detected spectra, unsymmetrical covariance can shorten the time required to obtain homonuclear correlations *from* the days or even *weeks* it can take to directly acquire such data *to* the *hours*, or possibly even less time, it takes to acquire two higher-sensitivity heteronuclear experiments.[2,18]

Quantifying the uncertainty of a covariance spectrum as a matrix of standard deviations in the covariance peak intensities facilitates setting a threshold in peak-picking covariance spectra. Element-wise division of an unsymmetrical indirect covariance spectrum by its matrix of standard deviations produces a Z-matrix that can be contoured starting from a statistically justified critical Z-score. This Z-matrix, or alternatively unsymmetrical indirect covariance spectra modified by related scaling techniques such as Maximum Ratio Scaling, may also produce easier to visualize and analyze results when applying covariance to spectra with high dynamic ranges and when combining spectra with very different sensitivities.[7]

Unlike Fourier transformation, which is linear, covariance processing can produce spectra with uneven noise floors, although additive Gaussian noise remains additive and Gaussian following covariance. In particular, unsymmetrical covariance spectra have increased noise in regions with chemical shifts in which there are peaks in the spectra input to covariance processing.[7] These "noise ridges" resemble similar phenomena seen when applying other non-linear processing schemes, such as maximum entropy, to reconstructing spectra from data collected using certain sampling schemes[19] and suggests that there may be deep underlying mathematical connections between the many non-linear processing and reconstruction schemes used to maximally leverage the information content of NMR data.

In fact, how noise propagates through non-linear processing schemes in general remains an open question, although careful statistical analyses are available for particular reconstruction schemes.[28,29] In particular, there seems to be no rigorous analysis of the propagation of uncertainty in GIC when $\lambda \neq 1$, where λ is the exponent applied to the covariance matrix in GIC: *e.g.*, unsymmetrical covariance is GIC with $\lambda = 1$. One possible avenue for performing such an analysis is calculation of the Fréchet derivative of the covariance transform, *e.g.* following the methods developed by Higham and collaborators.[30–32] *In lieu* of a closed formula for propagating uncertainty through a matrix square-root, assessment of the uncertainty of a covariance peak in a direct, indirect or GIC with $\lambda \neq 0.5$ spectrum can proceed by evaluation of the corresponding peak in the covariance spectrum prior to square-root calculation.

8.3.4 Artifact Detection

The main disadvantage of IC processed spectra is the potential presence of artifacts, extra peaks which are absent in the source spectra. Since peaks in a 2D spectrum generally reflect structural information (*i.e.* representing correlations between atoms (nuclei)), any artifact peak can represent "false" structural information, which may produce problems in terms of spectrum interpretation. The most important task in covariance processing therefore is removal of as many artifacts as possible.

The main source of artifacts in covariance spectra is the partial overlap of peaks in 1D slices (matrix rows or columns) of 2D spectra that are used in the covariance calculation. Figure 8.7 illustrates sources of legitimate peaks and artifacts for an IC HSQC-TOCSY spectrum.

Generally artifact peaks are less intense than legitimate correlation peaks because they are produced by partial overlap. But under some conditions (for example when one overlapping peak is very intense), artifact peaks may be comparable in intensity or even more intense than normal peaks.

There are several methods that have been proposed to reduce the number or intensity of artifact peaks. The first attempt is described by Blinov *et al.*[33] and describes a method of partial artifact removal in indirect covariance HSQC-TOCSY spectra. The IDR HSQC-TOCSY spectrum contains both types

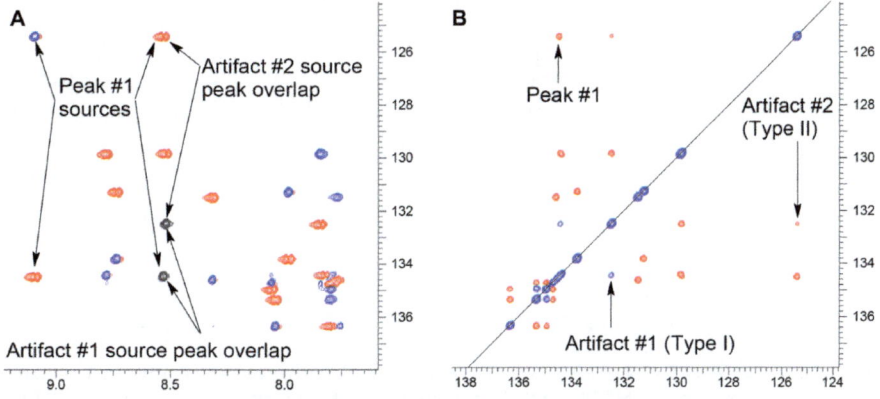

Figure 8.7 Partial peak overlap as a source of artifacts. (A) HSQC-TOCSY spectrum with sources of the legitimate peaks and the artifacts. (B) Indirect covariance HSQC-TOCSY spectrum with two types of artifacts produced by partial peak overlap of peaks with the same sign (Type I) and the different signs (Type II).

of correlations: direct C–H correlations are negative while long-range C–H correlations are positive. Each legitimate peak in IC spectrum should be produced by the pair of direct (negative) and long-range (positive) peaks and, therefore, should have a negative sign. Peaks that produce pairs of direct–direct or long-range–long-range peaks should have a positive sign. These peaks should be considered as artifacts and should be ignored in the IC spectrum. This type of artifact is called Type I. Another type of artifact (named Type II), is produced by partial or full overlap of direct C–H correlation with long-range C–H correlation from different hydrogens having close or equal chemical shifts. The Type II artifacts have the same negative sign as the legitimate peaks and it is impossible to distinguish the Type II artifacts from legitimate peaks. To solve this problem, the authors suggest dividing the original HSQC-TOCSY matrix into two matrices: one matrix containing only negative values from the original matrix, and the other matrix containing only positive values from the original matrix. The first matrix contains only direct responses while the second matrix contains only long-range responses. Eqn (8.6) is used to calculate the IC spectrum in this case. The resulting IC spectrum does not contain Type I artifacts at all. Type II artifacts appear only on one side of the diagonal while the legitimate peaks appear on both sides of the diagonal and can easily be distinguished from Type II artifacts (Figure 8.8). Figures 8.9–8.11 show the method in detail. The main idea of the described method is based on the feature that only correlations between CH carbons appear in the HSQC-TOCSY spectrum and each CH correlation has a paired correlation which produces a diagonally symmetrical peak in the IC spectrum. The probability that an artifact (overlapping) peak has a corresponding pair is very low; hence, the artifact peaks

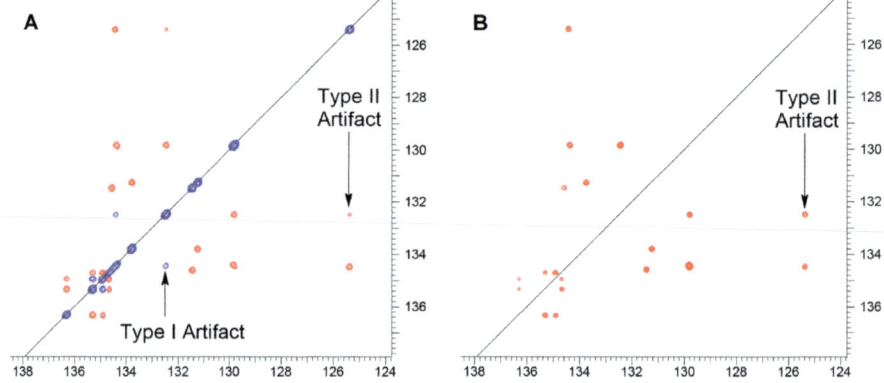

Figure 8.8   Indirect covariance *vs.* unsymmetrical indirect covariance. (A) Indirect covariance HSQC-TOCSY spectrum with two types of artifacts produced by partial peak overlap of peaks with the same sign (Type I) and different signs (Type II). (B) Unsymmetrical indirect covariance HSQC-TOCSY spectrum. The Type I artifacts are completely removed while the Type II artifacts become diagonally unsymmetrical and could be easily determined.

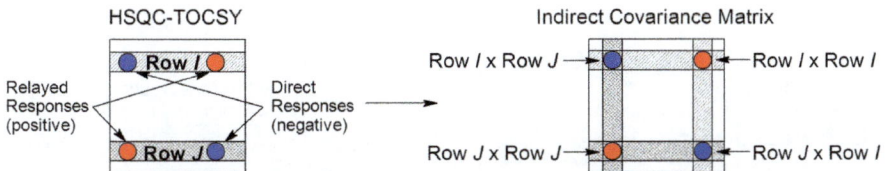

Figure 8.9 Schematic comparison between conventional and unsymmetrical indirect covariance: indirect covariance processing for an HSQC-TOCSY spectrum. This figure depicts origins of peaks in conventional indirect covariance. Each pair CH–CH has two direct and two relayed responses. These responses located on two rows corresponding to chemical shifts of carbons in pair. In IC process these rows produce two diagonal peaks when multiplied by itself and two non-diagonal peaks when multiplied by each other.

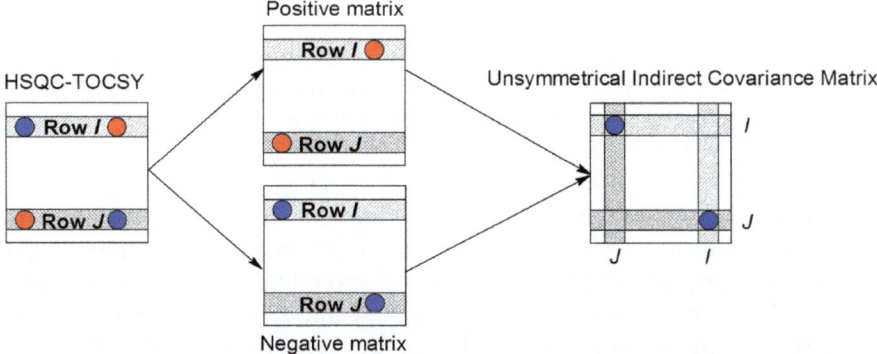

Figure 8.10 Schematic comparison between conventional and unsymmetrical indirect covariance: unsymmetrical indirect covariance processing for an HSQC-TOCSY spectrum. This figure depicts the origin of legitimate peaks in unsymmetrical indirect covariance. The original matrix is divided into two matrices: positive and negative. These matrices are multiplied by each other in the second step. Diagonal peaks are not present in the result matrix because rows do not multiply by themselves during multiplication of two different matrices. Non-diagonal correlations appear twice on both sides of the diagonal as the result of multiplication of Row I_{pos} to Row J_{neg} and Row J_{pos} to Row I_{neg}.

don't have a corresponding diagonally symmetrical pair and can be easily separated from the legitimate peaks.

The unsymmetrical indirect covariance logically follows from the described method of artifact removal and historically has grown from it. Unfortunately, UIC data is not symmetric relative to the diagonal in the general case, and the described method is not applicable to all UIC spectra.

A simple method to predict the potential presence of artifacts in the UIC spectra (but not remove them) has been described.[34] The method is based on the simple suggestion that peak projection overlap in HSQC spectra is a good sign of the presence of artifacts in IC HSQC-HMBC, HSQC-COSY and other

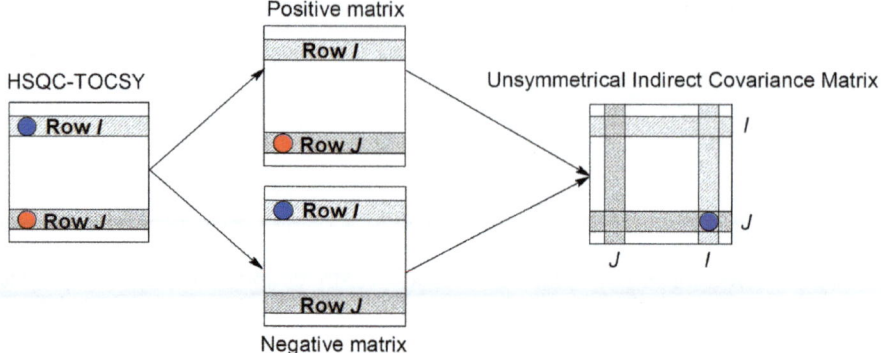

Figure 8.11   Schematic comparison between conventional and unsymmetrical indirect covariance: Type II artifacts in unsymmetrical covariance processing of an HSQC-TOCSY. This figure depicts the origin of Type II
artifacts produced by peak overlap. After unsymmetrical indirect covariance, artifact peaks do not have diagonally symmetrical pairs because
those are produced from only one pair of rows while legitimate peaks
are produced by two pairs of rows. In this case the artifact is produced
as the result of multiplication of Row J_{pos} to Row I_{neg} while multiplication of Row I_{pos} to Row J_{neg} produces zero.

forms of spectra. The main idea of the method is to calculate the indirect
covariance spectrum for the HSQC spectrum alone and then check the resultant spectrum for the presence of non-diagonal peaks. The non-diagonal
peaks in the indirect covariance HSQC are a result of peak projection overlap
along the ^{1}H (horizontal axis) and may be used to determine the areas
of potential locations of the artifact peaks. These are those areas with
carbon chemical shifts equal to the chemical shifts of non-diagonal peaks
in an IC HSQC spectrum. Unfortunately this approach is not able to
distinguish the artifact peaks in these areas. However, the absence of an
off-diagonal peak essentially guarantees the absence of artifacts in the
corresponding areas.

Another method of artifact removal has been described,[5] based on the
observation that the matrix square root procedure (matrix power = 0.5) reduces the intensity of artifact peaks more dramatically than the intensity of
"normal" peaks in the resulting covariance calculation. The square root
matrix power calculation is applicable only for a square matrix and is generally inapplicable to the product of unsymmetrical indirect covariance because this product is non-square in most cases. The authors developed
generalized indirect covariance to overcome this restriction. Figure 8.12
displays the results of conventional unsymmetrical indirect covariance and
generalized indirect covariance processing with a matrix power of 0.5 for an
artificial HMBC-TOCSY spectrum. It is clearly visible that some (but not all)
artifact peaks are removed by the matrix power operation.

The authors also showed that the intensity of the artifact peaks decreases
more rapidly than the intensity of the "real" peaks with decreasing matrix

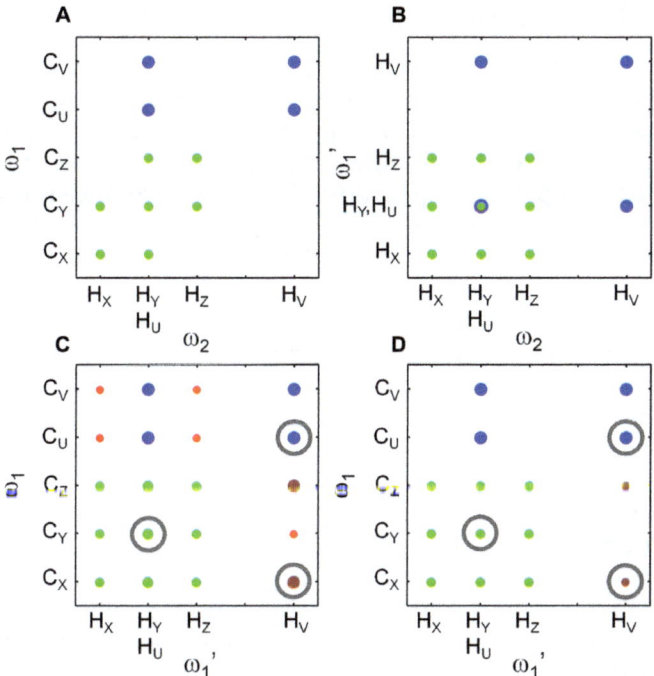

Figure 8.12 Schematic (A) HMBC, (B) TOCSY, (C) [HMBC*TOCSY],[1] and (D) [HMBC*TOCSY]$^{0.5}$ spectra for a model mixture containing one spin system with three connected ^{13}C–^{1}H pairs, X–Y–Z, and one spin system consisting of two ^{13}C–^{1}H pairs, U–V, where the carbon–carbon connectivities are among X–Y, Y–Z, and U–V. Note the degeneracy in chemical shift for the protons of ^{13}C–^{1}H pairs Y and U, which leads to false positive (red) signals in the [HMBC*TOCSY]$^{\lambda=1}$ spectrum. Application of the matrix-square root in the GIC formalism eliminates most false positives. The two most intense false positives (dark red) are not completely suppressed with decreasing λ. They can be identified as false positives because of their large slope as a function of λ.
Reprinted with permission from D. A. Snyder and R. Brüschweiler, Generalized Indirect Covariance NMR Formalism for Establishment of Multidimensional Spin Correlations, *J. Phys. Chem. A.* Copyright 2009 American Chemical Society.

power value used for the resulting covariance calculation. This observation could also be used to separate artifacts from real peaks.

There is currently no universal and robust method to avoid artifacts in covariance calculations, and moreover these artifacts are theoretically unavoidable in those cases when a structure has two proton signals with very close chemical shifts and peak shapes. In this case the "real" covariance signals and the artifacts are indistinguishable in a covariance spectrum. The presence of the artifacts is still the main reason why unsymmetrical indirect covariance spectra are not widely used for the purpose of structure elucidation.

8.3.5 Applications of Indirect Covariance NMR

The main application of indirect covariance is to the processing of HSQC-TOCSY data. As described above, the application of IC to homonuclear spectra is essentially useless. The application of IC to other heteronuclear spectra produces trivial or ambiguous results in terms of spectral interpretation. For example, the application of indirect covariance processing to a HSQC spectrum, in an ideal case with no peak overlap along F_2, produces a spectrum with only diagonal peaks. The application of IC to an HMBC spectrum[35] produces a spectrum which contains ambiguous carbon–carbon correlations (in terms of the distance between nuclei). However, the ambiguity in the covariance calculated spectrum is no different to the inherent ambiguity of an HMBC spectrum itself.

UIC and GIC have much more general applicability and can be used to produce some very convenient results that assist in structure identification, but which cannot be acquired by conventional 2D NMR. The most useful combination of spectra for UIC processing combines a "direct" CH correlation spectrum (HSQC) with a "long-range" CH or CC correlation spectrum (HMBC or ADEQUATE). Figure 8.13 displays the UIC HSQC-1,1-ADEQUATE spectrum, which contains information regarding the direct bonds between carbon atoms in a much more convenient manner for interpretation than as contained within the pair of source spectra.

The application of UIC is not limited to ^{13}C–^{1}H 2D NMR spectra. For example, combining a ^{13}C–^{1}H HSQC spectrum with a ^{15}N–^{1}H HMBC spectrum produces a spectrum containing ^{13}C–^{15}N correlations. While it has been feasible to acquire ^{13}C–^{13}C correlation spectra at natural abundance using either ADEQUATE or INADEQUATE experiments, the acquisition of ^{13}C–^{15}N spectra at natural abundance has only very recently become possible with the development of the HCNMBC[36] experiment. There is likely to be considerable potential for the HCNMBC experiment in fields such as alkaloid chemistry.

The application of UIC is also not restricted to a pair of *hetero*nuclear spectra. For example, a combination of HSQC and TOCSY[2] data produces an analog of a HSQC-TOCSY spectrum. In the case of a pair of ^{13}C–^{1}H HSQC and ^{1}H–^{1}H spectra, UIC can be applied twice using eqn (8.7):

$$C = S_{HSQC}S_{COSY}S_{HSQC}{}^{T} \tag{8.7}$$

to produce a ^{13}C–^{13}C spectrum. The combination of a H–H COSY and two C–H HSQC spectra (one spectrum used twice) produces a Doubly Indirect Covariance (DIC)[37] C–C spectrum analogous to a carbon–carbon COSY spectrum.

Doubly indirect covariance is not the only approach to obtain "short range" carbon–carbon correlations. In a series of papers, Martin and co-workers systematically combined HSQC and HMBC experiments with 1,1- and 1,*n*-ADEQUATE spectra to generate the covariance equivalent of 1,1- 1,*n*-, *n*,1- and *n*,*n*-ADEQUATE spectra.[38–42] While 1,1- and 1,*n*-ADEQUATE spectra

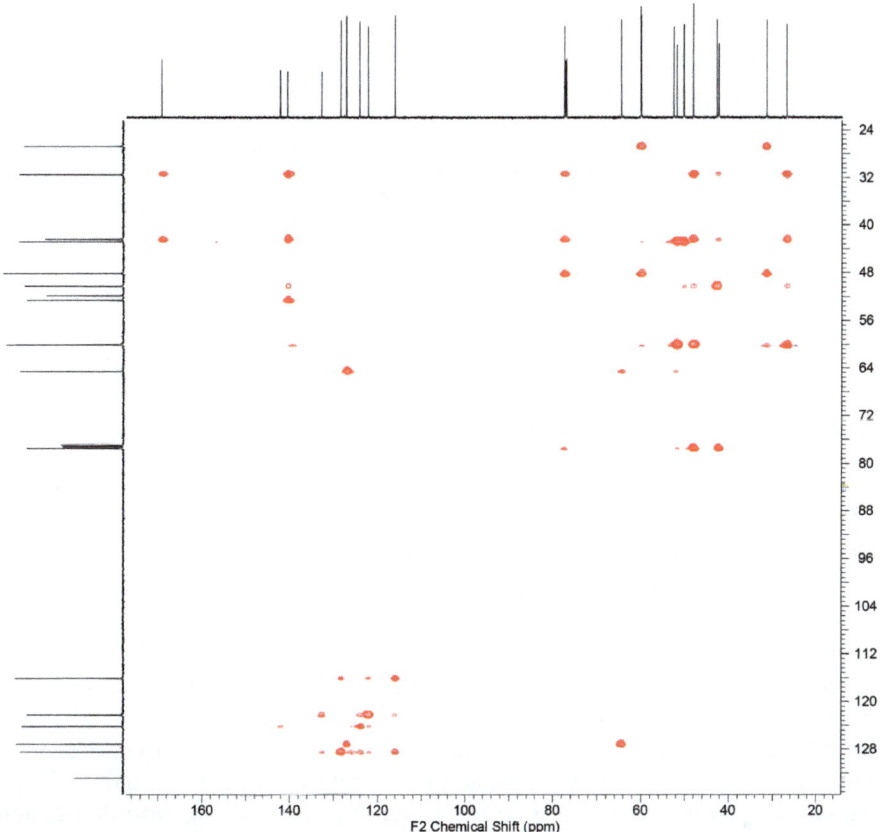

Figure 8.13 Unsymmetrical indirect covariance HSQC-1,1-ADEQUATE spectrum of strychnine.

can be readily acquired, the same cannot be said for the n,1- and n,n-ADEQUATE spectra.[34] While the 1,1- and 1,n-ADEQUATE experiments afford responses in the second frequency domain (F_1) at the intrinsic ^{13}C chemical shift, correlations are observed in the n,1- and n,n-ADEQUATE spectra at the double quantum frequency of the pair of correlated carbon resonances, rendering interpretation of the data extremely cumbersome. Furthermore, the sensitivity of the n,1- and n,n-ADEQUATE spectra is extremely low requiring significant amounts of sample even when the data are acquired with cryogenic NMR probes. In contrast, the HMBC-1,1-ADEQUATE and HMBC-1,n-ADEQUATE covariance spectra have much higher intrinsic sensitivity and correlations are observed in the F_1 frequency domain at the intrinsic ^{13}C chemical shifts of the correlated carbons.[43]

Perhaps one of the most promising recent developments in the covariance calculation of carbon–carbon correlation spectra is found in the covariance processing of an inverted $^1J_{CC}$ 1,n-ADEQUATE spectrum[17,44] with an unedited HSQC spectrum.[45] The carbon–carbon correlation spectrum that

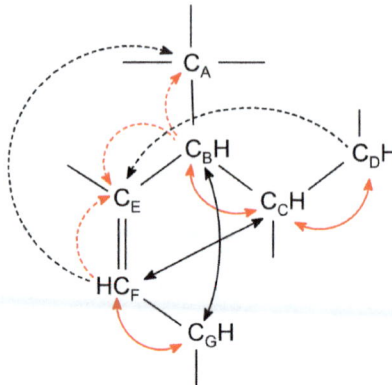

Scheme 8.1   Illustration of the four types of correlations possible in an inverted
$^{1}J_{CC}$ 1,*n*-ADEQUATE spectrum.

results from this covariance calculation can be depicted by starting from a
hypothetical structural fragment (Scheme 8.1). As illustrated by this hypothetical structure fragment, there are four types of correlations possible in
an inverted $^{1}J_{CC}$ 1,*n*-ADEQUATE spectrum.

Correlations shown in red are $^{1}J_{CC}$ correlations. They can occur either
between pairs of protonated carbons (double-headed red arrows) or between
a protonated carbon and an adjacent quaternary carbon (dashed, single-
headed red arrows). For $^{n}J_{CC}$ correlations, in a similar manner, there are
again two types of correlations possible. Correlations between pairs of $^{n}J_{CC}$
correlated protonated carbons are denoted by solid black, double-headed
arrows. Long-range carbon–carbon correlations between a protonated carbon and a non-protonated carbon are denoted by dashed, single-headed
black arrows.

Covariance processing of an inverted $^{1}J_{CC}$ 1,*n*-ADEQUATE spectrum and
an unedited HSQC spectrum will afford the result shown schematically
in Figure 8.14 for the hypothetical structural fragment shown above.
As shown in the figure, there are four different types of carbon–carbon
correlations differentiated in the covariance processed spectrum. $^{1}J_{CC}$
correlations between pairs of protonated carbons give rise to diagonally
symmetric correlation with negative phase. Long-range carbon–carbon
correlations between pairs of protonated carbons give rise to diagonally
symmetric correlations with positive phase. A $^{1}J_{CC}$ correlation between a
protonated carbon and an adjacent quaternary carbon give rise to a diagonally asymmetric correlation with negative phase. Finally a long-range
carbon–carbon correlation between a protonated carbon and a remote non-
protonated carbon give rise to a positively phased diagonally asymmetric
response.

The results obtained by covariance processing an unedited HSQC spectrum of strychnine with an inverted $^{1}J_{CC}$ 1,*n*-ADEQUATE spectrum are shown
in Figure 8.15. Figure 8.16 shows an expansion of the aliphatic region of the

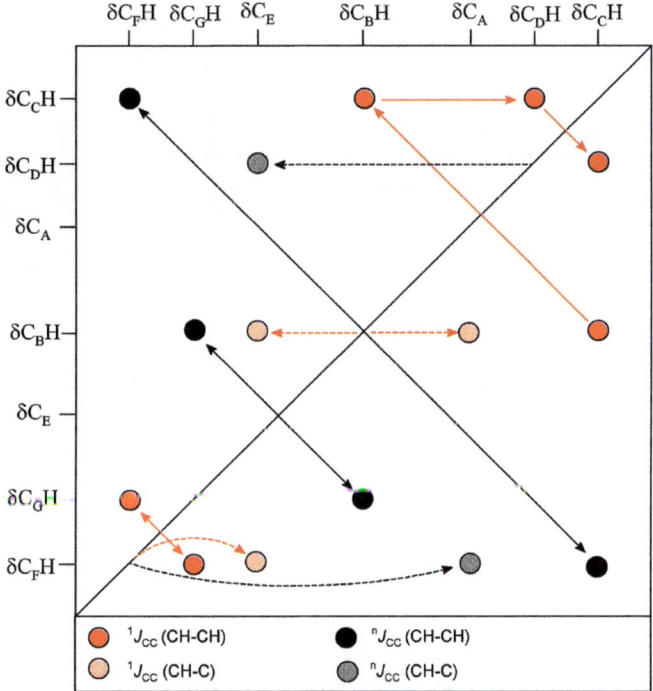

Figure 8.14 Schematic showing the four types of correlations possible in the covariance processing of an unedited HSQC spectrum with an inverted $^1J_{CC}$ 1,n-ADEQUATE spectrum.

spectrum shown in Figure 8.15 with correlations labeled as to their coupling pathways.

For the elucidation of complex structures, *e.g.* natural products, the covariance processing of an unedited HSQC spectrum with an inverted $^1J_{CC}$ 1,n-ADEQUATE spectrum offers perhaps the most promise of providing very useful structural data. The sole drawback of the method, of course, is the inherent difficultly of acquiring the inverted $^1J_{CC}$ 1,n-ADEQUATE data. It should be noted, however, that in a recent report on the application of advanced NMR techniques to the complex, proton-deficient alkaloid staurosporine, that the inverted $^1J_{CC}$ 1,n-ADEQUATE spectrum afforded all of the $^1J_{CC}$ correlations in the molecule while a conventional 1,1-ADEQUATE spectrum was missing some of the $^1J_{CC}$ correlations.[46]

8.3.6 Optimizing Spectra for Best Application to Covariance

A key advantage of covariance processing is that, although covariance processing is applicable to both uniformly (typically transformed into the frequency domain) and non-uniformly sampled (including non-deterministically sampled) data, it does not require specially acquired NMR

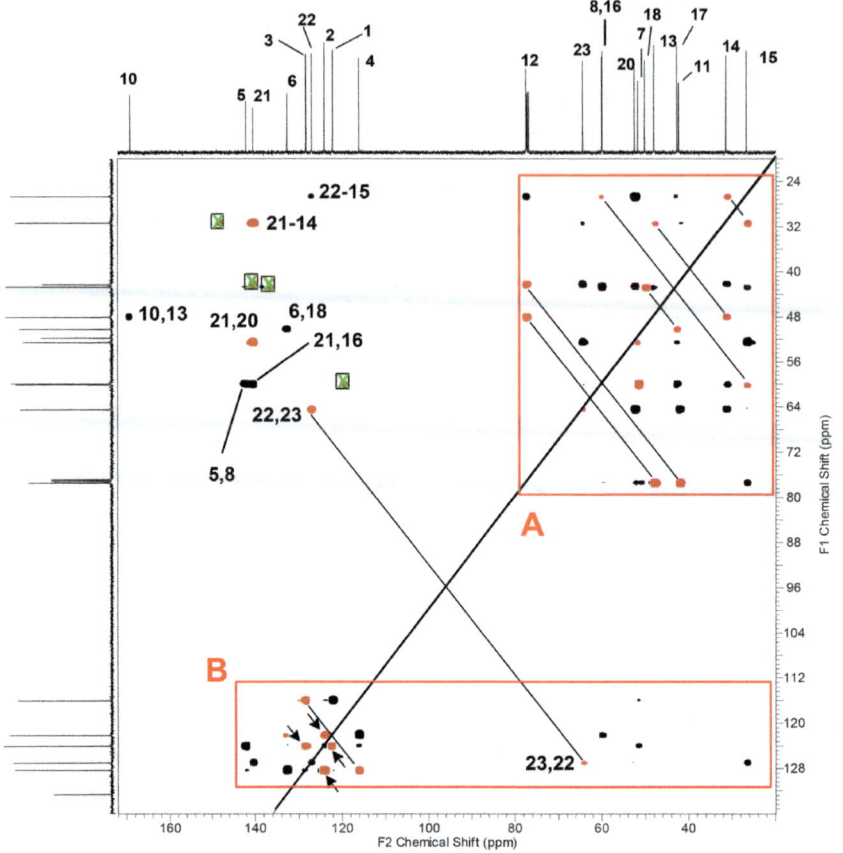

Figure 8.15 $^1J_{CC}$-Edited HSQC-1,*n*-ADEQUATE spectrum of strychnine. The spectrum was calculated using the generalized indirect covariance method from an unedited HSQC spectrum and an inverted $^1J_{CC}$ 1,*n*-ADEQUATE spectrum. It is important to note that the phase editing of this spectrum is based on $^1J_{CC}$ (negative phased, plotted in red) *vs.* $^nJ_{CC}$ (positive phase, plotted in black) carbon–carbon coupling rather than the more conventional multiplicity-editing of an HSQC spectrum based on CH_2 *vs.* CH/CH_3. The spectrum is diagonally symmetric. Correlations between pairs of protonated carbons *via* $^1J_{CC}$ coupling are diagonally symmetric, inverted, and plotted in red, *e.g.* the correlation between the 14-methine and 15-methylene carbons in the upper right corner of the spectrum. In contrast, in a more conventional HSQC-1,1-ADEQUATE spectrum calculated from multiplicity-edited HSQC and 1,1-ADEQUATE spectra, the same diagonally symmetric correlation would appear but the phase would be mixed since C14 is a methine carbon and C15 a methylene carbon. Correlations between a protonated carbon and a quaternary carbon *via* $^1J_{CC}$ are diagonally asymmetric, inverted, and plotted in red, *e.g.* the C14–C21 correlation. Correlations between a protonated carbon and a quaternary carbon *via* an $^nJ_{CC}$ coupling are diagonally asymmetric, positive, and plotted in black, *e.g.* the C6–C18 correlation.

Reprinted with permission from G. E. Martin *et al.*, $^1J_{CC}$ Edited HSQC-1,n-ADEQUATE: A New Paradigm for Simultaneous Direct and Long-Range Carbon-Carbon Correlation, *Magn. Reson. Chem.* Copyright © 2012 John Wiley & Sons, Ltd.

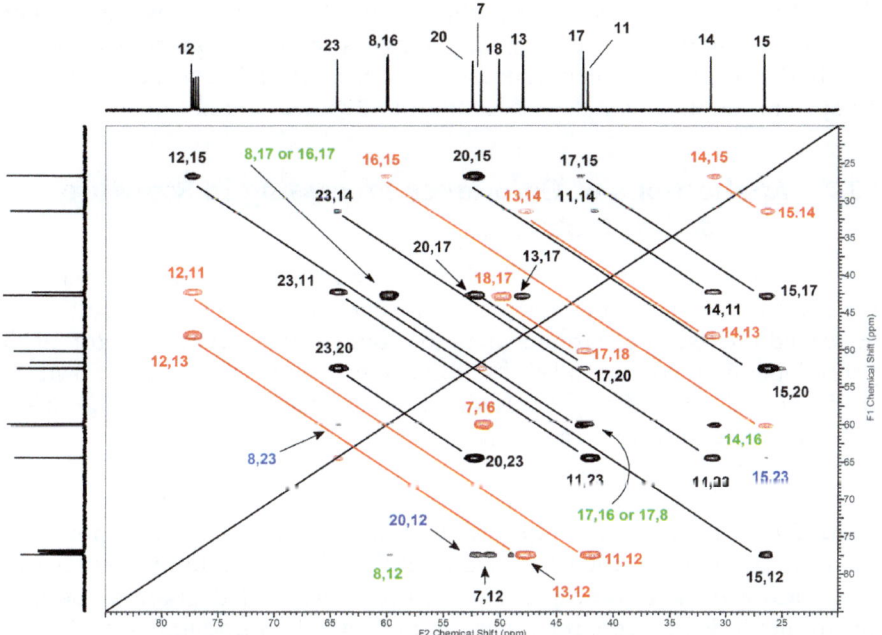

Figure 8.16 Expansion of region A of the $^1J_{CC}$ edited HSQC-1,*n*-ADEQUATE spectrum shown in Figure 8.15. Correlations are labelled according to the length of the coupling pathway. Correlations *via* $^1J_{CC}$ are inverted, plotted, and labelled in red. Long-range correlations *via* $^3J_{CC}$ are the most prevalent, and are positive, plotted and labelled in black. Correlations *via* $^2J_{CC}$ are weaker and less prevalent in the spectrum. These correlations are also positive, plotted in black, and labelled in green. Finally, correlations *via* $^4J_{CC}$ are generally quite weak, positive, plotted in black and labelled in blue.
Reprinted with permission from G. E. Martin *et al.*, $^1J_{CC}$ Edited HSQC-1,n-ADEQUATE: A New Paradigm for Simultaneous Direct and Long-Range Carbon-Carbon Correlation, *Magn. Reson. Chem.* Copyright © 2012 John Wiley & Sons, Ltd.

spectra. Nevertheless, the best covariance calculated spectra result when the initial NMR data are acquired with covariance processing in mind. Proton (F_2) spectral widths should be set identically. If, for example, an HSQC spectrum is to be processed with a 1,1-ADEQUATE spectrum, it should be kept in mind that the latter spectrum may have correlations, for example, to carbonyl resonances. In such cases, the F_1 spectral width of the HSQC spectrum should be set identically to the 1,1-ADEQUATE spectrum. While the HSQC spectrum will obviously have a significant region of the spectrum not containing any correlations, the sensitivity of the HSQC experiment is such that digitizing the "empty" space is of little consequence on acquisition time for the data. We have also found that better results are obtained when the higher sensitivity experiment is more highly digitized in the F_1 frequency domain. For example, it may be advantageous to acquire an HSQC spectrum with 512 t_1 increments, whereas perhaps 160–200 increments would be more normal. During data processing for the two-component spectra that will be used in the covariance calculation, we have also found it advisable to process the spectra identically, *e.g.* process both to 2K×1K data points.

8.3.7 Applications of Covariance Processing in Structure Elucidation Problems

To date there have been only limited examples of the use of covariance processed data in structure elucidation problems. Martin and Sunseri[47] have illustrated the use of covariance data to establish the carbon skeleton of the cyclin-dependent kinase inhibitor dinaciclib. Martin has also shown the application of covariance processed data in the establishment of the carbon skeleton of the antifungal agent posaconazole[48] (Scheme 8.2).

Those data led to the actual determination of the structure of an unprecedented degradation product of posaconazole.[49] While numerous oxidized degradants of the piperazine moiety in the structure of posaconazole have been determined,[50–53] the established degradation pathways were of no utility in determining the structure of a novel degradant. By mass spectrometry, it was apparent that the molecule had undergone the loss of the equivalent of a methyl group in the degradation process. Fragment ions in the MS/MS data localized the degradation to the "center" of the molecule. Covariance processing of an HSQC spectrum with a 1,1-ADEQUATE spectrum established the structure of the degradant as a unique benzodiazepine that involved first ring-opening of the piperazine moiety followed by ring closure involving one of the aromatic rings. Two alternative ring closures were possible as shown by the structures below (Schemes 8.3 and 8.4). From the HSQC-1,1-ADEQUATE spectrum (Figure 8.17), it was possible to assign the structure of the novel degradant as the structure shown in Scheme 8.4.

The application of covariance processing in structure elucidation is not limited to indirect covariance methods (IC, GIC, UIC and DIC); direct covariance has utility in structure elucidation beyond the resolution enhancement it provides. In particular, direct covariance converts an INADEQUATE

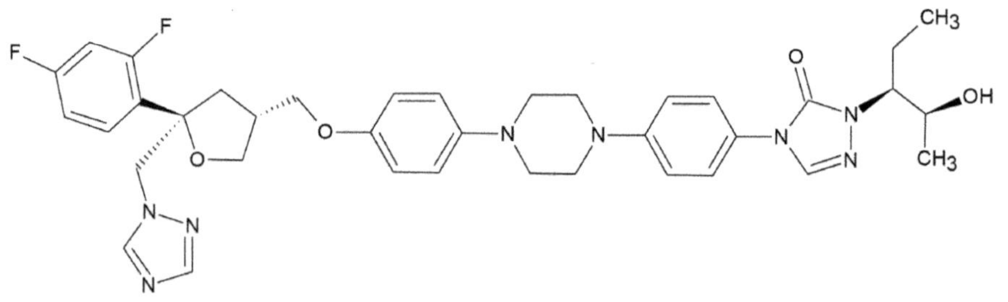

Scheme 8.2   The antifungal agent posaconazole.

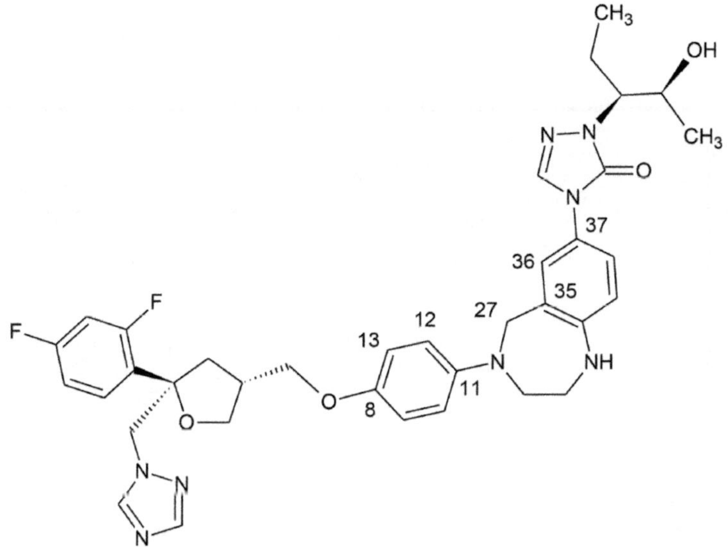

Scheme 8.3 One of the two possible degradation products of posaconazole.

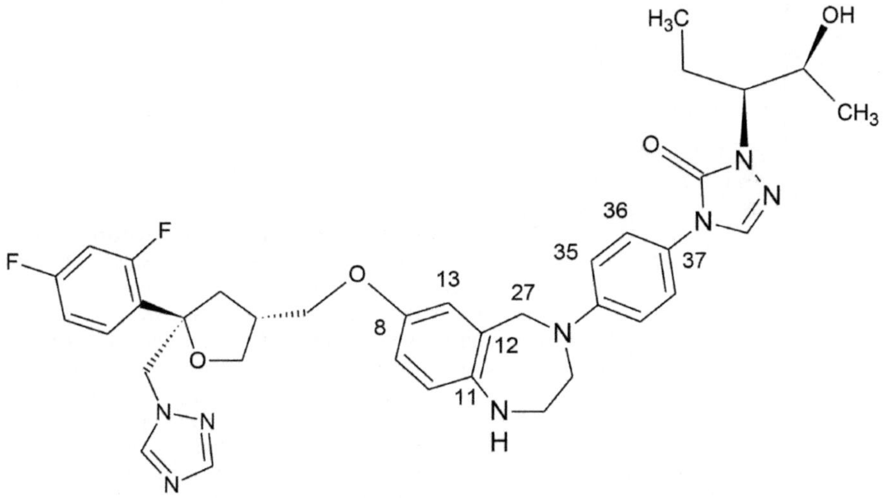

Scheme 8.4 The other possible degradation product of posaconazole. Covariance NMR identifies this structure the novel degradation product actually occurring.

spectrum from its conventional double quantum view to a diagonally symmetric view,[54] a view which is much more convenient for spectral interpretation. Figure 8.18 shows an example of applying direct covariance to an INADEQUATE spectrum.

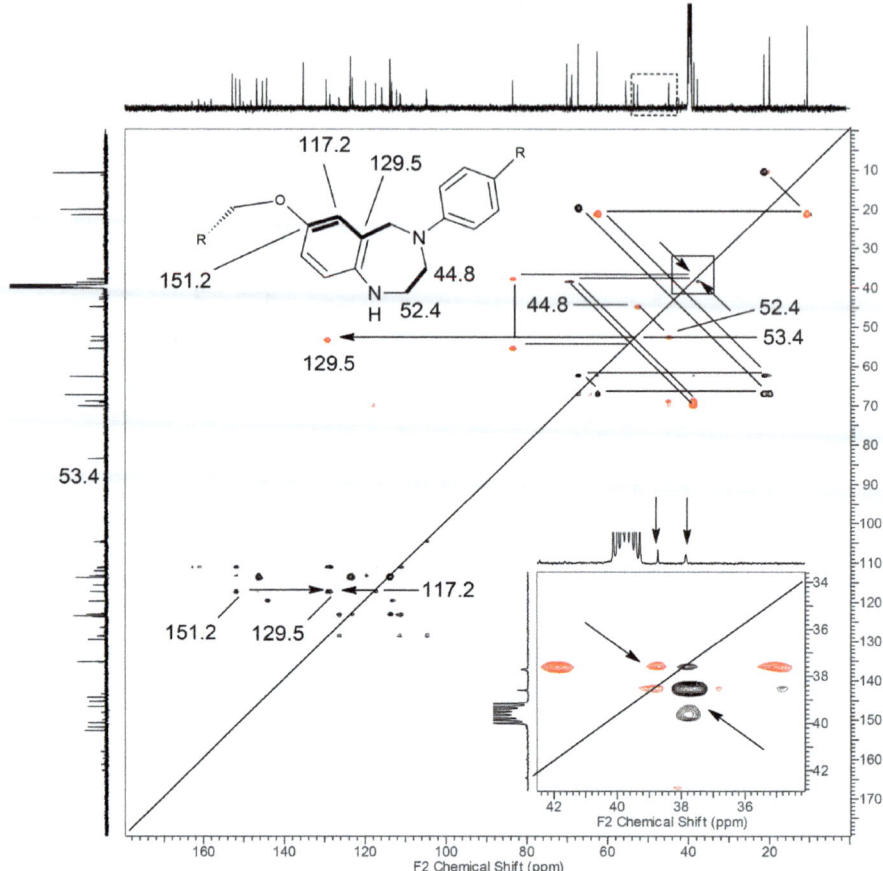

Figure 8.17 HSQC-1,1-ADEQUATE spectrum of a degradant of the antifungal agent posaconazole calculated from a multiplicity-edited HSQC spectrum and a 1,1-ADEQUATE spectrum using the generalized indirect covariance method. Correlations extracted from the spectrum are denoted by bold bonds on the inset structural segment. Correlations from the methylene carbon to the quaternary carbon resonating at 129.5 ppm and from the aromatic carbon resonating at 117.2 ppm to the quaternary carbon at 129.5 ppm defined the structure of the degradant as the benzo[1,4-]diazepine shown. (See also Scheme 8.4.) These data also provided the means of differentiating the structure of the degradant from the other structural possibility shown in Scheme 8.3. This report represents the first utilization of an HSQC-1,1-ADEQUATE in the elucidation of an unknown structure.

Reprinted from *J. Pharm. Biomed. Anal.*, **66**, W. Zhong, X. Yang, W. Tong and G. E. Martin, Structural characterization of a novel degradant of the antifungal agent posaconazole, 40, Copyright (2012), with permission from Elsevier.

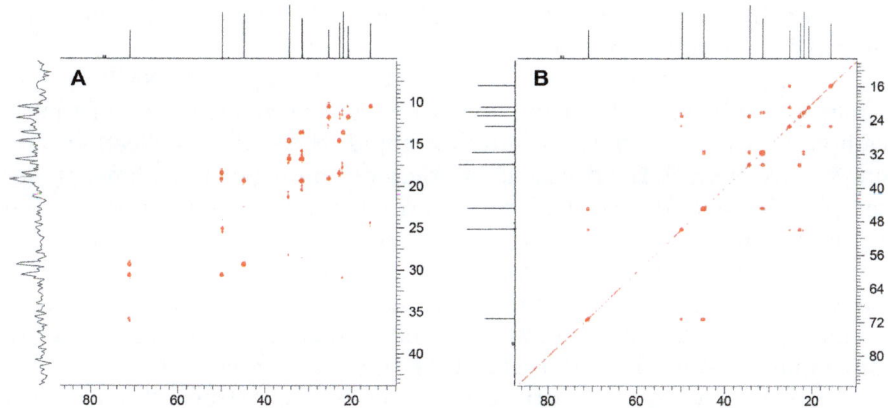

Figure 8.18 Example of application of direct covariance to INADEQUATE spectrum. (A) INADEQUATE spectrum of menthol. (B) Direct covariance INADEQUATE spectrum of menthol. The direct covariance processed spectra has a diagonal form that is more convenient for spectrum interpretation.

8.4 Related Methods

Direct covariance NMR of 2D spectra in the mixed time-frequency domain calculates the covariance matrix for a time series of 1D NMR spectra, *e.g.* a collection of 1D spectra parameterized by t_1. However, covariance analysis potentially applies to *any* collection of 1D NMR spectra. In particular, *Statistical Total Correlation Spectroscopy* (STOCSY) associates chemical shifts by correlating their associated intensities across 1D NMR spectra obtained from different samples: resonances arising from the same molecule will have correlated intensities. Thus STOCSY identifies molecules present across a variety of samples, such as metabolite samples from different tissues, individuals or disease states, and facilitates understanding of how molecule concentration varies across the analyzed samples. Similar techniques also apply when spatial coordinates parameterize the collection of NMR spectra, *i.e.* in magnetic resonance imaging (MRI).

In the analysis of spatially resolved spectra, however, non-negative matrix factorization[55] often replaces the singular value decomposition central to covariance NMR. Similarly in the analysis of TOCSY and STOCSY data recorded on mixtures, such as metabolite and natural product extracts, *Non-Negative Matrix Factorization* can extract traces representing the NMR spectra of each component of the sample.[56] Many matrix decompositions, such as non-negative matrix factorization, are actually equivalent to clustering algorithms[57] related to the clustering methods used in the analysis of TOCSY and similar NMR spectra include *NMR Consensus Trace Clustering*[58] and *DemixC*.[59] These methods are especially powerful when combined with doubly indirect covariance NMR and graph theoretic techniques (again related to clustering heuristics) for automatically tracing the carbon skeletons of unknown organic components in a complex mixture.[60]

The basis for covariance NMR is the ability to represent NMR spectra as matrices subject to the usual operations and decompositions of matrix algebra. Other techniques facilitating rapid NMR data collection and analysis also treat NMR spectra as matrices or, more generally, tensors. For example, the *multidimensional decomposition* techniques treat three and higher dimensional NMR spectra as tensors subject to standard tensor decomposition and reconstruction methods.[61] Similarly, *projection–reconstruction* applies matrix and tensor reconstruction methods pioneered in imaging and statistical analyses to the reconstruction of NMR spectra from rapidly acquired data. Techniques such as *Burrow-Owl*[62] use tensor analogues of covariance matrix calculation to reconstruct "higher-order spectra". The use of higher-order spectra obviates the need for expert-intense manual feature identification[63] by postponing peak picking until after inter-spectral correlations are computationally established, rendering correlation peaks within the resulting higher-order spectra self-evident.

8.5   Conclusions and Further Directions

The term "covariance NMR" describes a family of related techniques whose common thread is the treatment of NMR spectra as matrices subject to the usual operations of matrix algebra. By maximally leveraging the information present in even conventionally (uniformly sampled and subject to Fourier transformation) acquired NMR datasets, covariance facilitates the rapid acquisition of NMR spectra at resolutions and sensitivities for which direct experimental acquisition is otherwise infeasible. Direct covariance results in substantial resolution gains, allowing for up to 40-fold reductions in acquisition time, and also facilitates analysis of spectra such as INADEQUATE by endowing them with diagonal symmetry. Indirect covariance techniques, such as generalized indirect covariance and unsymmetrical indirect covariance, correlate relatively sensitive proton-detected spectra to reconstruct datasets whose experimental acquisition would require weeks of measurement time or isotope-labeled samples. Moreover, the matrix algebra underlying covariance NMR is closely related to the tensor algebra underlying many other useful techniques such as Burrow-Owl and multidimensional decomposition. Similarly, reconstruction of NMR datasets by correlating spectra is fundamentally similar to the peak-based correlation done by experts and used by expert systems. Thus, further analysis of the logic underlying covariance NMR techniques can facilitate the future development of expert systems for structure elucidation. In particular, as described in the following section, covariance processing complements computer-assisted structure elucidation (CASE)-based approaches, and assessment of covariance spectra can help identify the best inputs to CASE software.

8.6 Computer-assisted Structure Elucidation (CASE) and the Potential Influence of Covariance Processing

The primary applications of NMR are to either confirm a hypothetical structure or fully elucidate the molecular structure of an unknown. An ideal scenario is for a chemist to submit a sample to a walk-up NMR system and to suggest one or more hypothetical chemical structures for the purpose of "structure verification". Following the acquisition of one or more NMR spectra, the data would be passed, together with the structure(s), to a software system capable of predicting the spectra and utilizing the appropriate algorithms to verify the chemical structure by comparing the experimental and predicted pairs of spectra.[64,65] The preferred approach for structure verification is one of combined verification utilizing a triad of input data—a ^{1}H spectrum, a HSQC spectrum (multiplicity-edited preferred) and a hypothetical chemical structure.

Should spectral verification fail, then the potential to utilize complex algorithms to perform computer-assisted structure elucidation (CASE) exists. Williams first presented the concept of instrument-integrated structure elucidation in 2000[66] in relation to efforts regarding CASE systems. The last state-of-the-art review of CASE systems, by Elyashberg *et al.*[67] in 2008, provided a thorough analysis of work reported to date but progress in the capabilities of the performance of the structure elucidation software has continued unabated and a blind validation test was published recently.[68] Hardware manufacturers are now integrating automated structure elucidation software with their spectrometers, with Bruker leading the way with their complete molecular confidence (CMC-se) package.[69]

One potential and perhaps underappreciated or even unrealized advantage of covariance processing to the spectroscopist is the analogy of the information content in a covariance spectrum to the internal, reduced representation of 2D NMR data within the Structure Elucidator CASE program. Specifically, 2D correlation data input to Structure Elucidator are reduced to carbon–carbon and carbon–nitrogen correlation data. For example, the carbon–carbon connectivity information that one could derive from the covariance processing of an inverted direct response (IDR) HSQC-TOCSY spectrum is analogous to how those data would be reduced to connectivity information in the Structure Elucidator program. A similar argument can be made for the covariance processing of a ^{1}H–^{13}C HSQC spectrum with a ^{1}H–^{15}N HMBC spectrum to afford a ^{13}C–^{15}N correlation spectrum. Generally, as spectroscopists we would consider ^{1}H–^{15}N HMBC correlation data in terms of the correlation of a specific proton or protons to the nitrogen in question, rather than assimilating that information as the ^{13}C to which the proton is attached being long-range correlated to the ^{15}N resonance. Calculating the covariance spectrum eliminates the intervening step for the spectroscopist, allowing him/her to utilize the information in the same fashion that the Structure Elucidator program would use the data.

Consider the "hyphenated-ADEQUATE" spectra such as the HSQC-1,1-ADEQUATE spectrum shown in Figure 8.12. The covariance calculation establishes the direct ^{13}C–^{13}C correlation network for us, absolving us from the need to think in terms of "this CH proton is associated with a carbon at XX.X ppm which, in turn, in the 1,1-ADEQUATE spectrum is directly linked to a carbon resonating at YY.Y ppm". The covariance calculation removes a mental "processing" step. With more complicated combinations of HSQC or even HMBC and 1,*n*-ADEQUATE spectra, the mental gymnastics in considering spectral analysis become much more complicated and thus removing mental processing steps *via* covariance has further increased utility.

As stated earlier in this chapter, covariance processing does not produce any new information and only converts the data to a more convenient representation. Thus, subjecting NMR spectra to (indirect) covariance processing prior to CASE does little or nothing to improve CASE results or to reduce calculation times. However, by removing mental processing steps for scientists who may have to finely tune the selection of correlations that are input into the CASE system to perform elucidation, covariance processing complements CASE analysis.

Martin has shown, in a number of reports, that covariance processing has allowed him to identify specific long-range heteronuclear correlations that dramatically influence the elucidation process. For example, CASE methods have been successfully applied to resonance assignment in posaconazole.[70] However, very-long range heteronuclear correlation data[71] has a huge impact on CASE structure generation times[72] and the choice of experiments to use to best obtain such data oftentimes requires fine tuning. By presenting rapidly acquired heteronuclear data in terms of (experimentally difficult to acquire) heavy-atom correlations, covariance processing facilitates a scientist's ability to select appropriate datasets as inputs for CASE. Thus a key future direction for covariance processing will be to develop protocols to facilitate the use of covariance spectra in order to allow scientists, who need not even be expert spectroscopists, to select the best inputs for rapid and robust structural elucidation using CASE methods.

Finally, Parella and co-workers[73] have begun to explore the utilization of covariance processing techniques to construct "pure shift" versions of experiments for which pulse sequences have yet to be developed or which cannot be developed for physical experimental reasons using currently available methods. For example, by using covariance processing of a non-pure shift HSQC spectrum with a pure shift diagonal, it is possible *via* GIC processing to obtain a pure shift spectrum in which even the anisotropic methylene protons are collapsed to singlets rather than remaining as a doublet as in a conventional multiplicity-edited pure shift HSQC spectrum.[74] In a more recent report, Parella and co-workers[75] extended the use of covariance processing methods to produce pure shift versions of experiments such as HMBC, which currently have no experimental analog.

References

1. G. E. Martin, C. E. Hadden, D. J. Russell and M. Kramer, in *Handbook of Spectroscopy: Second, Enlarged Edition*, ed. G. Gauglitz and D. S. Moore, Wiley-VCH Verlag GmbH & Co. KGaA, Weinheim, Germany, 2014, pp. 209–268.
2. K. A. Blinov, N. I. Larin, A. J. Williams, K. A. Mills and G. E. Martin, *J. Heterocycl. Chem.*, 2006, **30**, 163.
3. F. Zhang and R. Brüschweiler, *J. Am. Chem. Soc.*, 2004, **126**, 13180.
4. R. Brüschweiler and F. Zhang, *J. Chem. Phys.*, 2004, **120**, 5253.
5. D. A. Snyder and R. Brüschweiler, *J. Phys. Chem. A*, 2009, **113**, 12898.
6. K. A. Blinov, N. I. Larin, A. J. Williams, M. Zell and G. E. Martin, *Magn. Reson. Chem.*, 2006, **44**, 107.
7. D. A. Snyder, A. Ghosh, F. Zhang, T. Szyperski and R. Brüschweiler, *J. Chem. Phys.*, 2008, **129**, 104511.
8. T. Short, L. Alzapiedi, R. Brüschweiler and D. A. Snyder, *J. Magn. Reson.*, 2011, **209**, 75.
9. ACD/Spectrus (Version 2014), Advanced Chemistry Development, Inc. Toronto, ON, Canada, 2014.
10. MNova (Version 9.0), Mestrelab Research, S.L, Santiago de Compostela, Spain, 2014.
11. R. Brüschweiler, *J. Chem. Phys.*, 2004, **121**, 409.
12. Y. Chen, F. Zhang, D. A. Snyder, Z. Gan, L. Brüschweiler-Li and R. Brüschweiler, *J. Biomol. NMR*, 2007, **38**, 73.
13. S. T. Smith, *IEEE Trans. Signal Process.*, 2005, **53**, 1610.
14. Y. Chen, F. Zhang, W. Bermel and R. Brüschweiler, *J. Am. Chem. Soc.*, 2006, **128**, 15564.
15. N. Trbovic, S. Smirnov, F. Zhang and R. Brüschweiler, *J. Magn. Reson.*, 2004, **171**, 277.
16. D. A. Snyder, Y. Xu, D. Yang and R. Brüschweiler, *J. Am. Chem. Soc.*, 2007, **129**, 14126.
17. M. Reibarkh, R. T. Williamson, G. E. Martin and W. Bermel, *J. Magn. Reson.*, 2013, **236**, 126.
18. G. E. Martin, B. D. Hilton, R. M. I. Willcott and K. A. Blinov, *Magn. Reson. Chem.*, 2011, **49**, 350.
19. M. Mobli, A. S. Stern and J. C. Hoch, *J. Magn. Reson.*, 2006, **182**, 96.
20. K. Zangger and H. Sterk, *J. Magn. Reson.*, 1997, **124**, 486.
21. P. Sakhaii, B. Haase and W. Bermel, *J. Magn. Reson.*, 2009, **199**, 192.
22. J. A. Aguilar, M. Nilsson and G. A. Morris, *Angew. Chem., Int. Ed.*, 2011, **50**, 9716.
23. L. Paudel, R. W. Adams, P. Király, J. A. Aguilar, M. Foroozandeh, M. J. Cliff, M. Nilsson, P. Sándor, J. P. Waltho and G. A. Morris, *Angew. Chem., Int. Ed.*, 2013, **52**, 11616.
24. N. H. Meyer and K. Zangger, *Angew. Chem., Int. Ed.*, 2013, **52**, 7143.
25. L. Castañar, J. Saurí, P. Nolis, A. Virgili and T. Parella, *J. Magn. Reson.*, 2014, **238**, 63.

26. I. Timári, L. Kaltschnee, A. Kolmer, R. W. Adams, M. Nilsson, C. M. Thiele, G. A. Morris and K. E. Kövér, *J. Magn. Reson.*, 2014, **239**, 130.
27. J. Ying, J. Roche and A. Bax, *J. Magn. Reson.*, 2014, **241**, 97.
28. A. S. Stern, K.-B. Li and J. C. Hoch, *J. Am. Chem. Soc.*, 2002, **124**, 1982.
29. S. G. Hyberts, S. A. Robson and G. Wagner, *J. Biomol. NMR*, 2013, **55**, 167.
30. A. H. Al-Mohy and N. J. Higham, *SIAM J. Matrix Anal. Appl.*, 2009, **30**, 1639.
31. N. J. Higham and L. Lin, *SIAM J. Matrix Anal. Appl.*, 2009, **32**, 1056.
32. N. J. Higham and L. Lin, *SIAM J. Matrix Anal. Appl.*, 2013, **34**, 1341.
33. K. A. Blinov, N. I. Larin, M. P. Kvasha, A. Moser, A. J. Williams and G. E. Martin, *Magn. Reson. Chem.*, 2005, **43**, 999.
34. G. E. Martin, B. D. Hilton, K. A. Blinov and A. J. Williams, *Magn. Reson. Chem.*, 2008, **46**, 138.
35. W. Schoefberger, V. Smrečki, D. Vikić-Topić and N. Müller, *Magn. Reson. Chem.*, 2007, **45**, 583.
36. S. Cheatham, P. Gierth, W. Bermel and Ē. Kupče, *J. Magn. Reson.*, 2014, **247**, 38.
37. F. Zhang, L. Brüschweiler-Li and R. Brüschweiler, *J. Am. Chem. Soc.*, 2010, **132**, 16922.
38. G. E. Martin, B. D. Hilton and K. A. Blinov, *Magn. Reson. Chem.*, 2011, **49**, 248.
39. G. E. Martin, B. D. Hilton, R. M. I. Willcott and K. A. Blinov, *Magn. Reson. Chem.*, 2011, **49**, 641.
40. G. E. Martin, B. D. Hilton and K. A. Blinov, *J. Nat. Prod.*, 2011, **74**, 2400.
41. G. E. Martin, B. D. Hilton, K. A. Blinov, C. G. Anklin and W. Bermel, *Magn. Reson. Chem.*, 2012, **50**, 691.
42. G. E. Martin, K. A. Blinov and T. R. Williamson, *Magn. Reson. Chem.*, 2013, **51**, 299.
43. G. E. Martin, *Annu. Rep. NMR Spectrosc.*, 2011, **74**, 215.
44. G. E. Martin, T. R. Williamson, P. G. Dormer and W. Bermel, *Magn. Reson. Chem.*, 2012, **50**, 563.
45. G. E. Martin, K. A. Blinov, M. Reibarkh and T. R. Williamson, *Magn. Reson. Chem.*, 2012, **50**, 722.
46. M. M. Senior, R. T. Williamson and G. E. Martin, *J. Nat. Prod.*, 2013, **76**, 2088.
47. G. E. Martin and D. Sunseri, *J. Pharm. Biomed. Anal.*, 2011, **55**, 895.
48. G. E. Martin, *J. Heterocycl. Chem.*, 2012, **49**, 716.
49. W. Zhong, X. Yang, W. Tong and G. E. Martin, *J. Pharm. Biomed. Anal.*, 2012, **66**, 40.
50. W. Feng, H. Liu, G. Chen, R. Malchow, F. Bennett, E. Lin, B. Pramanik and T.-M. Chan, *J. Pharm. Biomed. Anal.*, 2001, **25**, 545.
51. B. M. Warrack, A. K. Goodenough and G. Chen, *Am. Pharm. Rev.*, 2010, **13**, 20.
52. W. Zhong, J. Yang and X. Yang, *Rapid Commun. Mass Spectrom.*, 2011, **25**, 3651.

53. W. Zhong, B. D. Hilton, G. E. Martin, L. Wang and S. H. Yip, *J. Heterocycl. Chem.*, 2013, **50**, 281.

54. F. Zhang, N. Trbovic, J. Wang and R. Brüschweiler, *J. Magn. Reson.*, 2005, **174**, 219.

55. P. Sajda, S. Du, T. R. Brown, R. Stoyanova, D. C. Shungu, X. Mao and L. C. Parra, *IEEE Trans. Med. Imaging*, 2004, **23**, 1453.

56. D. A. Snyder, F. Zhang, S. L. Robinette, L. Bruschweiler-Li and R. Brüschweiler, *J. Chem. Phys.*, 2008, **128**, 052313.

57. H. Lu, Z. Fu and X. Shu, *Pattern Recogn.*, 2014, **47**, 418.

58. K. Bingol and R. Brüschweiler, *Anal. Chem.*, 2011, **83**, 7412.

59. F. Zhang and R. Brüschweiler, *Angew. Chem., Int. Ed.*, 2007, **46**, 2639.

60. K. Bingol, F. Zhang, L. Bruschweiler-Li and R. Brüschweiler, *J. Am. Chem. Soc.*, 2012, **134**, 9006.

61. V. Y. Orekhov, I. V. Ibraghimov and M. Billeter, *J. Biomol. NMR*, 2001, **20**, 49.

62. G. Benison, D. S. Berkholz and E. Barbar, *J. Magn. Reson.*, 2007, **189**, 173.

63. B. J. Harden, S. R. Nichols and D. P. Frueh, *J. Am. Chem. Soc.*, 2014, **136**, 13106.

64. S. S. Golotvin, E. Vodopianov, B. A. Lefebvre, A. J. Williams and T. D. Spitzer, *Magn. Reson. Chem.*, 2006, **44**, 524.

65. S. S. Golotvin, E. Vodopianov, R. Pol, B. A. Lefebvre, A. J. Williams, R. D. Rutkowske and T. D. Spitzer, *Magn. Reson. Chem.*, 2007, **45**, 803.

66. A. J. Williams, *Curr. Opin. Drug Discovery Dev.*, 2000, **3**, 298.

67. M. E. Elyashberg, A. J. Williams and G. E. Martin, *Prog. Nucl. Magn. Reson. Spectrosc.*, 2008, **53**, 1.

68. A. Moser, M. E. Elyashberg, A. J. Williams, K. A. Blinov and J. C. DiMartino, *J. Cheminf.*, 2012, **4**, 5.

69. CMC-se, (2014), Bruker BIospin, 2014.

70. B. D. Hilton, W. Feng and G. E. Martin, *J. Heterocycl. Chem.*, 2011, **48**, 948.

71. R. T. Williamson, A. V. Buevich, G. E. Martin and T. Parella, *J. Org. Chem.*, 2014, **79**, 3887.

72. K. A. Blinov, A. V. Buevich, R. T. Williamson and G. E. Martin, *Org. Biomol. Chem.*, 2014, **12**, 9505.

73. A. Fredi, P. Nolis, G. E. Martin and T. Parella, *J. Magn. Reson.*, 2016, **266**, 16–22.

74. R. W. Adams, Pure Shift NMR Spectroscopy, *eMagRes*, 2014, **3**, 1–15, DOI: 10.1002/9780470034590.emrstm1362.

75. A. Fredi, P. Nolis, C. Cobas and T. Parella, *J. Magn. Reson.*, 2016, **270**, 161–168.

76. M. Weingarth, P. Tekely, R. Brüschweiler and G. Bodenhausen, *Chem. Commun.*, 2010, **46**, 952–954.

CHAPTER 9

Maximum Entropy Reconstruction

MEHDI MOBLI,*[a] ALAN S. STERN[b] AND JEFFREY C. HOCH[c]

[a] Centre for Advanced Imaging, The University of Queensland, St Lucia, QLD 4072, Australia; [b] Rowland Institute at Harvard, 100 Edwin H. Land Blvd., Cambridge, MA 02142, USA, Email: stern@rowland.harvard.edu; [c] University of Connecticut Health Center, 263 Farmington Avenue, Farmington, CT 06030-3305, USA, Email: hoch@uchc.edu.au
*Email: m.mobli@uq.edu.au

9.1 Introduction

The relatively small magnitude of the nuclear magnetic moment leads to very weak polarisation in attainable magnetic fields, making NMR spectroscopy an inherently insensitive method. The introduction of pulsed NMR experiments by Richard Ernst revolutionised NMR spectroscopy, not only in ushering in the era of multidimensional experiments but also by improving the attainable sensitivity of NMR per unit time.[1,2] In pulsed NMR, an RF pulse excites all nuclei in a given bandwidth simultaneously, and as the nuclear magnetic moments return to equilibrium (through various relaxation mechanisms) the induction caused by the transverse component of the magnetic vectors is recorded as a discrete time series called a free induction decay (FID). The FID is then transformed using the discrete Fourier transform (DFT) to produce a frequency spectrum.

In the early applications of the DFT it was clear that more information could be extracted from the data if the FID was manipulated with an

New Developments in NMR No. 11
Fast NMR Data Acquisition: Beyond the Fourier Transform
Edited by Mehdi Mobli and Jeffrey C. Hoch
© The Royal Society of Chemistry 2017
Published by the Royal Society of Chemistry, www.rsc.org

apodisation function prior to DFT.[2] By multiplying the FID with an exponentially decaying function it was possible to improve the signal-to-noise ratio (SNR), albeit with a loss of resolution. Conversely, an exponentially increasing function was able to improve resolution but at the cost of decreased sensitivity. These limitations of DFT processing motivated a search for non-Fourier methods that do not exhibit this sensitivity–resolution trade-off. The earliest application of maximum entropy (MaxEnt) to NMR (in 1984) sought to increase the resolution of the NMR spectrum without introducing any artifacts or enhancing the noise in the spectrum.[3] The benefits of MaxEnt spectral reconstruction as an alternative to the DFT were particularly evident in cases where the signal had not decayed sufficiently during the acquisition time, and where an apodisation function would be required prior to zerofilling to avoid truncation artifacts that manifest as characteristic "sinc wiggles" owing to the step function at the tail of the signal and the start of the zerofill (Figure 9.1).

Much of the early work on MaxEnt reconstruction involved the evolution of the regularisation function that is central to the method, which in early implementations assumed all the signals to be positive absorption lines, *i.e.* that no phase correction was necessary. This assumption caused problems, in particular for multidimensional data where the phase modulation in the interferogram could not be reconstructed using MaxEnt. After some early attempts to overcome this issue by splitting the signal in different phase "channels",[4] an elegant solution was offered by Daniell and Hore, where MaxEnt was used to reconstruct the transverse magnetisation vectors (M_x and M_y) rather than the frequency spectrum.[5] In the same work, the authors derived an entropy functional that was suitable for spin 1/2 nuclei and proposed an efficient search algorithm, based on that earlier proposed by Skilling and Bryan, for computing the complex spectrum (see Section 9.2 for details). Indeed adaptations of these algorithms have been implemented in the majority of maximum entropy reconstruction software currently available (including the Rowland NMR Toolkit implementation used to generate the data in the illustrations of this chapter).

9.2 Theory

MaxEnt reconstruction finds the spectrum that maximizes the entropy while maintaining consistency with the measured data. The use of entropy as a measure of missing information originated with Shannon and is the foundation for information theory. The Shannon entropy is described by:[6]

$$S = -\sum_{i=0}^{N-1} p_i \log p_i \qquad (9.1)$$

where p_i is the ith element of a probability distribution describing the message or image, and S is a measure of the information capacity needed to transmit the message or image through a communication channel. Consistency with the measured data is defined by the condition:

$$C(\mathbf{f}, \mathbf{d}) \leq C_0 \tag{9.2}$$

where $C(\mathbf{f}, \mathbf{d})$ is the unweighted χ-squared statistic:

$$C(\mathbf{f}, \mathbf{d}) = \sum_{i=0}^{M-1} |m_i - d_i|^2 = \sum_{i=0}^{M-1} |\mathbf{k}_i \cdot \mathrm{iDFT}(\mathbf{f})_i - d_i|^2 \tag{9.3}$$

and C_0 is an estimate of the noise level; iDFT is the inverse DFT, and $\mathbf{m}$ is a "mock data" vector given by iDFT($\mathbf{f}$), $\mathbf{k}$ is a kernel to be deconvolved from the signal ($\mathbf{k}$ can be set to 1 everywhere if no deconvolution is desired. The constrained optimization problem is converted to an unconstrained optimization through introduction of a new objective function:

$$Q(\mathbf{f}, \mathbf{d}) = S(\mathbf{f}) - \lambda C(\mathbf{f}, \mathbf{d}) \tag{9.4}$$

where the value of the Lagrange multiplier λ is adjusted to obtain $C = C_0$.

The entropy defined by eqn (9.1) is not suitable for complex or hyper-complex data that occur in NMR. An entropy functional suitable for phase-sensitive NMR data is:[7,8]

$$S(\mathbf{f}) = -\sum_{n=0}^{N-1} \frac{|f_n|}{\mathrm{def}} \log\left(\frac{|f_n|/\mathrm{def} + \sqrt{4 + |f_n|^2/\mathrm{def}^2}}{2}\right) - \sqrt{4 + |f_n|^2/\mathrm{def}^2} \tag{9.5}$$

The parameter def is related to the sensitivity of the NMR spectrometer.[7] The constraint $C(\mathbf{f}, \mathbf{d})$ and the entropy $S(\mathbf{f})$, and thus the objective function $Q(\mathbf{f}, \mathbf{d})$, readily generalize to multiple dimensions.

In the most general case, MaxEnt reconstruction does not have an analytic solution, necessitating iterative numerical approaches. Because the entropy and constraint are both convex, in principle it should be possible to compute the MaxEnt reconstruction using methods such as steepest descent or conjugate gradients. In practice, however, these methods suffer from premature convergence, in which the step size shrinks to zero before the optimum is reached. The seminal development of the "Cambridge" algorithm by Skilling and Bryan,[9] which is both robust and highly efficient, effectively launched the modern application of the maximum entropy principle in NMR. In addition to searching within a subspace defined by gradients and second derivatives of the entropy and constraint, their algorithm incorporates concepts derived from variable metric methods to avoid premature convergence. Extensions to the Cambridge algorithm have provided additional performance gains and adapted it to the requirements of phase-sensitive NMR data.[10]

A schematic diagram for MaxEnt reconstruction is shown in Figure 9.1. The algorithm begins with a trial spectrum equal to zero everywhere. At each

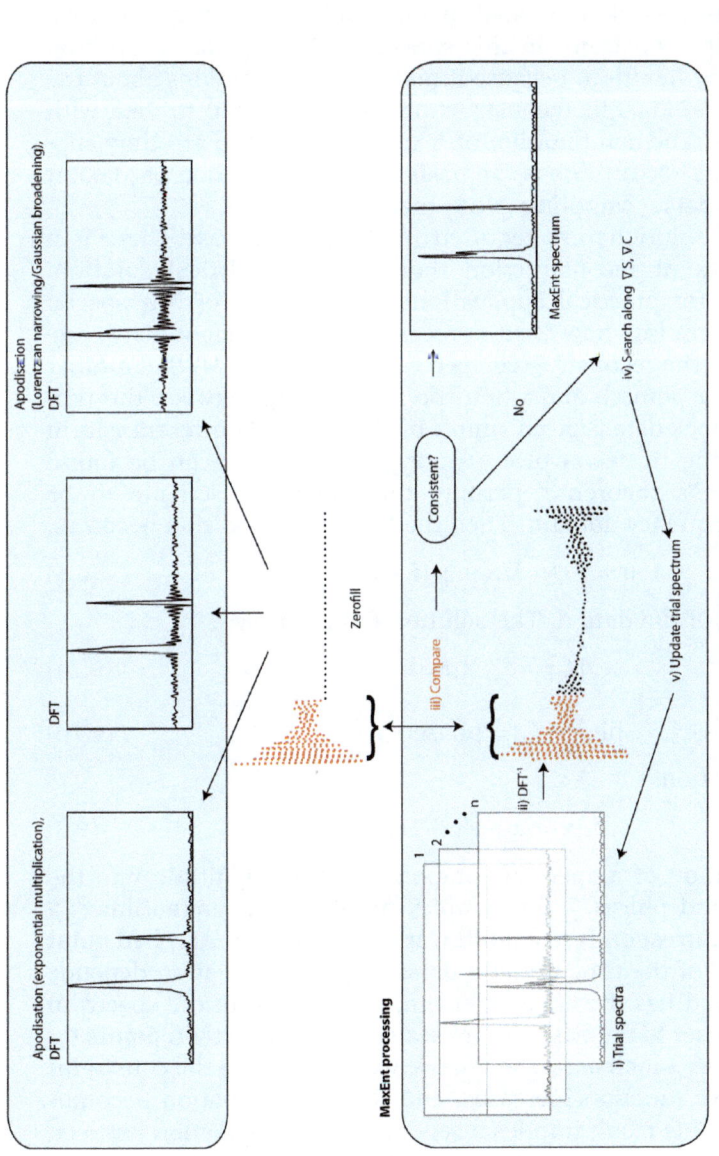

Figure 9.1 Schematic of the MaxEnt procedure and comparison with traditional DFT processing. The raw data consists of 64 complex points containing three exponentially decaying sinusoids representing Lorentzian lines one having a relative intensity of 0.5 and a linewidth of 2 Hz (central signal) and two having linewidths of 30 Hz (overlapped signals) with relative intensities of 0.7 and 1 (right to left). Gaussian noise is also added to the FID with a relative standard deviation of 0.05. **The central panel shows** the raw data with zerofilling to 256 complex points. The top panel marked **Traditional processing** shows, **left:** apodisation of the FID with an exponentially decaying function (Lorentzian broadening) followed by zerofilling to 256 complex points prior to DFT, **middle:** zerofilling to 256 complex points followed by DFT, **right:** apodisation of the FID using Lorentzian line-sharpening of 20 Hz and Gaussian line-broadening of 10 Hz followed by zerofilling to 256 complex points prior to DFT. The panel on the bottom labelled **MaxEnt processing** shows the inverse and iterative nature of the MaxEnt procedure; the sequential steps of the algorithm are indicated by ascending roman numerals (i–iv). Note that in the comparison of the DFT^{-1} of the trial spectrum to the FID only the first 64 points are compared (*i.e.* the measured data).

iteration, mock data is computed from the current value of the trial spectrum. The value of C is computed using eqn (9.3). The algorithm constructs a small set of direction vectors and computes a quadratic approximation to the entropy in the subspace spanned by these vectors. Since the constraint is itself quadratic, it is possible to analytically maximize the entropy approximation subject to the constraint in this subspace. This results in the trial spectrum for the next iteration. Because it makes no assumptions about the nature of the signals, MaxEnt reconstruction can be applied to data with arbitrary lineshapes. The computation of C can be limited to arbitrary subsets of the mock data vector; this is the basis for the application of MaxEnt with nonuniform (sparse) sampling (NUS) methods.

While numerical solutions are required in the general case, there is a special case of MaxEnt reconstruction that has an analytical solution. Though unrealistic for practical applications, examination of this special case gives some insight into how MaxEnt reconstruction works. When N (the number of points in the reconstructed spectrum) is equal to M (the number of experimental data points), and when the relationship between the trial spectrum and the mock data is given simply by the inverse Fourier transform (*i.e.* we are not trying to deconvolve a decay), the solution can be found analytically. Parseval's theorem[10] permits the constraint statistic to be computed in the frequency domain. Then the Lagrange condition becomes:

$$0 = \nabla Q = \nabla S - 2\lambda(\mathbf{f} - \mathbf{F}) \tag{9.6}$$

where $\mathbf{F}$ is the DFT of the data $\mathbf{d}$. The solution $\mathbf{f}$ is given by:

$$|f_n| = \delta_\lambda^{-1}\{|F_n|\} \tag{9.7a}$$

$$\text{phase}(f_n) = \text{phase}(F_n) \tag{9.7b}$$

where δ_λ is the function:

$$\delta_\lambda(x) = x - s'(x)/2\lambda, \tag{9.8}$$

$s(x)$ is the contribution of a spectral component with magnitude x to the overall entropy S, and phase(z), for complex numbers z is arctan(imag(z)/real(z)). This result corresponds to a nonlinear transformation, applied point by point to the DFT of the time domain data. The transformation depends on the value of λ, and has the effect of scaling every point in the spectrum down, but points closer to the baseline are scaled down more than points far above the baseline. As λ increases, the relative weight given to the constraint term in the objective function increases, and the transformation becomes more nearly linear. This result implies a very important distinction between signal-to-noise ratio (S/N) and sensitivity. Sensitivity is the ability to distinguish signal from noise. Applying the same transformation to both the signal and the noise cannot improve this ability, since peaks that are comparable in height to the noise level will be reduced by the same amount as the noise. The ratio between the highest signal peaks and the noise may increase, but small peaks will be just as difficult to distinguish as before. For linear operations (such as apodization), an improvement in S/N implies an improvement in sensitivity; for nonlinear operations (such as MaxEnt reconstruction) this is not so. In this special case, gains in S/N in the MaxEnt reconstruction are purely cosmetic. In the more general case, there may be real sensitivity gains.[11,12] It is, however, particularly relevant for NUS applications to question whether gains in S/N really correspond to gains in sensitivity.[13]

The nonlinearity of MaxEnt reconstruction is an inherent characteristic, and is responsible for the ability to achieve noise suppression without sacrificing resolution (in contrast to the apodized DFT). This nonlinearity has important implications whenever quantification of peak intensities or volumes is required, such as nuclear Overhauser effect measurements or difference spectroscopy. For difference spectroscopy, it is sufficient to compute the difference using the time domain data and compute the MaxEnt reconstruction of the difference. This assures that the same nonlinearity applies to both experiments. When measuring nuclear Overhauser effects, or otherwise quantifying peak intensities, there are two possible approaches. One is to tightly constrain the MaxEnt reconstruction to match the data, which forces the reconstruction to be nearly linear (although at the expense of noise suppression).[14,15] Another is to inject synthetic signals of known intensity into the time domain data prior to reconstruction. A calibration curve can then be constructed by comparing measured intensities or volumes to the known amplitudes of the signals.[16] In the following section we look at the effect of parameter selection on the linearity of the MaxEnt reconstruction in more detail.

9.3   Parameter Selection

In principle, multidimensional MaxEnt spectra can be reconstructed by computing the overall entropy of the full spectrum or by partitioning the reconstruction into a series of reconstructions computed for lower-dimensional subspectra. For example, a 2D MaxEnt spectrum can be computed *via* a series of 1D MaxEnt reconstructions in f_1 following DFT processing of f_2. If the constraint, C_0, is kept constant between rows then the weighting (λ) of the constraint relative to the entropy will vary. This will introduce small changes in the reconstruction between rows and can distort peak shapes. By using a constant value for λ, one can eliminate the variation of the nonlinearity between rows. Furthermore, the final result will be the same as if the full spectrum had been processed all at once using this value of λ. A good estimate of λ can be made by finding representative rows where the constraint statistic $C(\mathbf{f}) = C_0$ is satisfied and using the value of λ found for these rows to perform the complete reconstruction. The same strategy can be applied to higher dimensions, *i.e.* a 3D spectrum can be constructed as a series of 2D plane reconstructions. The approach of using a fixed value of λ, rather than a fixed value of C_0, is called the constant-λ algorithm.[17] An advantage of this approach is that the memory requirements for intermediate storage are significantly smaller than performing a full-dimensional MaxEnt reconstruction, leading to more efficient computation.

While the formal derivation of the MaxEnt algorithm specifies criteria for determining the appropriate values of def and C_0 (or λ), applying those criteria in practice is challenging and finding optimal values even more so. Fortunately the results of MaxEnt reconstruction are not terribly dependent on the precise values of the parameters, over a wide range. Using empirical rules of thumb, Mobli *et al.* described an automated procedure for determining values for the adjustable parameters.[18,19] While they cannot be said to be optimal, they have proven to be useful estimates for most applications. The key to the approach is to evaluate the noise level of the acquired data *in situ* in order to estimate the value of C_0. Using Parseval's theorem it is then possible to estimate the frequency domain def parameter. In practice, this approach tends to result in a very conservative estimate of def that yields faster and more linear (but noisier) reconstructions. Alternatively, the value of def can be systematically reduced to produce smoother reconstructions. This procedure can also be implemented using the constant-λ algorithm. In this approach, in addition to estimating the C_0 and def parameters, 10 multidimensional cross sections taken orthogonal to the direct dimension containing the largest signal components are each extracted and processed. The resulting λ values are averaged to produce a single value, which, along with the estimated def, are used to reconstruct the complete spectrum.

9.4   Linearity of MaxEnt Reconstruction

MaxEnt reconstruction manipulates the frequency-domain trial spectrum to maximize the entropy functional. The extent to which the algorithm allows the reconstructed FID to deviate from the measured FID is constrained by the unweighted χ-squared statistic (C_0, computed in the time domain). This procedure ensures that the mock FID corresponding to the output spectrum does not deviate from the measured FID beyond the uncertainty (noise) in the data.

Where the output spectrum is the same size as the FID, *i.e.* where all of the points in the mock FID are constrained against the FID, setting C_0 to a very small value (tightly constrained to match the experimental FID) the result will be very similar to the DFT of the data and no gains in spectral quality can be achieved by MaxEnt compared to DFT. If the data is undersampled, however, either by recording a truncated FID (where zerofilling is applied) or by non-uniform sampling (NUS), the MaxEnt procedure will effectively "fill in" the missing data points, be they at the end of the FID or between the acquired data samples, or both. Since these data points cannot be compared to any measured data, they are not constrained by the χ-squared statistic. Under these circumstances, making C_0 small allows significant improvements in the spectral quality without deviating from the measured data, leading to nearly linear reconstructions (see also above and Figure 9.2). Since there is a relationship between the value of C_0 and λ (reducing C_0 leads to an increase of the value of λ), it is possible to use the constant-λ algorithm to obtain nearly linear reconstructions by setting λ to a large value. It should be noted that the resulting reconstruction becomes computationally more demanding as the value of λ increases, to the point where it may become computationally unfeasible. Practically, an optimal value can be found by slowly increasing the value of λ until there is no significant change in the peak intensities while the reconstruction reaches convergence within a given timeframe.

Since using a high λ value results in a spectrum where the input data is essentially exactly matched and the entropy is maximised by adjusting the "missing" data values only, the approach is referred to as maximum entropy interpolation (MINT).[15] The approach is very similar to the Forward MaxEnt approach,[14] with the distinction that it does not alter the underlying Cambridge algorithm, and hence it retains the desirable convergence properties of the original method.[7,9]

9.5 Non-uniform Sampling

In the example provided in Figure 9.2, the signal is sampled uniformly and the truncated data is extended by MaxEnt. In this sense, what is achieved is similar to the application of linear prediction (LP) extrapolation (see Chapter 3 for details). The results, however, are distinctly different and detailed comparisons have shown that MaxEnt produces more accurate frequencies, fewer false positives, and higher resolution and sensitivity.[20] This comparison also revealed that further improvements could be achieved if the data were sampled non-uniformly. Indeed, the MaxEnt reconstruction of NUS data had been demonstrated much earlier and was shown to have great potential for speeding up NMR measurements.[21]

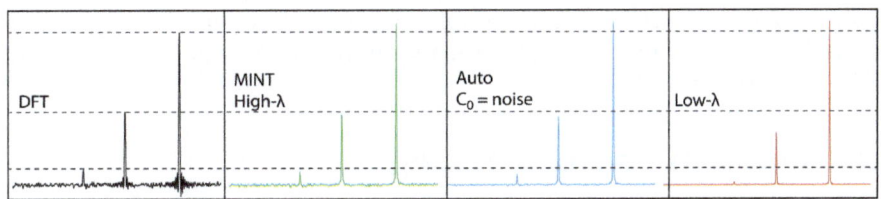

Figure 9.2 Synthetic data (128 points) processed using zerofilling (to 512 points) followed by DFT or by MaxEnt reconstruction (to 512 points) employing different parameter selection approaches. The MINT approach shows the highest linearity whilst effectively removing truncation artifacts and reducing the noise level. The Auto approach is less linear as the data is allowed to differ from the measured values by a level consistent with the uncertainty in the measurement introduced by the noise. Finally, using a low value for λ allows the data to differ from the measured values far more than is justified by the noise level in the FID. As the spectrum becomes more non-linear, the truncation artifacts and the noise are reduced.

Curiously, although the advantages of NUS were clear for many decades it wasn't until the introduction of back-projection reconstruction (BPR) that the NUS field gathered momentum in NMR spectroscopy (see also Chapters 5 and 6).[22] BPR demonstrated that by coupling the time-increments of two or more indirect dimensions it is possible to acquire radial cross-sections of the multidimensional object, and that these could be used to reconstruct the multidimensional spectrum using the inverse Radon transform.[22] However, it was soon demonstrated that sampling along radial projections is a special case of NUS, and that the deterministic nature of radial sampling results in significant artifacts compared to random sampling.[23] These developments have led to significant efforts in identifying and evaluating novel sparse sampling methods.[24–31]

A key tool in efforts to identify novel sampling methods is the point-spread-function (PSF) associated with the sampling function used. The PSF is the spectrum of the sampling function; the sampling function is a vector corresponding to the uniform sampling intervals spanning the measured FID, with values corresponding to sampled data set to one and values corresponding to data not sampled set to zero. The DFT of this function is the PSF and reveals the artifacts that will be convolved with the experimental data as a consequence of the incomplete sampling. For the case of uniform sampling (extended by zerofilling or through spectrum reconstruction), the sampling function is simply a box function, but for NUS the function takes a more complicated form. Considering zerofilling as a special case of NUS it is clear that the effect of the MaxEnt operation is to deconvolve the PSF from the spectrum of the under sampled signal (Figure 9.3).

The merits of various sampling approaches have been reviewed extensively,[32,33] and just as there are a myriad of window functions with different adjustable parameters, so it is likely to be with NUS, where the choice of the sampling method and the value of the different adjustable parameters will be dictated by the particular experimental setup, including sample-dependent parameters, such as sample concentration and stability, as well as the information being sought, *i.e.* requirement for high sensitivity, linearity or resolution.

9.6   Random Phase Sampling

As noted in the introduction of this chapter the NMR signal represents the transverse component of the magnetisation vector (M_x and M_y). If the detector only measures the Y component of this signal it will be impossible to discriminate the sign of the signal (clockwise or counter-clockwise rotation). If, however, both the X and Y components are detected, the sign of the frequency can be discriminated. The two components are the *cosine* and *sine* (or real and imaginary) components of the complex signal that is used to generate the frequency spectrum. Recording both components is referred to as "quadrature detection"; it can be achieved by use of two detectors, or by placing the receiver frequency at one extreme of the spectrum and

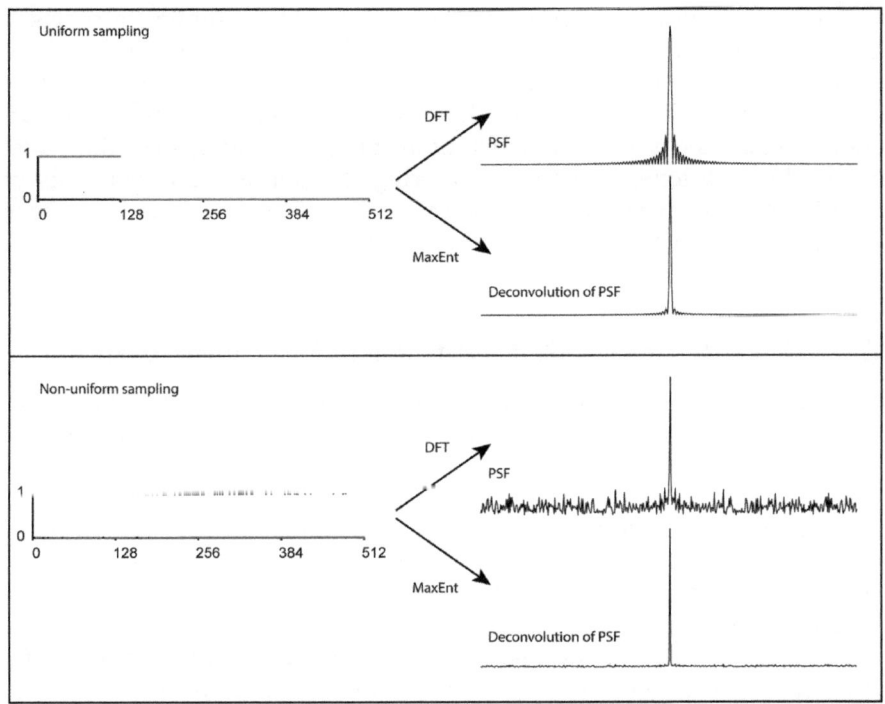

Figure 9.3 The DFT of the sampling function (containing 0 where points are omitted, and 1 where data is sampled) results in a PSF. The non-zero (outside central signal) components of the PSF are convoluted with all signals in any spectrum generated using the corresponding sampling function. The top panel shows that the DFT of a truncated signal leads to prominent artifacts about the central component, commonly referred to as sinc-wiggles. Where the data are sampled randomly the (bottom panel) the artifacts are distributed across the spectral window. In both cases MaxEnt reconstruction leads to the deconvolution of the PSF outside the central component.

oversampling the signal. In both instances, the sampling requirement is increased by a factor of two. This is not a significant burden in the acquisition dimension where the signal is recorded in real time. However, for the indirect dimensions of a multidimensional NMR experiment the real and imaginary components must be recorded as separate experiments, effectively doubling the experiment time for a two-dimensional experiment. An additional factor of two is introduced for each added dimension, increasing the total experiment time by a factor of four and eight for three- and four-dimensional experiments, respectively.

The phase components can be considered as an additional sampling dimension and by randomly sampling the phase components of a signal it is possible to use MaxEnt reconstruction to recover the correct phase of the signal and discriminate the signs of frequency components.[34,35] Indeed, the added phase dimension allows further randomization of the sampling distribution, resulting in reduced coherence of the sampling artifacts.[36] This is analogous to what had been observed for deterministic *versus* non-deterministic sampling when comparing random sampling with sampling along projections, *i.e.*, increasing the randomness of the sampling distribution reduced the associated coherent artifacts.[23,37] There does, therefore, seem to be additional benefit to exploring the phase sub-dimensions of multidimensional experiments. However, the application of random phase detection has yet to be demonstrated broadly and is likely to be the object of future investigations.

9.7   MaxEnt Reconstruction and Deconvolution

The inverse nature of MaxEnt reconstruction provides some additional advantages in that it enables stable deconvolution.[38] The convolution theorem states that the multiplication of two functions in the time domain corresponds to their convolution in the frequency domain. Thus, typically the procedure for deconvolving a spectral feature such as linewidth or *J*-modulation requires the point-by-point division of the FID by the function that describes the feature being deconvolved. This can lead to instabilities where the function approaches zero. The inverse nature of maximum entropy allows for deconvolution [see eqn (9.3)] where the function *multiplies* the inverse DFT of the mock spectrum, prior to comparison with the experimental data (see also Figure 9.1), thus avoiding divide-by-zero instabilities or noise amplification.

9.7.1   *J*-coupling

Spin–spin couplings (also called *J*-couplings or scalar couplings) between two covalently bound nuclei results in the splitting of the nuclear resonance. For all but the simplest molecules these splittings lead to increased complexity of the analysis, thus, there have been numerous methods proposed for removing these splittings. The removal of the splittings has the added benefit of collapsing the signal to a single component containing the intensity of the entire multiplet. In practice, decoupling of heteronuclear spins is most effectively achieved through broadband decoupling, where the nucleus not being observed is irradiated with a broadband pulse, hence effectively removing its contribution to the coupling network of the observed nuclei. However, where the coupled nuclei are of the same type and therefore have similar nuclear frequencies, the problem becomes more complex as the decoupling pulse cannot be applied to one nucleus without irradiating the other. Indeed, removing homonuclear couplings is an active area of research and several experimental approaches have been developed that achieve this with varied levels of success.[39] From a signal processing perspective, deconvolution can also achieve decoupling. However, as the deconvolution kernel is applied to all signals it would require that all of the observed nuclei contain the same splitting. This is rarely the case, and in the general case where multiple different splitting patterns are present for different nuclei in the same spectrum a very complex computational approach of multichannel decoupling[40] would be required that is not feasible in practice. There are, however, scenarios where a constant spin–spin coupling must be decoupled and where broadband decoupling cannot be applied. This is a general problem for ^{13}C detection of multidimensional NMR experiments applied to proteins in both solution and solid states. Typically the signal detected contains one or two one-bond couplings of a relatively fixed value. Often one of the two *J*-couplings can be removed through selective excitation, leaving the second coupling active. In these cases stable deconvolution through MaxEnt reconstruction has proven to be a very useful approach in removing the remaining splitting.[41,42]

9.7.2 Linewidths

In the case of linewidth deconvolution the situation is far less complex. Indeed, the procedure has a very similar effect on the resulting spectrum as application of a window function, with the distinct difference that it does not amplify the noise. Thus, it is neither crucial to know the exact linewidth of all the lines in the spectrum nor are the lines required to have the same linewidth (see Figure 9.4). The procedure has a similar effect on the line intensity as *J*-deconvolution, in that reducing the linewidth leads to an increase in the signal intensity. In both cases, the apparent gain in sensitivity is real and is a consequence of utilising *a priori* knowledge in the processing step. Although clearly a powerful method for improving spectral quality, linewidth deconvolution has not been explored in NMR applications[43] and remains an area for future exploration.

9.8 Perspective and Future Applications

The inherent low sensitivity of NMR coupled with the limitations of the DFT have motivated spectroscopists to explore alternative methods for signal processing. MaxEnt reconstruction emerged early as a robust and reliable alternative that was capable of resolving one of the most vexing limitations of the DFT, namely to avoid artifacts associated with truncated data without introducing line-broadening or reducing sensitivity. Although MaxEnt has been demonstrated to outperform commonly applied methods of resolution enhancement, including apodisation and linear prediction, it remained largely underutilised for decades. With the introduction of NUS for multidimensional NMR experiments there has been renewed interest in non-Fourier methods of signal processing, and together with other methods covered in this volume, such as compressed-sensing (CS, see Chapter 10) and iterative soft thresholding (see Chapters 7 and 10), MaxEnt reconstruction is becoming more broadly applied by the NMR community. This trend is likely to continue with the introduction of higher fields and more sensitive electronics in NMR instruments, and with the recent increase

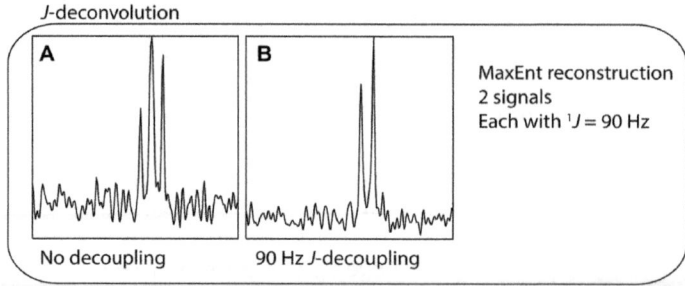

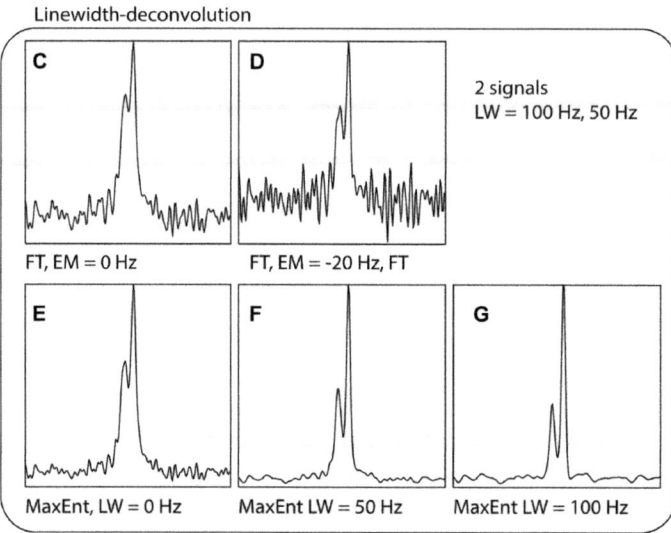

Figure 9.4 *J*- and linewidth deconvolution using MaxEnt reconstruction. Panels A and B show the reconstruction of two overlapped doublets in the presence of noise. (A) In the absence of *J*-deconvolution the two doublets appear as a triplet. (B) When *J*-decoupling is applied the doublets collapse into two singlets with twice the intensity, leading to an increase in S/N. (C, D) Two overlapping signals with different linewidths are processed using either zerofilling followed by DFT (C) or after application of a line-narrowing window function (multiplication with an exponential rising function) followed by zerofilling and DFT (D). (E–G) The same data is processed using MaxEnt reconstruction applying 0, 50 and 100 Hz linewidths in the deconvolution kernel in E, F and G, respectively. The stronger line-narrowing leads to improved resolution and sensitivity.

in solid-state NMR applications. Indeed, the incentive for utilizing NUS methods increases with magnetic field strength, as the smaller sampling intervals required in the indirect dimensions that are needed because of the increased spectral dispersion make the sampling requirements for achieving high resolution in the indirect dimensions more onerous.

There is still much to be explored in the application of MaxEnt in NMR. In particular, the areas of random phase detection (RPD) and linewidth deconvolution have shown great potential to improve spectral quality. Future studies are likely to explore these avenues for different types of NMR experiments and may indeed include the combination of one or several of NUS, RPD and linewidth decoupling to further enhance the spectral quality achievable from the inherently noise NMR signals.

References

1. R. R. Ernst and W. A. Anderson, *Rev. Sci. Instrum.*, 1966, **37**, 93–102.
2. R. R. Ernst, *Adv. Magn. Reson.*, 1966, **2**, 1–135.
3. S. Sibisi, J. Skilling, R. G. Brereton, E. D. Laue and J. Staunton, *Nature*, 1984, **311**, 446–447.
4. E. D. Laue, M. R. Mayger, J. Skilling and J. Staunton, *J. Magn. Reson. (1969)*, 1986, **68**, 14–29.
5. G. J. Daniell and P. J. Hore, *J. Magn. Reson. (1969)*, 1989, **84**, 515–536.
6. C. E. Shannon, *Bell Syst. Tech. J.*, 1948, **27**, 379–423.
7. G. J. Daniell and P. J. Hore, *J. Magn. Reson.*, 1989, **84**, 515–536.
8. J. C. Hoch, A. S. Stern, D. L. Donoho and I. M. Johnstone, *J. Magn. Reson.*, 1990, **86**, 236–246.
9. J. Skilling and R. Bryan, *Mon. Not. R. Astron. Soc.*, 1984, **211**, 111–124.
10. J. C. Hoch and A. S. Stern, *NMR Data Processing*, Wiley-Liss, New York, 1996.
11. J. A. Jones and P. J. Hore, *J. Magn. Reson.*, 1991, **92**, 363–376.
12. J. A. Jones and P. J. Hore, *J. Magn. Reson.*, 1991, **92**, 276–292.
13. D. L. Donoho, I. M. Johnstone, A. S. Stern and J. C. Hoch, *Proc. Natl. Acad. Sci. U. S. A.*, 1990, **87**, 5066–5068.
14. S. G. Hyberts, G. J. Heffron, N. G. Tarragona, K. Solanky, K. A. Edmonds, H. Luithardt, J. Fejzo, M. Chorev, H. Aktas, K. Colson, K. H. Falchuk, J. A. Halperin and G. Wagner, *J. Am. Chem. Soc.*, 2007, **129**, 5108–5116.
15. S. Paramasivam, C. L. Suiter, G. Hou, S. Sun, M. Palmer, J. C. Hoch, D. Rovnyak and T. Polenova, *J. Phys. Chem. B*, 2012, **116**, 7416–7427.
16. P. Schmieder, A. S. Stern, G. Wagner and J. C. Hoch, *J. Magn. Reson.*, 1997, **125**, 332–339.
17. P. Schmieder, A. S. Stern, G. Wagner and J. C. Hoch, *J. Magn. Reson.*, 1997, **125**, 332–339.
18. M. Mobli, M. W. Maciejewski, M. R. Gryk and J. C. Hoch, *J. Biomol. NMR*, 2007, **39**, 133–139.
19. M. Mobli, M. W. Maciejewski, M. R. Gryk and J. C. Hoch, *Nat. Methods*, 2007, **4**, 467–468.
20. A. S. Stern, K.-B. Li and J. C. Hoch, *J. Am. Chem. Soc.*, 2002, **124**, 1982–1993.
21. J. C. J. Barna, E. D. Laue, M. R. Mayger, J. Skilling and S. J. P. Worrall, *J. Magn. Reson. (1969)*, 1987, **73**, 69–77.
22. E. Kupce and R. Freeman, *J. Am. Chem. Soc.*, 2003, **125**, 13958–13959.
23. J. C. Hoch, M. W. Maciejewski and B. Filipovic, *J. Magn. Reson.*, 2008, **193**, 317–320.

24. D. Rovnyak, M. Sarcone and Z. Jiang, *Magn. Reson. Chem.*, 2011, **49**, 483–491.
25. M. Palmer, B. Wenrich, P. Stahlfeld and D. Rovnyak, *J. Biomol. NMR*, 2014, **58**, 303–314.
26. M. Bostock, D. Holland and D. Nietlispach, *J. Biomol. NMR*, 2012, **54**, 15–32.
27. P. C. Aoto, R. B. Fenwick, G. J. A. Kroon and P. E. Wright, *J. Magn. Reson.*, 2014, **246**, 31–35.
28. S. G. Hyberts, K. Takeuchi and G. Wagner, *J. Am. Chem. Soc.*, 2010, **132**, 2145–2147.
29. A. D. Schuyler, M. W. Maciejewski, H. Arthanari and J. C. Hoch, *J. Biomol. NMR*, 2011, **50**, 247–262.
30. K. Kazimierczuk, A. Zawadzka and W. Koźmiński, *J. Magn. Reson.*, 2008, **192**, 123–130.
31. M. Mobli, *J. Magn. Reson.*, 2015, **256**, 60–69.
32. M. Mobli and J. C. Hoch, *Prog. Nucl. Magn. Reson. Spectrosc.*, 2014, **83**, 21–41.
33. M. Mobli, M. W. Maciejewski, A. D. Schuyler, A. S. Stern and J. C. Hoch, *Phys. Chem. Chem. Phys.*, 2012, **14**, 10835–10843.
34. M. Maciejewski, M. Mobli, A. Schuyler, A. Stern and J. Hoch, in *Novel Sampling Approaches in Higher Dimensional NMR*, ed. M. Billeter and V. Orekhov, Springer, Berlin Heidelberg, 2012, vol. 316, ch. 185, pp. 49–77.
35. M. W. Maciejewski, M. Fenwick, A. D. Schuyler, A. S. Stern, V. Gorbatyuk and J. C. Hoch, *Proc. Natl. Acad. Sci. U. S. A.*, 2011, **108**, 16640–16644.
36. A. D. Schuyler, M. W. Maciejewski, A. S. Stern and J. C. Hoch, *J. Magn. Reson.*, 2015, **254**, 121–130.
37. M. Mobli, A. S. Stern and J. C. Hoch, *J. Magn. Reson.*, 2006, **182**, 96–105.
38. J. A. Jones, D. S. Grainger, P. J. Hore and G. J. Daniell, *J. Magn. Reson., Ser. A*, 1993, **101**, 162–169.
39. L. Castañar and T. Parella, *Magn. Reson. Chem.*, 2015, **53**, 399–426.
40. V. Stoven, J. P. Annereau, M. A. Delsuc and J. Y. Lallemand, *J. Chem. Inf. Comput. Sci.*, 1997, **37**, 265–272.
41. N. Shimba, A. S. Stern, C. S. Craik, J. C. Hoch and V. Dötsch, *J. Am. Chem. Soc.*, 2003, **125**, 2382–2383.
42. J. B. Jordan, H. Kovacs, Y. Wang, M. Mobli, R. Luo, C. Anklin, J. C. Hoch and R. W. Kriwacki, *J. Am. Chem. Soc.*, 2006, **128**, 9119–9128.
43. F. Ni, G. C. Levy and H. A. Scheraga, *J. Magn. Reson. (1969)*, 1986, **66**, 385–390.

CHAPTER 10

Compressed Sensing ℓ_1-Norm Minimisation in Multidimensional NMR Spectroscopy

MARK J. BOSTOCK,[a] DANIEL J. HOLLAND[b] AND
DANIEL NIETLISPACH*[a]

[a] Department of Biochemistry, University of Cambridge, 80 Tennis Court
Road, Cambridge, UK; [b] Department of Chemical Engineering, University
of Canterbury, Christchurch, New Zealand
*Email: dn206@cam.ac.uk

10.1 Introduction

Along with X-ray crystallography, NMR spectroscopy is the only atomic resolution technique available to study molecular structure. Over the years, many developments have significantly increased the size-limit to which biomolecular NMR studies can be applied, including in particular the TROSY technique,[1] advanced labelling strategies including selective methyl protonation[2] and methyl-directed experiments,[3] as well as continuing improvements in NMR hardware. Such advances have enabled NMR spectroscopy to probe increasingly demanding applications, such as protein complexes[4,5] and larger membrane proteins.[6,7]

In the majority of cases, multidimensional NMR experiments have proved indispensable to achieve backbone and side chain assignments, to acquire

New Developments in NMR No. 11
Fast NMR Data Acquisition: Beyond the Fourier Transform
Edited by Mehdi Mobli and Jeffrey C. Hoch
© The Royal Society of Chemistry 2017
Published by the Royal Society of Chemistry, www.rsc.org

sufficient structural restraints required for 3D structure determination, or to obtain data that provides insight into the functionality of biomolecules. Nevertheless, an outstanding limitation of multidimensional NMR is the long recording times required to complete the data acquisition of multidimensional experiments, particularly those with three or more dimensions. The frequently excessive experiment times arise from the use of the Fourier transform to convert the time-domain signal into a frequency domain spectrum which, according to the Nyquist theorem, requires data points to be sampled equally spaced on a grid with point spacing determined by the spectral width. As a result for an N-dimensional spectrum with k_n points in the nth-dimension, $2^{N-1} \times k_1 \times k_2 \times k_3 \times \cdots \times k_n$ points must be recorded leading to an exponentially increasing number of points, and by extension experiment times, as additional dimensions are recorded. For high-dimensional experiments, this typically leads to a trade-off between achievable signal-to-noise ratio (SNR) and resolution to ensure acceptable experiment times. For low concentration samples or large proteins, which typically show relatively low sensitivity, the available experiment time will be weighted in favour of improved SNR while sacrificing resolution, thus partially negating the benefit of separating peaks into additional dimensions. As a result, the complete potential of higher dimensional experiments *e.g.* 3D, 4D and 5D cannot be realised with full Nyquist sampling. In contrast, in situations where the NMR signal is plentiful the requirement for sampling of the full Nyquist grid results in experiment times that far exceed the demands based on sample SNR.

A proposed solution is to use a non-uniform, undersampling or sparse sampling strategy where only a small part of the full Nyquist matrix is collected. This approach was demonstrated many years ago[8-10] with its potential to increase resolution or SNR helping to realise the benefits of higher dimensional experiments.[11,12] However, undersampled or non-uniformly sampled datasets cannot be reconstructed using the Fourier transform. In order to overcome the problem described above, a range of different processing methods have been developed, all of which aim to reconstruct a spectrum from undersampled time-domain data, while attempting to reduce the artefact level incurred by the reduced sampling scheme.[13]

Several of these approaches are discussed elsewhere in this book. In this chapter, we discuss the relatively recent introduction of methods broadly termed 'compressed sensing' (CS), based on ℓ_1-norm minimisation. ℓ_1-Norm minimisation has been available for several decades,[14] however the notion of 'compressed sensing' was formulated more recently in the literature of information theory[15,16] and presents theoretical guidelines that can be used to explain how and why ℓ_1-norm minimisation is so effective. CS has become increasingly popular in a number of signal processing fields, including image reconstruction,[17-20] and more recently its potential for reconstruction of highly undersampled NMR spectra was demonstrated by ourselves and others.[21-23] To date, CS has shown considerable promise for a wide-range of NMR applications and looks set to become a considerable asset for the study of large biomolecular systems. In this chapter, the theory of CS and examples of the algorithms used are outlined in Sections 10.2 and 10.3; some of the challenges of implementing the methods are discussed in Section 10.4 and a discussion of the terminology in use is covered in Section 10.5; an overview of the current state of the field is given in Section 10.6; applications to 4-dimensional spectra are discussed in Section 10.7 and future directions are given in Section 10.8.

10.2   Theory

Central to the problem of reconstructing a spectrum from undersampled data is the requirement to solve a system of underdetermined linear equations. Consider the system of linear equations:

$$\mathbf{Ax} = \mathbf{b} \tag{10.1}$$

where $\mathbf{A}$ is an $M \times N$ matrix and $\mathbf{x}$ is a vector of length N that is to be recovered from measurements $\mathbf{b}$, where $\mathbf{b}$ is a vector of length M. Considering this for the case of NMR spectroscopy, $\mathbf{x}$ corresponds to the spectrum in the frequency domain, $\mathbf{b}$ to the measurements in the time domain, and $\mathbf{A}$ is the inverse Fourier transform. Conventionally $M = N$ and this system of equations has a unique solution. However, for non-uniformly sampled data, $M < N$ and eqn (10.1) has infinitely many solutions. CS theory states that $\mathbf{x}$ can be exactly reconstructed from $O(k)$ random projections assuming that $\mathbf{x}$ is k-sparse, *i.e.* containing no more than k nonzero components, by solving:

$$\min_{x} \|\mathbf{x}\|_0 \quad \text{subject to } \mathbf{Ax} = \mathbf{b} \tag{10.2}$$

where $\|\mathbf{x}_0\|$ denotes the ℓ_0-"norm" of $\mathbf{x}$, which is a pseudo-norm defined by:

$$\|\mathbf{x}\|_0 = \sum_i |x_i|^0 \tag{10.3}$$

where[15] $0^0 = 0$ and x_i is the ith element of $\mathbf{x}$. In other words, the ℓ_0-"norm" is equivalent to the number of non-zero entries in the spectrum $\mathbf{x}$, and thus solving eqn (10.2) is equivalent to finding the sparsest solution, *i.e.* that with the fewest non-zero elements, that is consistent with the measured data; in effect minimizing artefacts generated by the reconstruction from an undersampled dataset. The minimization described by eqn (10.2) is a combinatorial problem and therefore for realistic M and N it has been shown that it is not possible to solve eqn (10.2) computationally.[24] The achievement of CS is to demonstrate that where $\mathbf{x}$ is sufficiently sparse, minimising the ℓ_1-norm:

$$\min_{x} \|\mathbf{x}\|_1 \quad \text{subject to } \mathbf{Ax} = \mathbf{b} \tag{10.4}$$

Where:

$$\|\mathbf{x}\|_1 = \sum_i |x_i|^1 \tag{10.5}$$

returns the same solution and can be solved using standard linear programming. It is thus regarded as returning the exact, artefact-free reconstruction.[15,22,25,26] CS theory states that the solution to the ℓ_1-norm minimization problem will be equivalent to the ℓ_0 minimum solution for a k-sparse signal, provided that the number of samples is:

$$M \geq C\mu k \, \log(N) \qquad (10.6)$$

where C is a small constant and is independent of k, M and N whilst μ characterises the incoherence of the chosen sampling scheme. This result suggests that the number of measurements required to reconstruct a spectrum is determined primarily by the number of non-zero elements, and is only weakly dependent on the resolution of the spectrum.

The success of the ℓ_1-norm approach has resulted in generalisation of the ideas of compressed sensing to the ℓ_p-norm:

$$||\mathbf{x}||_p = \left(\sum_i |x_i|^p \right)^{1/p} \qquad (10.7)$$

where $p > 0$. It should be noted that a true norm is only defined on the interval $1 \leq p \leq \infty$; for the interval $0 < p \leq 1$, eqn (10.7) may be thought of as a quasi-norm. By solving norms for $p < 1$, it should be possible to further reduce the number of measurements required.

For NMR spectra, as in other signal processing applications, $\mathbf{b}$ may contain noise, or may be compressible rather than truly sparse. In such cases, the constraint in eqn (10.4) can be relaxed to:

$$\min_\mathbf{x} ||\mathbf{x}||_1 \quad \text{subject to } ||\mathbf{Ax} - \mathbf{b}||_2 \leq \delta \qquad (10.8)$$

Various algorithms are available to solve eqn (10.8), several of which will be discussed in more detail in Section 10.3.

Compressed sensing requires: (i) sparse representation of the desired signal in a particular basis and (ii) incoherent sampling with respect to that basis, *i.e.* the sampling schedule should be randomised to avoid artefacts arising owing to coherent sampling. The incoherence is characterised by the parameter μ, which is small for random samples from the Fourier matrix; systematic or coherent samples typically lead to an increase in μ and hence, from eqn (10.6), an increase in the number of samples is required. The strong theoretical basis underpinning CS provides a theoretical guarantee of recovery as well as a theoretical basis for the type of algorithm needed for the best reconstruction. More recent developments have explored the theoretical basis for the selection of sampled points, providing a framework to underpin the data acquisition scheme.[27] Adcock *et al.*[27] introduced the notion of asymptotic coherence and asymptotic sparsity and showed that sampling should be concentrated in the points with the most energy. The mathematical framework for these concepts reinforces empirical results showing the success of weighted sampling strategies, which have been used for many years in NUS-NMR, such as exponential sampling[8] and more recently weighted Poisson sampling.[28,29]

In the case of NMR spectra, sparsity is generated on transformation to the frequency domain; in other cases, sparsity may be generated by transforming to a wavelet domain[26] or by minimising differences in the spectra to yield the Total Variation (TV), determined by calculating the derivatives at all pixel positions, typically measured as the finite difference between neighbouring pixels,[30] allowing applications to a wide range of systems, including, for example, solid state spectra where much broader lineshapes are encountered. Importantly, an initial estimate of the signal location is not required for CS, only the knowledge that the spectrum (or some transform of the spectrum) is sparse.

Many NMR spectra are sparse in the frequency domain, and furthermore sparsity increases as the number of dimensions increases, of particular benefit when considering applications to high-dimensional spectra. Crucially, CS theory states that the number of measurements required to recover a signal is $M \geq C\mu k \log(N)$ (eqn (10.6)) which, for a given sparsity (k), has only a weak dependence on the number of points in the final spectrum, N. Therefore, the dimensionality of the experiment can be increased with only a slight increase in the number of measurements. For example, to add a fourth dimension to a 3D experiment containing n_2 points in the first indirect dimension, n_3 points in the second indirect dimension and an additional n_4 points in the third indirect dimension would conventionally require a factor of n_4 additional measurements; in the CS framework only a factor of $(1 + \log n_4/(\log n_2 + \log n_3))$ additional measurements are required. Assuming say 128 points in each indirect dimension, conventional Nyquist sampling would require a factor of 128 extra points (16 384 to 2 097 152 for 3D to 4D); in the CS framework a factor of only 1.5 extra points would be required equivalent to a sampling fraction of 0.3% for the 4D (assuming 25% sampling for the equivalent 3D), representing a considerable time saving over conventional Nyquist sampling and making this extremely powerful for application to high-dimensional spectra.

10.3 Algorithms

With the rising popularity of the CS approach, a range of different algorithms have been proposed to implement the methodology. Before discussing the different algorithms in more detail, a schematic of a typical iterative CS reconstruction for a 1D spectrum is shown in Figure 10.1, illustrating the selection of a signal from a noisy frequency domain spectrum (d), storage of this signal (f), and identification of the contribution of the stored signal to the original FID (c). This contribution is then removed from the original FID (a) before repeating iteratively; as a result, over a certain number of iterations, a spectrum is reconstructed (e) that compares extremely well with the fully-sampled spectrum (g). The algorithms used for CS reconstructions can be grouped into five main classes: greedy pursuit; convex relaxation; Bayesian methods; non-convex optimisation and brute force approaches.[31] To date, CS applications to solution NMR spectroscopy have focussed on greedy pursuit, convex and nonconvex optimisation approaches, which are discussed in more detail below.

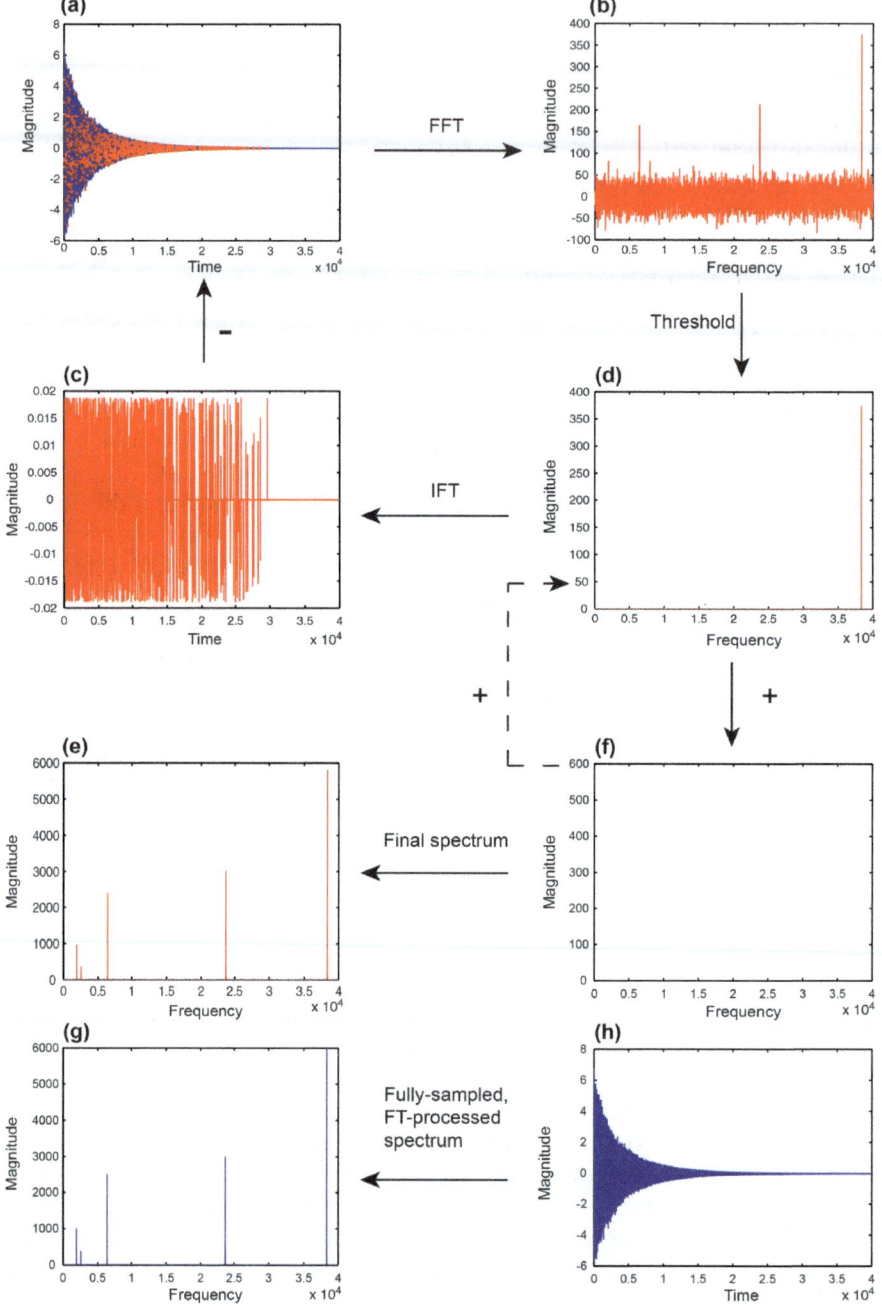

10.3.1 Greedy Pursuit

Greedy pursuit algorithms take an initial estimate of the sparse solution and iteratively refine this by modifying one or more components that lead to the greatest improvement in the sparse solution.[31] Algorithms that implement this approach include orthogonal matching pursuit (OMP)[32] as well as more modern implementations, such as compressive sampling matching pursuit (CoSaMP).[33] The greedy algorithms that have found favour in NMR applications are the iterative thresholding methods—iterative hard thresholding and iterative soft thresholding—which are discussed in more detail below. The iterative thresholding algorithms have the advantage of being fast, important when considering the large datasets typically acquired in NMR experiments, and are effective in both empirical and theoretical tests.[31]

10.3.1.1 Iterative Hard Thresholding (IHT)

This algorithm is based on work by Bredies and Lorenz.[34] The basic approach is to calculate an initial estimate of the solution by replacing missing data points with zeros followed by Fourier transform (FT) (equivalent to minimising the ℓ_2-norm[35]). A user-defined threshold is then set in the frequency domain with data points above the threshold stored. The inverse Fourier transform (IFT) of this thresholded spectrum is then calculated and the result subtracted from the original dataset before repeating this procedure iteratively. In this way, the largest points and the noise they generate can be subtracted from the raw data and on successive iterations it becomes easier to extract the weakest data points. This operation can be represented as:

$$\mathbf{x}^{[n+1]} = \mathbf{x}^{[n]} - H_\lambda(\mathbf{A}^T(\mathbf{A}\mathbf{x}^{[n]} - \mathbf{b})) \tag{10.9}$$

Figure 10.1 Illustration of the iterative process of compressed sensing (CS) for a one-dimensional spectrum. The loop starts with the undersampled/non-uniformly sampled data (red), in this case sampled according to an exponential distribution (a); the fully sampled data (blue) is shown for comparison (a and h). The Fourier transform (FT) is used to compute an estimate of the spectrum (b), and signals above a threshold are selected (d) and stored in an empty spectrum initiated with zeros (f). The thresholded spectrum is transformed back to the time-domain and subtracted from the original undersampled FID, thus removing both the contribution from the signal in the thresholded spectrum as well as any artefacts from that signal owing to the undersampling schedule (c); the sampling mask is reapplied such that only the recorded data are compared with the reconstructed data. The process is repeated iteratively, at each stage adding the new thresholded spectrum to (f), until after a sufficient number of iterations the final spectrum is reconstructed (e). At this point, the residual (b) should contain only noise. The FT-processed, fully sampled spectrum is shown in (g). In the 1D example shown, the fully sampled signal contains 19 998 points, of which 580 points (2.9%) were selected for the undersampled dataset. The spectrum was reconstructed with 100 iterations.

where **A**, **b** and **x** are defined as for eqn (10.1), $\mathbf{x}^{[0]} = 0$, and H_λ is a non-linear operator that sets the elements of $\mathbf{a}(a_i)$ that are below the user-defined threshold λ to zero:

$$H_\lambda(\mathbf{a}) = \begin{cases} 0 & |a_i| < \lambda \\ a_i & |a_i| \geq \lambda \end{cases} \tag{10.10}$$

The parameter λ effectively controls the steepness (gradient) of the reconstruction and has an effect on the speed and quality of the reconstruction. In conventional hard thresholding, λ is chosen such that only the k largest elements of the spectrum are retained. In our implementation, we modify the traditional hard thresholding approach and choose $\lambda = \alpha \max_i a_i$, where α is a constant between 0 and 1. As the number of iterations, n, increases, $\mathbf{A}^T(\mathbf{Ax}^{[n]} - \mathbf{b}))$ tends to a spectrum containing only noise. Bredies and Lorenz[34] reported strong convergence with rates defined by $O(n^{-1/2})$, which may be faster than for the iterative soft thresholding approach (IST), although it has been observed that whilst IST gives good descent in every step, IHT has some steps which reduce functional values drastically and others where the functional may remain unchanged for a number of iterations.[34]

10.3.1.2 Iterative Soft Thresholding (IST)

IST involves soft thresholding or non-linear shrinkage.[36] In contrast to the hard thresholding approach, in IST data points below a designated threshold are set to zero, whilst all data points above the threshold have the value of the threshold subtracted. IFT is then calculated and the result used to fill in missing data points. It can be represented as:

$$\mathbf{x}^{[n+1]} = S_\lambda(\mathbf{x}^{[n]} - \mathbf{A}^T(\mathbf{Ax}^{[n]} - \mathbf{b})) \tag{10.11}$$

where:

$$S_\lambda(\mathbf{a}) = \begin{cases} 0 & |a_i| < \lambda \\ (|a_i| - \omega\lambda) \cdot \text{sign}(a_i) & |a_i| \geq \lambda \end{cases}$$

λ is a user-defined threshold parameter, and ω is a scaling factor. A consideration for the IST approach is that the regulariser $\|\mathbf{Ax} - \mathbf{b}\|_2$ will converge to a constant value for a given value of λ regardless of the number of iterations. In contrast to our IHT approach, the regulariser will continue to decrease with increasing number of iterations (although at different rates depending upon the spectrum). Thus, in the IST approach the choice of the parameter λ has a direct influence on the tightness of the reconstruction to the original data, whilst for IHT this is not a concern, but in contrast it may prove harder to know when to stop the iterations. The IST approach was previously discussed in the context of extending truncated data and more generally in the context of NMR data processing, although the convergence properties were questioned.[37,38] This algorithm has been employed by Hyberts *et al.*[23] and has also been used in the context of NMR data processing by Kazimierczuk and Orekhov.[22,35] Hyberts *et al.*[23] use an implementation (istHMS) where the threshold is calculated as a proportion (75%) of the maximum peak height and will thus change on each iteration, in contrast to the fixed threshold used in other implementations of IST, including in our work.[39]

10.3.2 Convex Relaxation Methods

These methods "relax" the ℓ_0-norm to the ℓ_1-norm yielding a convex optimisation problem, solving problems of the form given in eqn (10.4). Whilst the iterative approaches can be shown to be equivalent to ℓ_1-norm minimisation,[37] the convex relaxation approaches set out to solve the ℓ_1-norm approach directly. Gradient methods constitute a group of convex relaxation algorithms, typically solving a problem of the form:

$$\min_x \frac{1}{2} \, ||\mathbf{Ax} - \mathbf{b}||_2^2 + \lambda \, ||\mathbf{x}||_1 \tag{10.12}$$

The major operation involves calculating the gradient of the current iteration, which can make such approaches slower than the iterative thresholding methods or other greedy algorithms, particularly when the solution is very sparse. However, if the solution is not very sparse, convex relaxation methods may be preferable.[40] Various methods have been presented to solve eqn (10.12), including conjugate gradient descent, the homotopy method, and various primal-dual methods.[40] Here we consider conjugate gradient descent and the primal-dual algorithm of Yang and Zhang,[41] which have been used to reconstruct NMR data.

10.3.2.1 Conjugate Gradient Descent

A seminal paper in the application of compressed sensing to medical imaging was presented by Lustig *et al.*[19] In this work, the authors use nonlinear conjugate gradient descent with backtracking line-search, similar to approaches in Block *et al.*[30] and Bronstein *et al.*[42] As discussed by Lustig *et al.*,[19] this process involves finding the value of **x** that minimises the function:

$$f(\mathbf{x}) = \min_x ||\mathbf{Ax} - \mathbf{b}||_2^2 + \lambda ||\mathbf{\Psi x}||_1 \tag{10.13}$$

where λ is a regularisation parameter determining the weighting towards ℓ_1-minimisation (sparsity) or consistency with the raw data (the ℓ_2 term) and $\mathbf{\Psi}$ is an appropriate sparsifying transform *e.g.* the identity transform, wavelet transform or spatial finite differences transform (for total variation).

The solution to eqn (10.13) requires calculation of the gradient of $f(\nabla f)$, which is defined as:

$$\nabla f(\mathbf{x}) = 2\mathbf{A}^T(\mathbf{A}\mathbf{x} - \mathbf{b}) + \lambda \nabla \|\Psi \mathbf{x}\|_1 \tag{10.14}$$

Since the gradient of the ℓ_1-norm is not well defined for all values of $\mathbf{x}$, the approximation $x_i \approx \sqrt{x_i {}^* x_i + \varepsilon}$ is used giving $\dfrac{d|x_i|}{dx_i} \approx \dfrac{x_i}{\sqrt{x_i {}^* x_i + \varepsilon}}$. Eqn (10.14) can then be solved by using a diagonal matrix, $\mathbf{w}$, with diagonal elements $w_i = \sqrt{(\Psi x)_i {}^* (\Psi x)_i + \epsilon}$, which gives:

$$\nabla f(\mathbf{x}) = 2\mathbf{A}^T(\mathbf{A}\mathbf{x} - \mathbf{b}) + \lambda \psi {}^* \mathbf{W}^{-1}\psi x \tag{10.15}$$

Since the ℓ_1-norm is a convex problem, the algorithm seeks a solution where $\nabla f(\mathbf{x}) = 0$, signifying convergence. The approximation $x_i \approx \sqrt{x_i {}^* x_i + \varepsilon}$ is crucial in determining whether the solution converges. If ε is chosen to be too small, the algorithm may be unable to converge as $|\mathbf{x}|$ approaches zero, however if ε is too large, the algorithm will never reach the true minimum of eqn (10.14). Therefore, it is essential that ε is chosen appropriately based on the magnitude of $\mathbf{x}$.

10.3.2.2 YALL1 (Your Algorithm for ℓ_1)

Another convex-relaxation method applied to NMR in our work[39] is YALL1 presented by Yang and Zhang,[41] which uses an alternating direction method (ADM) to solve a range of ℓ_1-minimisation models, including the model appropriate for NMR data, the ℓ_1/ℓ_2 convex minimisation model (eqn (10.8)). The ADM approach solves problems of the form:

$$\min_{\mathbf{x},\mathbf{y}}(f(\mathbf{x}) + g(\mathbf{y})) \quad \text{subject to } \mathbf{A}\mathbf{x} + \mathbf{B}\mathbf{y} = \mathbf{b} \tag{10.16}$$

which can be converted into an augmented Lagrangian function of the form:

$$\mathscr{L}(\mathbf{x},\mathbf{y},\lambda) = f(\mathbf{x}) + g(\mathbf{y}) - \lambda^T(\mathbf{A}\mathbf{x} + \mathbf{B}\mathbf{y} - \mathbf{b}) + \frac{\beta}{2}\|\mathbf{A}\mathbf{x} + \mathbf{B}\mathbf{y} - \mathbf{b}\|_2 \tag{10.17}$$

Using the augmented Lagrangian approach, λ and β are updated at each step and, as a result, it is not necessary to search as many values of λ to solve the original constrained problem (eqn (10.13)). In order to use the augmented Lagrangian scheme it is necessary to introduce the second variable, $g(\mathbf{y})$. The authors report that the algorithm reduces the relative error between a fully sampled data set and an undersampled reconstruction faster than other available solvers under conditions of noisy data, that the algorithm is largely insensitive to the choice of starting point and initial parameters, and does not rely on a line-search technique.[41]

10.3.3 Non-convex Minimisation

Non-convex minimisation also involves relaxation of the ℓ_0-norm, however in contrast to the convex-relaxation methods, an ℓ_p-norm is used as the objective function where $0 < p < 1$ with $p \to 0$ in some algorithms; such functions do not have a global minimum and hence are described as non-convex. Exact reconstruction can be achieved by finding a local minimiser requiring far fewer signal measurements than for the case where $p = 1$.[43] The iteratively re-weighted least squares algorithm, discussed below, is an example of both convex and non-convex minimisation techniques for $p = 1$ and $p \to 0$, respectively.

10.3.3.1 Iteratively Re-weighted Least Squares (IRLS)

As discussed by Kazimierczuk and Orekhov,[22,35] IRLS aims to minimise the pth power of an ℓ_p-norm, which is defined as:

$$||\mathbf{x}||_p^p = \sum_i |x_i|^p \tag{10.18}$$

where x_i is the ith element of the vector $\mathbf{x}$. The ℓ_p-norm can be written in the form of a weighted ℓ_2-norm as follows:

$$||\mathbf{x}||_p^p = \sum_i w_i |x_i|^2 \tag{10.19}$$

where:

$$w_i = |x_i|^{p-2} = \frac{1}{|x_i|^{2-p}}$$

Eqn (10.19) allows eqn (10.8) to be reformulated as a least squares minimisation problem as follows:

$$f(\mathbf{x}) = \min_{\mathbf{x}} ||\mathbf{A}\mathbf{x} - \mathbf{b}||_2^2 + ||\mathbf{W} \cdot \mathbf{x}||_2^2 \tag{10.20}$$

where:

$$W_{ii} = \lambda w_i$$

This can be solved by:

$$\mathbf{x} = \mathbf{W}^{-1} \mathbf{A}^\dagger \boldsymbol{\psi} \tag{10.21}$$

with $\mathbf{A}^\dagger$ (the conjugate transpose of $\mathbf{A}$) being the FT matrix and $\boldsymbol{\psi}$ the solution to the system of linear equations:

$$(\mathbf{A}\mathbf{W}^{-1}\mathbf{A}^\dagger + \mathbf{E})\boldsymbol{\psi} = \mathbf{b}$$

where $\mathbf{E}$ is the identity matrix. Weights are set by solving eqn (10.21) iteratively using the weights from iteration $n - 1$ to set w_i for the next iteration. To avoid division by zero, weights are regularised according to Candès *et al.*:[44]

$$w_i^{n-1} = \frac{1}{|x_i^n|^{2-p} + \varepsilon} \tag{10.22}$$

with $\varepsilon > 0$. Minimisation of norms <1 is achieved by decreasing the value of p in the weights (eqn (10.19)) from 1 down to 0 ($\ell_{p \to 0}$-norm).[45] Since the ℓ_p-minimum solution is non-convex, the reduction in p must be done in such a way as to ensure that the solution remains close to the global solution of the ℓ_0-norm problem. Furthermore, ε must be chosen in an appropriate manner, analogous to Section 10.3.2.1.

10.3.4   Other Approaches

The methods discussed above represent the most common implementations for NMR spectroscopy. Bayesian approaches convert the standard CS problem into a linear regression problem where the constraint (prior) is a distribution of signal coefficients that in some basis is sparse. Posterior probabilities that incorporate the prior assumptions can be calculated, resulting in a range of models with estimates for the uncertainty amongst the various model solutions.[46,47] Brute force methods search through all possible support sets, and hence are typically very slow and not widely used.

10.4   Implementation and Choice of Stopping Criteria

It is clear from the discussion above and from the present implementations in the literature that a range of possible algorithms are available to reconstruct spectra *via* the CS method. Different implementations may be more suitable for different applications. For example, conjugate gradient approaches are typically slow, so are more suitable for 2D and some 3D spectra, whilst IRLS may be more suitable in situations of high dynamic range and low sampling fraction, although as the sampling fraction increases, IRLS reconstruction time increases exponentially whilst IST is constant.[35] Nevertheless, despite these differences, we have demonstrated extremely comparable reconstructions between IHT, IST and YALL1 reconstructions of 3D ^{15}N NOESY data.[39] In making comparisons between different methods, it is important to ensure the comparison is performed under comparable conditions; different algorithms vary the weighting between the sparsity term ($\|\mathbf{x}\|_1$) and the data consistency term ($\|\mathbf{Ax} - \mathbf{b}\|_2$). In our hands, in the case of a 3D ^{15}N NOESY, we found that forcing the solution towards a smaller value for the data consistency term (*i.e.* increased weighting towards $\|\mathbf{Ax} - \mathbf{b}\|_2$), typically led to a decrease in sparsity (increase in $\|\mathbf{x}\|_1$) owing to the introduction of many small 'noise-like' peaks that improved the fit to the raw data (Figure 10.2). In the case of our 3D NOESY

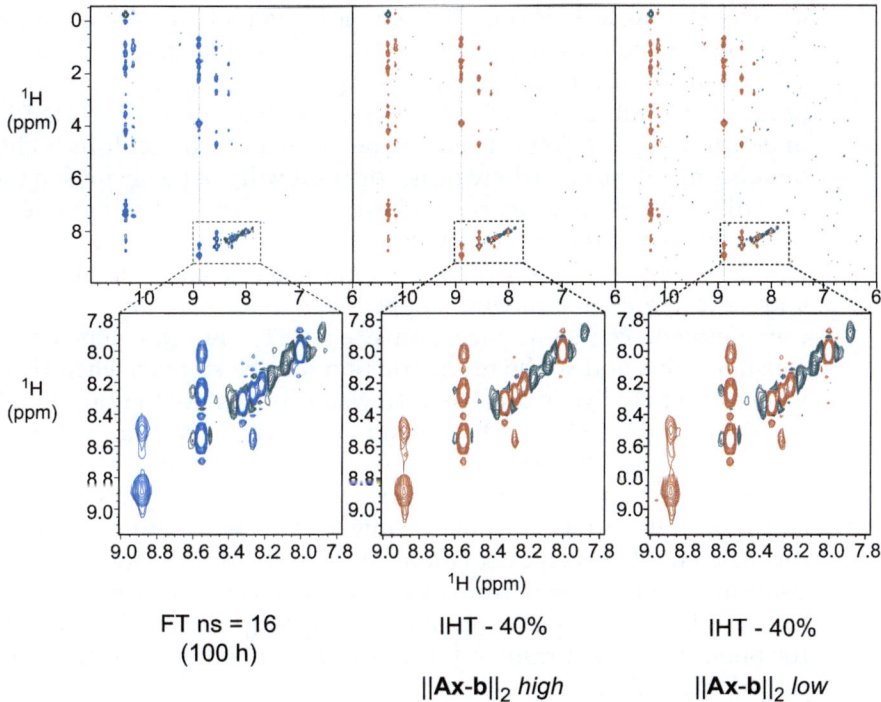

Figure 10.2 Comparison of a [^{1}H–^{1}H] strip plot from a 3D NOESY ^{15}N HSQC experiment at ^{15}N position 127.58 ppm (residue Ala 64) recorded on a ^{15}N labelled sample of sensory rhodopsin II, processed with FT (blue) or CS (red) using either a high ($\sim 2 \times 10^4$) or low (~ 2) value for the data-matching constraint $\|\mathbf{Ax} - \mathbf{b}\|_2$. The higher level of noise-like artefacts is apparent in the reconstruction with low $\|\mathbf{Ax} - \mathbf{b}\|_2$. Negative peaks are shown in dark blue. IHT 40% indicates reconstruction of a 40% sampled data set (equivalent to 40 h experiment time) using the CS-iterative hard thresholding algorithm. Both FT and CS spectra were recorded with 16 scans.

Adapted from *J. Biomol. NMR*, Compressed sensing reconstruction of undersampled 3D NOESY spectra: application to large membrane proteins, **54**, 2012, 15–32, M. J. Bostock, D. J. Holland and D. Nietlispach, (© Springer Science + Business Media B.V. 2012) with permission of Springer.[39]

^{15}N HSQC reconstructions, typically the best reconstructions were observed when $\mathbf{Ax} - \mathbf{b}_2$ was $\sim 2 \times 10^4$. Given the potential to introduce noise-like peaks and for reconstruction times to increase without any improvement in spectral quality, it is important to choose appropriate stopping criteria. A number of possibilities are available:

(i) As the iterations proceed, the term $(\mathbf{A}^T(\mathbf{Ax} - \mathbf{b}))$ *i.e.* the FFT of the residual, tends to a spectrum containing only noise. Given an estimate of the expected noise level in the spectrum to be reconstructed, the reconstruction may be stopped when the standard deviation of $(\mathbf{A}^T(\mathbf{Ax} - \mathbf{b}))$ is equivalent to the expected noise level. In order to determine a noise estimate, the standard deviation of a region of the zero-filled FT reconstructed spectrum containing no peaks should be calculated. However, this requires a region of the spectrum with no peaks to be found, otherwise the noise estimate will be distorted by undersampling artefacts. This may be particularly difficult in highly undersampled spectra where peak positions will not be obvious in the zero-filled FT reconstruction and where extensive aliasing is employed to reduce the spectral width.

(ii) The data consistency term $\|\mathbf{Ax} - \mathbf{b}\|_2$ may be calculated at each iteration and the reconstruction stopped when $\|\mathbf{Ax} - \mathbf{b}\|_2 \leq \delta$ where δ is user defined. For some algorithms, *e.g.* IST, $\|\mathbf{Ax} - \mathbf{b}\|_2$ tends to a constant value and so the reconstruction may be stopped when there is no further change in $\|\mathbf{Ax} - \mathbf{b}\|_2$. However, this is not the case for all algorithms, *e.g.* IHT, which, depending on the spectrum, will continue to decrease the value of $\|\mathbf{Ax} - \mathbf{b}\|_2$ as iteration number increases. δ may be based on the expected noise level in the spectrum. Thus (i) and (ii) effectively become equivalent, taking into account the difference between calculations of the standard deviation and the ℓ_2-norm as well as the need to estimate the noise in the time domain for $\|\mathbf{Ax} - \mathbf{b}\|_2 \leq \delta$. This is a limitation of stopping criterion (ii) owing to the potential for distortion by water signals and other artefacts in the time-domain data.

(iii) The change in the spectrum (or the residual $(\mathbf{Ax} - \mathbf{b})$) may be detected and the reconstruction stopped when no further change is detected, or in the case of the residual, when it falls below a user-defined value. The properties of these terms may be assessed by standard deviation or by the ℓ_2-norm.

Hyberts *et al.*[23] use (iii), stopping the reconstruction when the ℓ_2-norm of the residual falls below a user-defined value, whilst Kazimierczuk and Orekhov[35] stop reconstructions when the ℓ_2-norm of the residual falls below 10% of the ℓ_2-norm of the fully sampled spectrum, used as a measure of the noise (in this case the full dataset was also recorded). In our work on 3D NOESY spectra with the IHT implementation, we favour stopping criteria based on an estimate of the noise as in (i)[39] and this approach was also used for the 4D spectra discussed in Section 10.7.

Figure 10.3 demonstrates these different criteria for a 3D cube extracted from a methyl-selective 4D HCCH NOESY dataset reconstructed using our IHT implementation. The 3D subspace extracted at position 1.26 ppm in the acquisition dimension is reconstructed in the three indirect dimensions. A representative 2D ^{13}C–^{13}C plane from the 3D cube extracted at 1.0 ppm in the ^1H indirect dimension is shown. Figure 10.3(a) shows the reconstruction after 400 iterations; the reconstruction is clearly incomplete showing distorted peak shapes and compared to the reconstructions with more iterations in (b) and (c) is missing some of the weaker peaks. Running for too many iterations as is the case in (c) (∼2500 iterations) leads to the generation

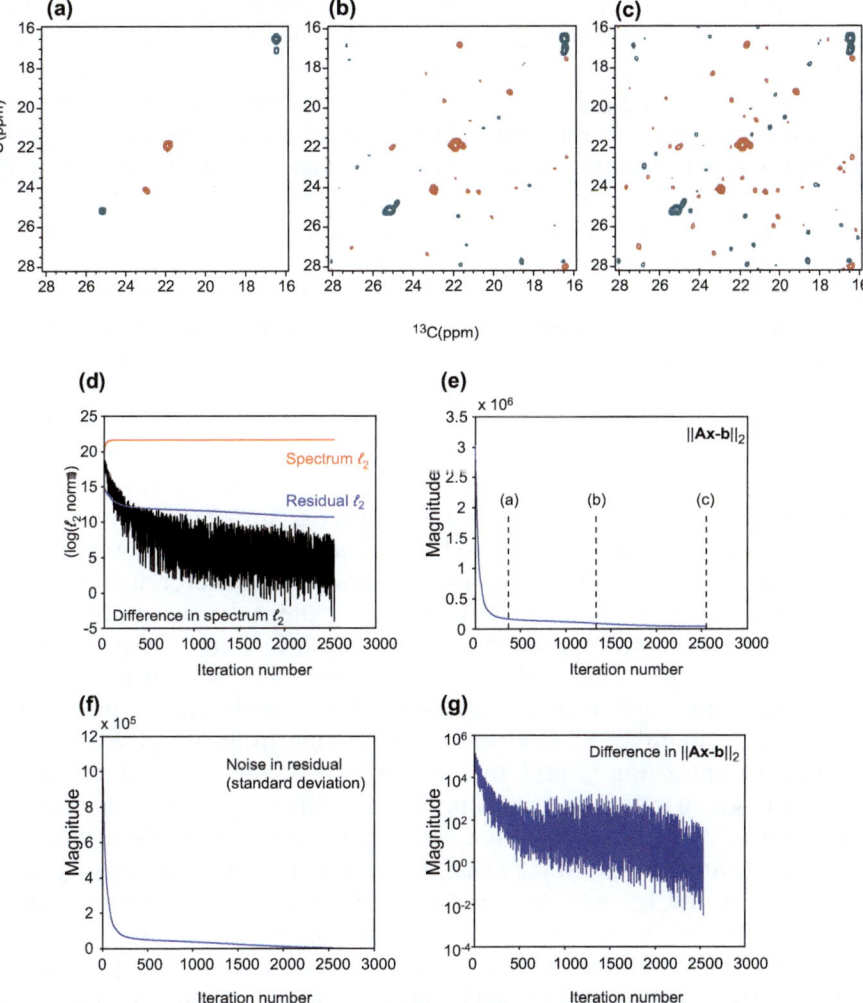

Figure 10.3   2D [^{13}C–^{13}C] plane extracted at the proton positions F_1 1.26 ppm/F_2 1.0 ppm of a methyl-selective 4D HCCH NOESY dataset reconstructed using the CS IHT implementation. The data was recorded at 35 °C and 800 MHz using a 0.5 mM sample of pSRII, selectively methyl-protonated as described in the text. (a) the reconstruction after 400 iterations; (b) after 1362 iterations and (c) after 2545 iterations. The end-points of the reconstructions shown in (a)–(c) are marked with dashed lines in (e) in terms of iteration number. Statistics for the reconstruction are shown in the graphs (d)–(g). (d) the spectrum ℓ_2 (red line), residual ℓ_2 (blue line) and the difference in the spectrum ℓ_2 from iteration to iteration (black line). (e) the data matching constraint, $\|\mathbf{Ax} - \mathbf{b}\|_2$, (f) the noise in the FT transformed residual spectrum (calculated as the standard deviation) and (g) the difference in the constraint in (e) from iteration to iteration.

of significant amounts of noise with no noticeable improvement in peak reconstruction. Clearly for large datasets such as a 4D experiment this is also computationally very expensive. Figure 10.3(b) was reconstructed with ~1400 iterations; the peaks are faithfully reconstructed and the noise does not interfere with the recognition of genuine signals. Statistical reports during the progress of the reconstructions are shown in the graphs below the spectra. Dashed lines in (e) represent the points at which the different spectral reconstructions (a, b, c) were terminated. The stopping criterion used for the best reconstruction shown in (b) was based on an estimated noise level of 3×10^4 in the data, following criterion (i) above.

Terminating the reconstruction when the norm of the residual is equivalent to the noise-level in the data (approach (ii)) is known as Morozov's discrepancy principle[48] and has been widely used as a stopping criterion for reconstruction of spectra from noisy data.[49,50] However, it is limited by the accuracy of estimating the noise from the time-domain signal. From the graphs in Figure 10.3(d), it can be seen that selecting a stopping criterion based on method (iii) is less appropriate. Any significant changes in the spectrum or residual ℓ_2 (red and blue lines in Figure 10.3(d), respectively) are not seen after approximately 250–300 iterations. However, it is clear from (a) that the reconstruction is incomplete at this point. Furthermore, after the initial sharp drop in the gradient of the residual ℓ_2-norm, the residual ℓ_2 continues to decrease gradually and so it is not clear at what point the optimum reconstruction is achieved. Meanwhile, the difference in the spectrum ℓ_2 from iteration to iteration is oscillatory, indicating the continued increase in spectrum ℓ_2 and the difficulty in using the gradient of the spectrum ℓ_2 as a stopping criterion. Clearly addition of noise-like peaks to the spectrum will continue to increase the spectrum ℓ_2 while at the same time decreasing the residual after the reconstruction has converged, as shown in (e) and (g), and so these parameters are a less useful stopping criterion. Furthermore, the magnitude of the spectrum or residual ℓ_2-norm will also depend on the spectra and samples, such that the stopping criteria will need to be readjusted for each case. Hence we favour the noise-based stopping criterion (i).

10.5   Terminology

Whilst application of CS to NMR spectra has been popularised relatively recently[21–23,39] and the literature on this field is still relatively sparse, there nevertheless exist a range of different implementations and terminologies and we provide an overview in this section as well as clarifying some of the language.

As mentioned in the introduction to this review, the approach of ℓ_1-norm minimisation has existed for a number of decades.[14,51] However, the term 'compressed sensing' (CS) was coined most recently in the field of information theory in an attempt to expand the ideas behind 'lossy' compression formats, *e.g.* in image or video storage, to a more general theory for sparse data acquisition[15] and covers application of the ℓ_p-norm (eqn (10.7)) for $0 < p \leq 1$ to the solution of underdetermined systems of linear equations (eqn (10.1)).[15,16,43] Hyberts *et al.*[23] define the term compressed sensing to apply to the general field of reconstructing nonuniformly sampled data. Since the various other approaches in this field typically have their own descriptions, we prefer to keep the terminology consistent with its initial introduction in the context of ℓ_1-norm minimisation[15,16] and, by extension, application to solving the more general situation of ℓ_p-norms where $0 < p \leq 1$, a definition also used by Kazimierczuk and Orekhov.[35]

Using the definition of CS above, a number of reconstruction approaches can be classified as similar to the CS method. For example, the forward maximum entropy reconstruction procedure (FM) has been extended to include ℓ_1-norm minimisation as a target function, although negative values of the Shannon, Skilling or Hoch/Stern entropies may also be used[52] and original implementations employed the Shannon entropy.[53] The CLEAN procedure[54–58] relies on iterative signal selection and artefact removal comparable to the iterative thresholding CS implementations;[59] these algorithms were developed empirically without the theoretical basis of the CS framework. Depending on the implementation, there are some differences, for example signal removal may be carried out in the frequency rather than time domain,[54] at each iteration selection may include only a single resonance and some implementations may use line-shape fitting.[56] A more recent extension of the CLEAN approach is SCRUB (Scrupulous CLEANing to Remove Unwanted Baseline Artifacts); this has its basis in the CLEAN-type algorithms and thus has a number of comparisons to CS-iterative thresholding implementations, but instead of processing the whole spectrum, looks at groups of peaks above a threshold and then carries out a CLEAN-based algorithm on each group of peaks, again identifying only the strongest signal within a group at each iteration.[60] Another algorithm for iterative artefact removal is SIFT (spectroscopy by integration of frequency and time domain information),[61] which also relies on an iterative reconstruction procedure based on the Gerchberg–Papoulis algorithms;[62,63] nevertheless, the SIFT approach is designed to rely on information about 'dark' (information-free) points in the spectra. Whilst CS can readily incorporate such prior knowledge, the CS theorem only requires a spectrum to be sparse, without knowledge of the signal location. Furthermore, unlike the iterative thresholding approaches used for CS reconstruction, there is no iterative signal removal in SIFT. Consequently, these approaches will not be discussed further.

10.6 Current Applications

An early application of CS methodology to NMR was the use of the iterative soft thresholding (IST) algorithm, which is shown to be equivalent to ℓ_1-norm minimisation, to extend a truncated time-domain signal.[37,38] In this application, truncated time-domain data was initially extended with zeros, followed by discrete Fourier transformation (DFT), thresholding, inverse DFT (IFT) and then substitution of the new points in place of the zero-filled points. This process was repeated iteratively until no further change was seen in the spectrum. Whilst extension to non-uniformly sampled data was not undertaken in this work, the potential for this was discussed as well as the similarity with the maximum entropy approach.

Application to non-uniformly sampled NMR data was presented in two papers in 2011.[21,22] Work by ourselves[21] demonstrated application of the CS methodology to undersampled 2D spectra of ubiquitin allowing assessment of the error relative to fully-sampled FT spectra. 30% sampling allowed a reduction of recording time from 2 h 45 min to 50 min (128 to 30 complex pairs), whilst the root mean squared (RMS) error for peak positions and intensities remained low at 0.02 ppm and 3%, respectively. Extension to 3D backbone assignment spectra was also discussed in the context of HNCA and HN(CO)CA spectra; successful reconstruction with 22.8% and 16% under-sampling was achieved. In comparison with maximum entropy (ME) reconstruction it was noted that CS appears to perform more linearly, and did not bias weaker peaks to lower intensities, although alternative versions of ME reconstruction are available, *e.g.* MINT, where the nonlinearity of the reconstruction can be varied.[64] This suggested that CS is more appropriate for situations where the quantitative interpretation of peak intensities is critical, such as when deriving distance information from NOESY cross peaks and in situations of large dynamic range. It should be noted, however, that in our study the ME implementation in Azara (W. Boucher, un-published) was used and a systematic comparison of the different ME im-plementations was not undertaken. Spectra were reconstructed using the conjugate gradient descent with back-tracking line search approach of Lustig *et al.*[19] to solve eqn (10.4):

$$\min_x \|\mathbf{Ax} - \mathbf{b}\|_2^2 + \lambda\|\mathbf{x}\|_1 \tag{10.23}$$

where λ is a parameter that allows variation of the weighting between data consistency ($\|\mathbf{Ax} - \mathbf{b}\|_2^2$) and sparsity ($\mathbf{x}_1$).

In contrast, Kazimierczuk and Orekhov[22] introduced CS using an alter-native approach (IRLS), minimising the function shown below:

$$\min_x \|\mathbf{Ax} - \mathbf{b}\|_2^2 + \lambda\|\mathbf{x}\|_p \tag{10.24}$$

where if $p = 1$, the penalty function is convex, *i.e.* has a global minimum, whilst if $0 < p < 1$, the function is non-convex. Whilst the non-convex mini-misation may have more than one local minimum, typically the solution is as good as those obtained with ℓ_1-norm minimisation and, furthermore, exact reconstruction is possible with substantially fewer measurements than for the ℓ_1-norm.[43] Kazimierczuk and Orekhov[22] use the iterative soft thresholding (IST)[63] and iterative re-weighted least squares (IRLS) imple-mentations to solve the convex and non-convex minimisation, problems respectively.[44,65] The authors demonstrate results for 2D HSQC, 2D NOESY and 2D DQF COSY spectra, suggesting that the IRLS approach allows faster convergence in situations of higher dynamic range, although for the IRLS approach with $p = 1$, the convergence and results are similar to IST, but IRLS is slower. Details of the stopping criteria, however, are not discussed.

More recently, the popularity of the iterative thresholding implementations has continued with applications to 3D and 4D NOESY spectra.[23,39] Hyberts *et al.*[23] used an IST variant, istHMS, which is run on a large CPU cluster. In the istHMS implementation data points above the threshold, calculated as a proportion of the tallest peak, rather than using a fixed threshold, are truncated at the level of the threshold; inverse FT to re-generate time-domain data at the sampled points is then carried out on the truncated spectrum, before iterating, reducing the threshold at each step. This is equivalent to a shrinkage operation on the peaks above the threshold,[37] whilst setting points not recorded to zero ensures artefacts from these positions are minimised. Since the IST implementation relies only on FFT and IFT operations, it is orders of magnitude faster than other methods of NUS reconstruction, *e.g.* forward maximum entropy (FM), which rely on matrix multiplication.[23] High quality reconstructions of both 3D and 4D datasets are demonstrated with excellent fidelity in recovery of peak intensities.

Analysis of the strengths and limitations of the CS approach when applied to crowded and high dynamic range data have been discussed by ourselves.[39] We investigated the reconstruction of a 3D ¹⁵N NOESY HSQC spectrum of the 7TM α-helical membrane protein pSRII. The molecular weight of pSRII is 26 kDa, but at 35 °C, the protein tumbles like an 80 kDa protein. Severe overlap and the wide range of peak intensities make this a particularly challenging application to test the accuracy and robustness of the CS approach. Highly accurate reconstructions using the IHT approach were demonstrated for both crowded and less crowded regions of the spectra. In less crowded regions a sampling fraction as low as 20% was possible, whilst in the more crowded regions of the 3D NOESY 30–40% sampling was advised. Some examples of strip plots for different amino acid residues for the CS and FT reconstructions of the 3D ¹⁵N NOESY are shown in Figure 10.4. RMS errors in intensity between the CS and FT versions are estimated as ∼10% for the weakest peaks (SNR ∼1.2–6), falling close to 0% for SNR > ∼10². With greater undersampling, a much slower decay of intensity errors as SNR increases is observed; consequently, at lower sampling fractions even the more intense peaks may show quite significant errors. Whilst errors in peak position (mean chemical shift deviation) also decrease with increasing SNR, there is much less variation between different sampling fractions. Generally, all errors fall within the digital resolution of the spectra.

At the time of writing, our preferred algorithm is a modified version of the IHT procedure.[34] Unlike the soft thresholding approach where shrinkage of peaks occurs, in the IHT approach, a threshold is set as a proportion of the maximum peak intensity; peaks above this threshold are selected and stored in a separate spectrum initiated with zeros and the IFT of this is used to remove the contribution from the largest peaks and any noise resulting from

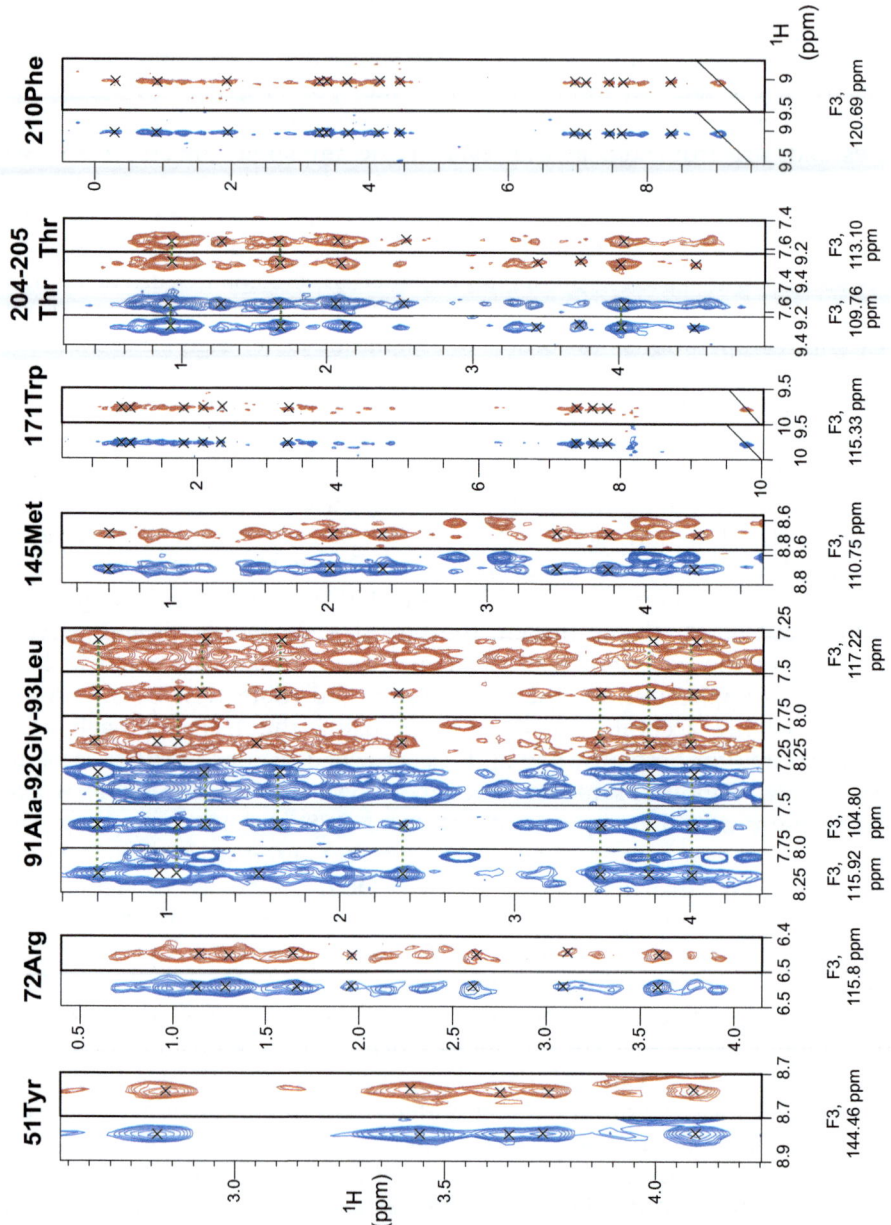

truncation of their time-domain FIDs, before repeating iteratively. However, despite success with this approach, as discussed above, a number of other algorithms are available and we used the challenging case of the 3D ^{15}N NOESY recorded on pSRII to test a range of different algorithms. Gradient optimisation approaches tend to be quite slow[23] making them less suitable for application to high dimensional spectra, such as 3D and 4D datasets. Consequently, we did not continue with application of the conjugate gradient descent procedure from our initial work.[21] We compared the IHT, IST and YALL1 approaches. Overall we found that all three approaches produce very comparable results, suggesting an inherent robustness of the CS approach to the algorithm used for the ℓ_1-norm minimisation. In our hands we observed slightly improved performance of the IHT and IST approaches, shorter reconstruction times, and slightly higher quality spectra at lower sampling factors; however, overall the performance of the three approaches was very comparable. In comparing the different CS implementations, it was important to compare situations of comparable weighting between the data consistency ($\|\mathbf{Ax} - \mathbf{b}\|_2$) and sparsity ($\|\mathbf{x}\|_1$) terms. In our implementations, enforcing stricter data matching (lower value for $\|\mathbf{Ax} - \mathbf{b}\|_2$) led to higher artefact levels in the spectra and consequently we prefer to weight the reconstruction towards enforcing sparsity (discussed further in Section 10.4).

Whilst in many cases application of CS to 2D spectra is of less significance for practical applications to proteins and has been used primarily as an example of the methodology,[21] application of CS to determination of coupling constants has been demonstrated;[66] here it is necessary to record both a decoupled and coupled experiment leading to longer recording times. Thiele and Bermel[66] investigated measurement of one-bond scalar and RDC couplings using the IRLS CS implementation with $\ell_{p \to 0}$-norm. The authors observed highly accurate reconstructions and measurements of coupling constants. In line with CS theory, they observed varying minimum possible sampling fractions (below which peaks are lost and spectral quality and accuracy of coupling constants may decline) depending on the number of peaks observed using the different acquisition schemes *i.e.* sparsity. For the

---

Figure 10.4   [^{1}H–^{1}H] (F_1, F_2) strip plots from 3D NOESY ^{15}N HSQC spectra recorded on pSRII (0.5 mM ^{15}N V17C pSRII, 800 MHz, 35 °C). The spectra were reconstructed using either CS iterative hard thresholding (IHT) with 40% sampling (ns = 16, experiment time: 40 h) (red) or Fourier transformation of the fully sampled data (ns = 16, experiment time: 100 h) (blue). Selected crosspeaks are marked with black crosses; crosspeak assignments are omitted for clarity. Dashed lines (green) between strip plots indicate sequential NOEs. Black diagonal lines indicate the positions of the diagonal auto-correlations.
Adapted from *J. Biomol. NMR*, Compressed sensing reconstruction of undersampled 3D NOESY spectra: application to large membrane proteins, **54**, 2012, 15–32, M. J. Bostock, D. J. Holland and D. Nietlispach, (© Springer Science + Business Media B.V. 2012) with permission of Springer.[39]

different sparsity levels, sampling fractions may vary from ~20–30% of 4k complex points down to 5% of 4k complex points. In line with other applications of CS, resolution enhancement may also be achieved, in this case increasing resolution to 16k complex points with sampling at 1.25%, or for particularly sparse situations, a combination of resolution improvements and time-savings may be achieved.

A further application to 2D spectra has come from assignment of near-symmetric β-cyclodextran derivatives. The CS approach using the IRLS implementation provided considerable time savings for acquisition of 2D ^{13}C HSQC (5% sampling), 2D ^{13}C HSQC-TOCSY (20% sampling) and 2D ^{13}C HSQC-NOESY (20% sampling) spectra allowing significantly improved resolution in the same time for the HSQC (1 h) and allowing reduced experiment times of 9 h and 10 h for the TOCSY and NOESY experiments respectively.[67] This demonstrates the potential impact of CS for a wide-range of NMR applications.

A more recent application of CS methodology to NMR applications involves the use of iterative thresholding for fitting pulsed gradient spin echo (PGSE) data to determine diffusion coefficients.[68] Typically such data has been analysed by methods such as CONTIN,[69] maximum entropy[70] and non-negative least-squares (NNLS).[71] Both CONTIN and ME are forms of Tikhonov regularization[72] solving a least-squares problem of the form:

$$\|\Phi\mathbf{x} - \mathbf{b}\|_2^2 + \lambda\Theta(\mathbf{x}) \tag{10.25}$$

where Φ is a Laplace Transform Matrix, $\Theta(\mathbf{x})$ is a regularisation function and τ adjusts the balance between experimental data and regularisation. For ME, $\Theta(\mathbf{x})$ uses a form of the Shannon entropy, whilst for CONTIN, an ℓ_2-norm is used. Urbańczyk *et al.*[68] suggest the use of the ℓ_1-norm as the regularisation parameter, implementing an algorithm based on FISTA.[73] The authors report improved stability compared to CONTIN and NNLS and it appears most successful at recovering the weakest peaks, consistent with other reports that the CS methods are more 'linear'.[23,39] Regularisation with the ℓ_1-norm will enforce a sparse solution, which in the case of diffusion data will not always be appropriate. It may be possible to overcome this limitation by using model-based approaches, such as has been demonstrated for T_1 and T_2 estimation in imaging applications.[74] Nevertheless, the potential for overfitting in the case of polydisperse samples means the method should be used with care, making it most suitable for monodisperse samples.

The flexibility of the CS approach allows implementation in a variety of situations, provided the conditions of sparsity and randomness of sampled points can be assumed. Shrot and Frydman[75] applied the CS approach in the context of single-scan ultrafast 2D NMR. Here the objective is to reduce the amplitude of the decoding gradient, which for FT acquisition is applied as an oscillating square-wave gradient during the acquisition period; this can ease hardware demands and increase SNR. Refocusing of the gradient-encoded magnetisation to give the t_1 points occurs during the t_2 period; in order to ensure incoherent sampling, the square-wave oscillating gradient for the FT version is replaced with a pseudo-random oscillating gradient. Consequently, refocusing of the encoded t_1 time-points occurs in a non-uniform manner, effectively giving a pseudo-NUS sampling schedule. The authors use the IST approach with 250 iterations to reconstruct a 2D spectrum and demonstrate that the refocusing gradient amplitude can be reduced from 38 G cm^{-1} to 7.5 G cm^{-1}, allowing a reduction in the receiver bandwidth and an increase in SNR.

A broad range of CS applications from the literature has been discussed above. A key question in this field is the relative benefits of the convex and non-convex CS reconstruction approaches. To date, to our knowledge, the IRLS approach in the context of NMR spectra has only been used by Kazimierczuk and Orekhov,[22] Thiele and Bermel[66] and Urbańczyk *et al.*[68] An alternative non-convex minimisation has been proposed by Qu *et al.*,[76] who used minimisation of the $\ell_{0.5}$-norm using the '*p*-shrinkage operator with continuation algorithm' (PSOCA), showing lower reconstruction errors with the $\ell_{0.5}$-norm minimisation approach compared to reconstruction *via* ℓ_1-norm minimisation. To date, the relative benefits of IRLS *versus* standard ℓ_1-norm minimisation have not been widely discussed. Some work to address this has been carried out by Kazimierczuk and Orekhov[35] looking at simulations as well as some real spectra with the IRLS and IST approaches. Reconstructions for different sparsity levels are compared by taking the average ℓ_2-norm of the residual over 50 sampling schedules with different seed numbers for each sparsity level. A reconstruction is deemed to be 'exact' when the average ℓ_2-norm of the residual falls below the noise level in the FT spectrum, where the noise is calculated as 10% of the ℓ_2-norm of the fully-sampled FT-processed spectrum. Using simulations of 1D spectra, the authors demonstrate that for spectra with a fast-decaying distribution of non-zeros (narrow distribution of spectral magnitude), IRLS may outperform IST achieving 'exact reconstructions' at a lower sparseness than for IST, whilst when the distribution of spectral magnitudes is broader, IST and IRLS are comparable. IRLS also appears to outperform IST in a situation of high dynamic range (again a narrow distribution of spectral magnitudes), achieving an 'exact' reconstruction at a lower sparsity than CS. This appears to be borne out for real spectra with a 2D ^{15}N-HSQC and 3D HNCO of ubiquitin (76 amino acid residues) showing comparable results for IRLS and IST (IRLS slightly outperforms IST for the 3D HNCO), whilst for a 3D ^{15}N HSQC-NOESY of azurin (128 amino acid residues), IRLS appears to outperform IST allowing a sampling sparsity of around 25% compared to around 50% for IST. These observations appear to be in line with CS theory which indicates that for non-convex minimisation of the ℓ_p-norm where $p \in [0,1]$ sampling requirements are between that for the ℓ_0-norm ($2k$ sampling points where k is the number of non-zeros, with $k \ll N$ and N is the length of the signal vector) and the ℓ_1-norm ($O(k\log(N/k))$). This suggests that the IRLS implementation may be a serious contender for reconstruction of NMR spectra. However, it should be noted that whilst the average ℓ_2-norm of the residual provides a good measure of convergence to the 'correct' FT spectrum, this provides no indication of the accuracy of the reconstruction in terms of peak positions and intensities. In fact the ℓ_2-norm can be distorted by large signals owing to the squared dependence. In our own study on 3D NOESY data, we found that enforcing greater data consistency by forcing a lower ℓ_2-norm between the raw and reconstructed data led to higher artefact level in the spectra without any appreciable improvement in the quality of the spectra.[39] Furthermore, whilst Kazimierczuk and Orekhov[35] found that a sampling sparseness of 50% (11 200 out of 22 400 complex points in the indirect dimensions) is required for the average noise level in the IST reconstruction to fall below the FT-noise level, we found high quality spectra of a much larger protein than azurin (pSRII, 241 residues) using the IHT approach at 30–40% sampling (1287–1716 out of 4290 complex points in the indirect dimensions); we also tested IST at 40% sampling and found comparable error levels in peak position and intensity to the IHT reconstruction. Thus, whilst there is clearly evidence that non-convex minimisation may outperform IST (and may allow faster computational times at low sampling sparseness[35]), further tests will be needed to assess the impact of IRLS in comparison with other algorithms on the final spectral reconstruction quality.

As discussed earlier, a key criterion for application of the CS approach is sparsity in a particular domain. Many of the papers discussed so far use the sparsity of the frequency domain (*i.e.* after Fourier transform).[21–23,39] Nevertheless, other transforms are possible, including difference methods, such as total variation,[30] or differences between multiple spectra,[77] which have been used in imaging applications as well as wavelet transforms. The wavelet transform was first applied to ℓ_1-norm minimisation of simulated NMR data by Drori[26] solving the following optimisation problem *via* the iterative soft-thresholding algorithm:

$$\min_x \|x\|_1 \quad \text{subject to} \quad \|y - SAW^T x\|_2 \tag{10.26}$$

where y is an $n \times 1$ observation vector, S is a random sampling operator, A is the inverse Fourier transform, W is an orthogonal 2D wavelet transform and x is the frequency domain spectrum. More recently, Qu *et al.*[76] compared the sparsity of NMR spectra in the wavelet domain with the frequency domain for ^1H–^1H and ^1H–^{13}C COSY spectra. They observed a faster decay of coefficients in the real part of the spectrum in the frequency rather than wavelet domain (using a symmlet wavelet transform), indicating greater sparsity in the frequency domain. Furthermore, the authors observed loss of some peaks as well as reduced peak height when reconstruction was carried out from selected wavelet coefficients, suggesting a slightly lower reconstruction quality. In our own experience (data not published), we also found no notable improvement in reconstruction quality when using the Daubechies wavelet transform. However, with many possible sparsifying wavelet transforms available, it is possible that for other cases improvements may be seen. Furthermore, for spectra that are not represented sparsely in the frequency domain, as are often encountered in solid state NMR spectroscopy, the use of the wavelet transform may be beneficial. It should be noted that Qu *et al.*[76] describe reconstruction of spectra from sparse frequency domain coefficients as *'self-sparse'* spectra, indicating that no further transforms, *e.g.* wavelet or TV, are required to enter a sparse domain. However, we prefer to describe this as sparsity in a particular transform domain, in this case the frequency domain.

10.7 Applications to Higher Dimensional Spectroscopy

Whilst a number of applications of the CS methodology in the literature have already been reviewed, we would particularly like to focus on the potential for CS to expand the appeal of high-dimensional experiments, in particular 4D datasets. Separation of peaks into an additional dimension in 4D spectra can significantly reduce spectral crowding, allowing extraction of more data compared to a 3D experiment where many peaks may be overlapped. Furthermore, additional resonance correlations may be elucidated and the process of spectral assignment significantly simplified. However, whilst such experiments provide extremely useful structural information, achieving sufficient resolution and SNR within a viable experiment time is very challenging without the use of undersampling approaches. Using conventional processing methods that are based on full sampling, the extremely long experiment times and losses in SNR incurred owing to additional frequency discrimination and longer coherence transfer pathways typically outweigh the benefits of 4D spectroscopy, precluding its widespread use to date. The combination of 4D experiments with non-uniform sampling (NUS) and CS reconstruction can overcome these limitations, allowing the benefits of such 4D experiments to be realised within competitive time periods and with good SNR, offering a significant advantage when assigning large proteins.

A number of 4D experiments have been proposed in the literature in combination with various reconstruction methods, including: 4D methyl HCCH NOESY spectra combined with multi-dimensional decomposition (MDD) reconstruction;[3] 4D methyl–methyl HCCH and amide–methyl HNCH NOESY spectra with coupled-MDD (Co-MDD);[78] a 4D HCC(CO)NH-TOCSY used to replace two complementary 3D experiments, the H(CCCO)NH-TOCSY and (H)C(CCO)NH-TOCSY, which produce proton and carbon correlations, respectively, and are typically used in small- to medium-sized proteins to assign correlations between amide residues and side chain atoms of the previous residue;[79] a 4D HCCH-TOCSY experiment using random concentric shell sampling with reconstruction *via* the FFT and artefact removal *via* the CLEAN procedure;[54] a 4D time-shared NOESY experiment reconstructed using SCRUB, a variant of the CLEAN procedure, which acquires amide–amide, methyl–methyl, amide–methyl and methyl–amide correlations,[60] and approaches for 4D backbone assignment experiments, *e.g.* 4D HNCACO, 4D HNCACACB, 4D HB/HACB/CANH and 4D HN(CA)NH.[80] However, many of the available reconstruction methods are restricted to situations of high SNR ratio,[79] suffer from poor reconstruction quality in situations of high signal dynamic range,[53] or are computationally expensive.[23] The high-quality reconstructions using the CS approach, including situations of high dynamic range, as well as shorter reconstruction times make the combination of CS with 4D experiments an extremely attractive prospect. Hyberts *et al.*[23] demonstrated the use of 4D ^{13}C-dispersed methyl–methyl NOESY experiments on a 1 mM sample of MED25 in complex with the VP16 transactivation domain using 17.5% sampling (20 698 of 118 272 complex points in the indirect dimensions) and four scans resulting in a recording time of 7.5 days. Using a 0.9 mM sample of the 10 kDa B1 domain of protein G a 4D HCCH NOESY spectrum was obtained with 0.8% sampling (10 800 of 1 350 000 indirect complex points) and four scans within 5 days.

We recorded a methyl-selective 4D HCCH NOESY experiment with 7.3% exponential sampling (7000 of 95 680 complex points) in 7 days with eight scans on a 0.5 mM highly deuterated, selectively methyl-protonated U-[^{2}H,^{13}C,^{15}N] Ileδ1-[^{13}CH$_3$] Leu,Val-[^{13}CH$_3$,^{12}CD$_3$], Ala-[^{13}CH$_3$] labelled sample of the 26 kDa seven transmembrane α-helical membrane protein, pSRII. At 35 °C the protein–detergent complex tumbles with a correlation time corresponding to a molecular weight of 80 kDa. Spectrum reconstruction was carried out using the IHT procedure, with the threshold set to 90% of the maximum peak intensity. Reconstruction took one week in MATLAB on an AMD Phenom II, Quad Core 3.0 GHz processor with 8 GB memory based on an algorithm not optimised for speed and using a machine with shared usage resulting in a variable load. Consequently, it is clear that significant improvements in reconstruction time can be achieved.

An example of a representative 2D plane from the full CS reconstruction of a NUS experiment run in 7 days is shown in Figure 10.5 (red) together with the equivalent plane from an FT reconstruction (blue) from an experiment recorded over 16 days; the benefits of the NUS approach with CS reconstruction are immediately obvious. The NUS-CS spectrum shows considerably higher resolution compared to the FT spectrum as well as increased SNR, demonstrated, for example, by the peaks at the bottom left and right of the spectrum (93 Leu, 102 Leu, 137 Leu). From the overlay, it is clear that it would not be possible to resolve the peak positions found in the NUS-CS spectrum, using the FT spectrum. The FT spectrum was recorded with spectral widths of 1200, 3012, and 3012 Hz in the indirect dimensions ^{1}H, ^{13}C, and ^{13}C, respectively, at 600 MHz. The CS-reconstructed spectrum was recorded with spectral widths of 1600, 2480, and 2480 Hz in the indirect dimensions at 800 MHz. Taking into account the difference in spectrometer frequency, the indirect proton dimension is properly scaled for both experiments, whilst the spectral widths for the carbon dimensions at 800 MHz are smaller (equivalent spectral width at 800 MHz would be 4016 Hz). In part, the dramatic difference in peak widths results from the difference in spectrometer frequency at which the two experiments were recorded, together with the representation in Figure 10.5 using a ppm scale and also the

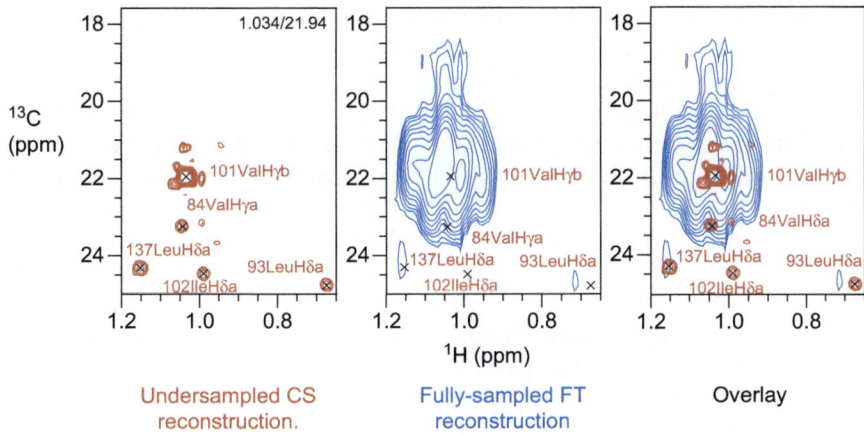

Undersampled CS reconstruction. Fully-sampled FT reconstruction Overlay

Figure 10.5 Comparison of a CS reconstruction of an undersampled methyl-selective 4D HCCH NOESY experiment (red), using 7% sampling (recording time one week) with the fully sampled FT-processed experiment (blue) (recording time 16 days). For comparison, peak positions from the CS-processed experiment (black crosses) are shown on the FT spectrum. The recording time for the CS data was only 44% of the FT experiment, illustrating the dramatic improvements in signal-to-noise achievable through CS. Chemical shift positions in the indirect proton and carbon dimensions (ppm) are shown in the top right of the first panel.

Table 10.1 Maximum evolution times in the indirect dimensions of the FT- and CS-processed versions of the 4D HCCH experiment.

	t_{1max} (ms) (^{1}H)	t_{2max} (ms) (^{13}C)	t_{3max} (ms) (^{13}C)
FT (full sampling)	13.33	6.64	6.64
CS (NUS)	28.75	20.96	16.13

reduced ^{13}C spectral widths employed at 800 MHz, leading to a resolution benefit per sampled data point. However, the overwhelming improvement in peak shapes is the result of the powerful combination of NUS and CS. The use of extensive undersampling allows a considerable increase in spectral resolution, as evidenced by Figure 10.5. The t_{1max} values for both experiments are shown in Table 10.1; a two-fold increase in the indirect proton dimension and an approximately three-fold increase in the indirect carbon dimensions compared to the FT experiment are achieved.

The maximum evolution times for the CS spectrum (Table 10.1) are comparable to the ones used by Tugarinov *et al.*[3] When compared to the fully sampled case, this represents a considerable increase in resolution, but still falls far short of sampling to $1.26 \times T_2$,[81] assuming $t_2 \sim 40$ ms. However, our comprehensive analysis of the 4D spectrum shows that for this entirely α-helical 7TM protein, the resolution is more than sufficient to effectively exploit the available distance information to a maximum. For illustration, the 2D plane shown in Figure 10.5 indicates that the CS spectrum is able to distinguish NOEs to four additional methyl groups, while with the resolution in the FT spectrum, this is clearly not possible.

The benefits inherent to NUS approaches can be used to increase the resolution or by collecting more scans per increment in the earlier regions of the evolution periods where the signal is more abundant. In our case the returns from sampling only 7% of the full-matrix was used entirely to increase the resolution. The fully-sampled spectrum was recorded on a room-temperature probe with 24 scans, whilst the CS-processed spectrum was recorded on a cryoprobe with eight scans, preventing a fair comparison in terms of absolute SNR. However, based on other data sets (data not shown) it is clear that with NUS-CS high quality 4D data can be recorded in a fraction of the time required for full sampling and with dramatically increased resolution.

A key advantage of high-dimensional spectra is the simplicity of the assignment procedures, providing significant benefits when carrying out structural studies of large proteins such as membrane proteins. In the case of the 4D HCCH NOESY experiment shown in Figure 10.6, a 2D ^{1}H–^{13}C correlation experiment is used to navigate to a particular diagonal peak position and a 2D ^{1}H–^{13}C strip plot is extracted in this case at the Hγb/Cγb methyl group frequencies of Val 101 (orange arrow); the other peaks in this plane represent NOEs from the diagonal peak (purple arrows), which can be easily assigned (blue arrows) by referring back to the 2D ^{1}H–^{13}C correlation experiment (green). Comparing the 4D HCCH spectrum with that acquired

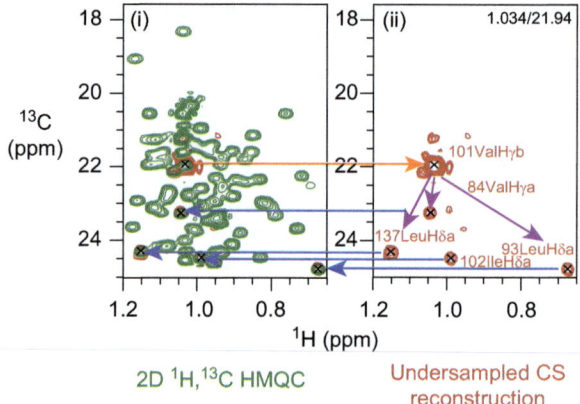

2D ^{1}H,^{13}C HMQC   Undersampled CS reconstruction

Figure 10.6   Illustration of the assignment procedure for the CS-reconstructed methyl-selective 4D HCCH NOESY experiment (7% sampling, recording time one week). A representative 2D ^{1}H–^{13}C strip plot (red) extracted at the Hγb/Cγb methyl group frequencies of Valine 101 (orange arrow) (ii) reveals long-range NOEs from methyl groups of neighbouring residues (purple arrows), which can be easily assigned (blue arrows) by direct comparison with a readily recorded and assigned 2D ^{1}H–^{13}C correlation experiment (green), by simply overlaying the spectra (i), providing direct information on the tertiary structure of the protein. Chemical shift positions in the indirect proton and carbon dimensions (ppm) are shown in the top right of panel (ii).

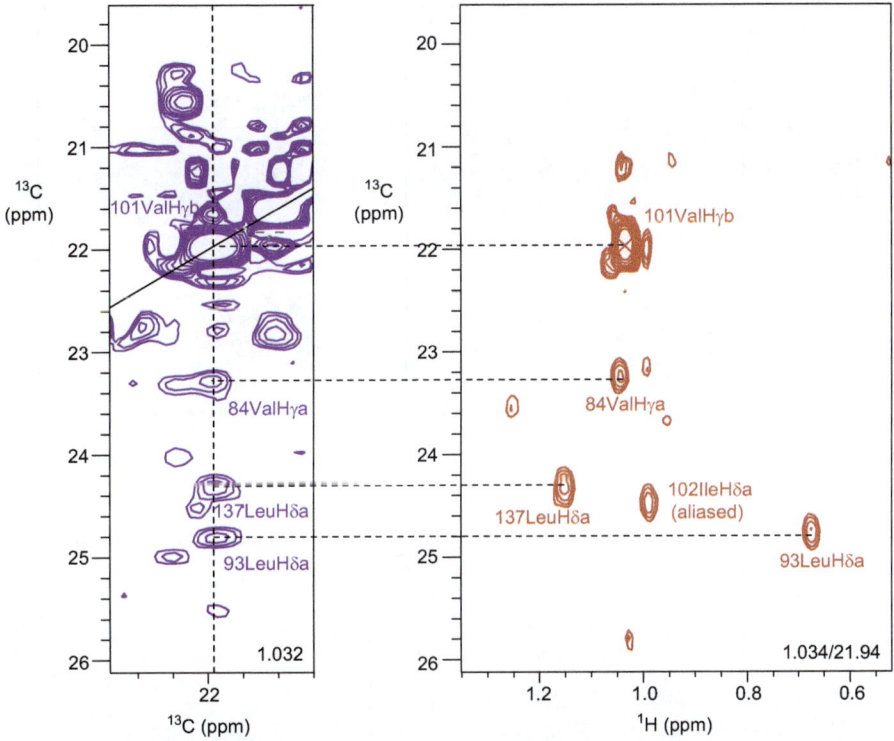

Figure 10.7   Comparison of NOEs from Val 101Hγb using a 3D (H)CCH experiment (purple), where NOEs are indicated by peaks on the vertical dashed line or using a 4D HCCH experiment (red). The diagonal peak is 101ValHγb, indicated by the diagonal line in the 3D experiment and a red cross for the 4D experiment. Shifts in the indirect dimension(s) are shown in the bottom left of each strip. Unassigned peaks are owing to peaks not centred in this plane.

from an equivalent 3D (H)CCH experiment (Figure 10.7), it can be seen immediately that the NOEs observed in the CC plane of the 3D experiment are also detected in the 4D experiment. However, the plane from the 4D is considerably less crowded, making assignment easier and furthermore the NOE peaks show comparable intensity, indicating no detrimental loss of SNR despite the addition of an extra dimension. In contrast to the 4D HCCH, the assignment procedure for a 3D ^{13}C NOESY is considerably more time consuming involving matching forward and return cross-peaks mostly between different planes. A further limitation of the 3D NOESY assignment procedure is shown in Figure 10.8. Here two cases are displayed where in the 3D the detection of the cross peak is challenging due to the presence of a very intense diagonal signal nearby, which compromises both exact peak positions and intensities of the cross peaks. In addition, since the NOE has the opposite sign to the diagonal, owing to aliasing of the peak, the NOE could easily be confused as a truncation artefact. In contrast, as can be seen

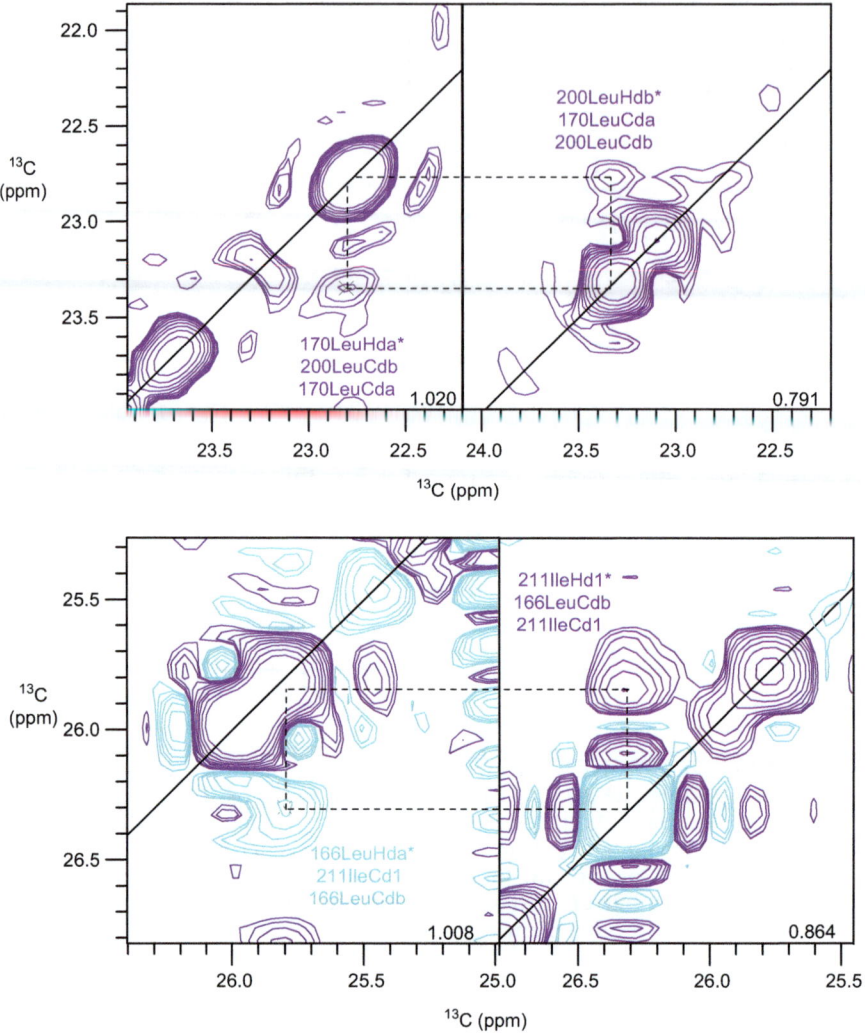

Figure 10.8 Two sections from a methyl-selective 3D (H)CCH NOESY showing the difficulties of NOE assignment when the cross peak is close to an intense diagonal signal. Blue peaks are negative in intensity and purple peaks positive. Negative cross peaks result from the (90, −180) phase setting in F_3. The ^{1}H shift for the strips (ppm) is shown in the bottom right-hand corner of each strip. Diagonal peaks fall along the diagonal line.

in Figure 10.9, in the 4D HCCH the cross-peaks are clearly distinguishable and not hidden by the diagonal peaks, making it less likely that important interhelical NOEs will be missed owing to their being obscured by the diagonal signals.

NOE-derived distances are a key restraint for protein structure determination, and in the original structure determination of pSRII, long-range

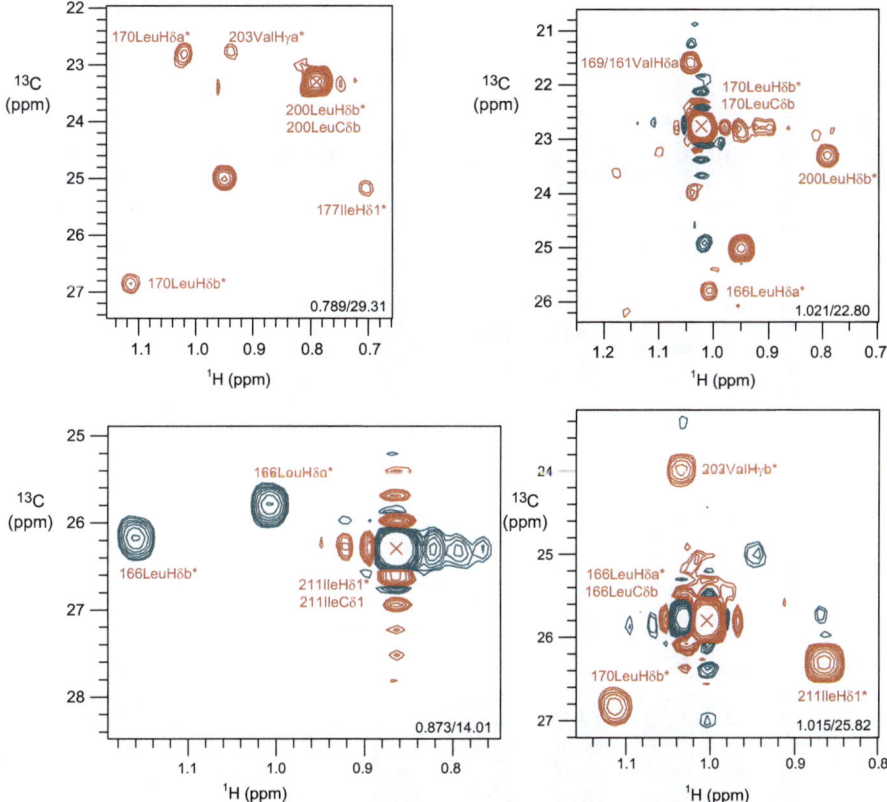

Figure 10.9 2D planes from a methyl-selective 4D HCCH NOESY for the residues shown in the 3D (H)CCH NOESY in Figure 10.8, demonstrating that the crosspeaks in the 4D are more easily distinguished from the diagonal peak compared with the 3D NOESY. Diagonal peaks are distinguished by a red cross. Negative peaks are shown in dark blue.

interhelical NOE restraints were critical for the orientation of the secondary structure.[82] In order to assess the quality of the 4D HCCH spectra obtained, the number of experimentally observed NOEs was compared with the number of theoretically expected NOEs derived from the 30 lowest energy structures for pSRII (PDB 2KSY). Distances were measured using the software MolMol,[83] averaging according to $\langle r^{-6} \rangle^{-1/6}$, and a set of expected distance restraints produced. NOEs in the 3D (H)CCH (fully sampled, DFT reconstructed) and 4D HCCH (undersampled, CS reconstructed) spectra were assigned using Analysis[84] and a subset of unidirectional and unambiguous NOE restraints was generated from the experimental restraint lists. For consistency distances were assigned to the restraints from the 3D and 4D spectra based on the distances measured in MolMol. Since a given pair of residues may show multiple restraints between the different combinations of methyl groups, distances were assigned in order of magnitude,

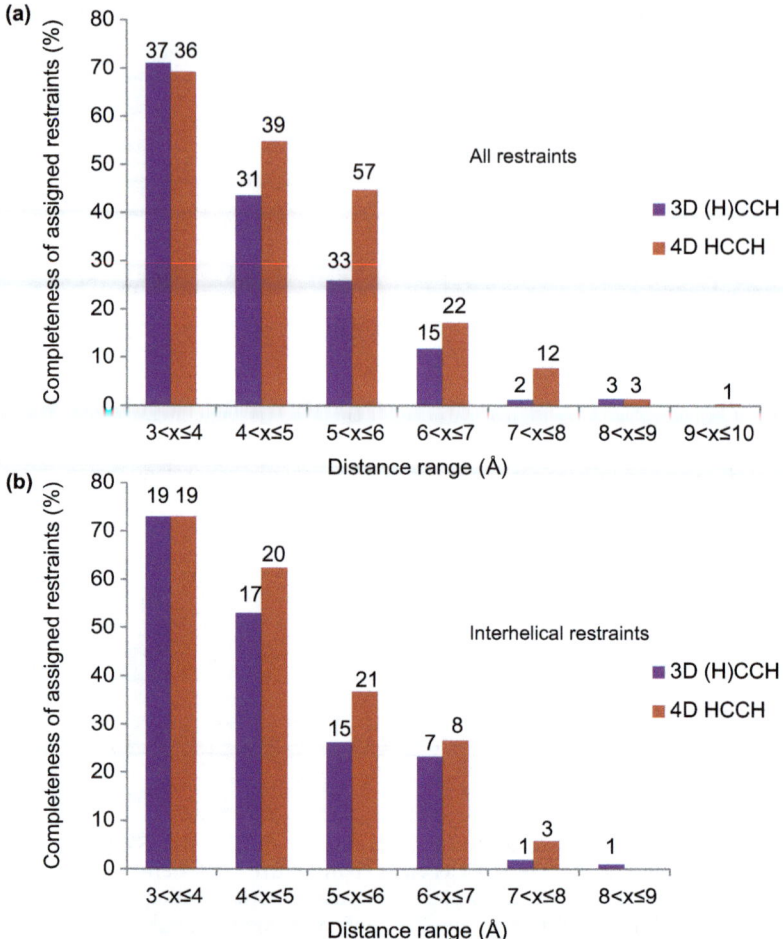

Figure 10.10   Percentage completeness of inter-methyl NOE restraints from 3D (H)CCH (purple) or 4D HCCH (red) NOESY methyl-selective spectra in different distance categories. The absolute number of restraints is shown above the bars. (a) all restraints, whilst (b) shows only restraints between different helices of the 7TM bundle.

i.e. the first restraint was assigned the shortest distance for that pair of residues. The results are shown graphically in Figure 10.10 as percentage completeness. Figure 10.10(a) shows all assigned restraints and it can be seen that apart from the first category, $3 < x \leq 4$ Å, in all cases more NOEs are observed in the 4D HCCH experiment (red) than in the 3D (H)CCH (purple), indicating that provided there is adequate resolution, separating peaks into a fourth dimension significantly increases the number of NOE peaks assigned and that CS processing of the 4D NUS data dramatically offsets the loss in SNR anticipated by the introduction of the additional fourth dimension to a level comparable to the 3D. The proportion of restraints observed declines with increasing distance for both 3D and 4D datasets; this is consistent with the r^{-6} distance dependence for the NOE, which results in a significantly slower NOE crosspeak build-up time for longer distances, such that at the mixing time used, 200 ms, longer distance restraints are much less likely to be observed. Nevertheless significant numbers of restraints are still observed up to 9 Å. The absence of short distance restraints ≤ 3 Å reflects the nature of the labelling scheme where only the pro(R) or pro(S) methyl group on leucine and valine residues is protonated. Consequently, NOEs between geminal methyl groups are not observed in the methyl–methyl NOESY, limiting the number of short distance restraints. Furthermore, proportions do not reach 100% in the lowest categories since for valine and leucine residues; the labelling scheme statistically reduces the magnetisation transfer between these amino acid residues to 25%.

Whilst the absolute number of NOE restraints assigned is important, of more significance for the tertiary structure of α-helical proteins is the number of interhelical restraints; typically for α-helical proteins the individual helices are well-restrained using hydrogen bond restraints, interamide NOEs and dihedral angle information,[7] however packing the helices together is more difficult owing to the shortage of long-range restraints between helices. Figure 10.10(b) shows the proportion of interhelical restraints assigned in the 3D (H)CCH and 4D HCCH spectra. Apart from the lowest distance category, the 4D HCCH detects more interhelical restraints than the 3D (H)CCH, although the numbers of restraints detected are closer in this case, indicating that the 3D (H)CCH detects most of the interhelical restraints available from the 4D. Nevertheless, the ease of assignment of the 4D HCCH still makes this an extremely valuable experiment and even the extra 11 interhelical restraints could make a significant contribution to structure calculations. Furthermore, the 4D HCCH experiment was run for 1 week; the SNR could be improved by recording more scans either in the same time (by increasing the undersampling fraction) or by increasing the total recording time and maintaining the same undersampling fraction. This could lead to identification of more of the weaker NOEs which were not observed in this experiment.

10.8 Future Perspectives

The examples discussed in this chapter, as well as the application of CS to a wide range of NMR experiments in the literature, indicate that CS shows great potential for reconstruction of undersampled NMR spectra, allowing improvements in SNR and/or resolution as well as yielding other benefits in less conventional NMR applications, for example single-scan NMR.[75] In the field of multidimensional NMR, CS has already been demonstrated as successful for reconstruction of 3D spectra, including 3D NOESY datasets, which are typically less well reconstructed by other methods. Data shown in this review and also by others[23] indicate that CS has the potential to make 4D spectra more accessible by significantly reducing recording times while providing high resolution and SNR. We demonstrate the benefits of 4D approaches combined with CS reconstruction, in particular the ease of assignment and improved resolution of a 4D HCCH NOESY. Bringing experiment times for higher dimensional spectra into an acceptable range, whilst ensuring sufficient resolution and SNR, should encourage application and development of a wide range of new experiments, allowing improved assignment strategies for large proteins. Combined with hardware changes such as ^{13}C detection to exploit favourable relaxation properties of other nuclei, 5D experiments could also become a more widely used technique.[85] Owing to the large disk-space requirements for such high-dimensional spectra, and consequently long read and write times, alternative approaches allowing reconstruction of only the relevant parts of the spectra will be necessary.[85,86] Nevertheless, we feel that reconstruction of frequency domain spectra (as opposed to peak lists) will still be crucial as, particularly in the case of large proteins, there is no substitute for visual interpretation of spectra to assess data quality and to resolve any remaining overlapped regions.

Faster reconstruction times by implementation of new algorithms, such as FISTA,[73] as well as exploiting the potential of graphical processing units to parallelise the reconstruction procedure[87] will help to bring processing times for the CS approach in line with the FFT. Reconstruction times for 2D and 3D experiments, with appropriate computer hardware are already comparable, however, longer reconstruction times for 4D and higher dimensional experiments make interactive processing, for example for phasing or apodization currently impossible.

As well as application to new higher dimensional experiments, the examples discussed in this chapter, *e.g. J*-coupling determination,[66] indicate that the accuracy of CS also makes it extremely valuable for time-consuming 2D experiments. Applications to relaxation experiments may also be possible, given the extremely long experiment times required to monitor spin relaxation.

10.9   Conclusion

In a relatively short time since its initial introduction to the field of protein NMR, compressed sensing has been demonstrated as suitable for an extremely wide-range of NMR applications. Fast processing times and highly accurate reconstructions, as well as simple implementation and a strong theoretical basis, make CS an extremely valuable reconstruction approach for undersampled experiments. This opens the possibility for new experimental techniques to simplify assignment and acquisition of structural restraints as well as for improvement of existing approaches.

References

1. K. Pervushin, R. Riek, G. Wider and K. Wüthrich, *Proc. Natl. Acad. Sci. U. S. A.*, 1997, **94**, 12366.
2. V. Tugarinov, V. Kanelis and L. E. Kay, *Nat. Protoc.*, 2006, **1**, 749.
3. V. Tugarinov, L. E. Kay, I. V. Ibraghimov and V. Y. Orekhov, *J. Am. Chem. Soc.*, 2005, **127**, 2767.

4. J. Fiaux, E. B. Bertelsen, A. L. Horwich and K. Wüthrich, *Nature*, 2002, **418**, 207.
5. R. Sprangers, A. Velyvis and L. E. Kay, *Nat. Methods*, 2007, **4**, 697.
6. H. J. Kim, S. C. Howell, W. D. Van Horn, Y. H. Jeon and C. R. Sanders, *Prog. Nucl. Magn. Reson. Spectrosc.*, 2009, **55**, 335.
7. D. Nietlispach and A. Gautier, *Curr. Opin. Struct. Biol.*, 2011, **21**, 497.
8. J. C. J. Barna, E. D. Laue, M. R. Mayger, J. Skilling and S. J. P. Worrall, *J. Magn. Reson.*, 1987, **73**, 69.
9. P. Schmieder, A. Stern, G. Wagner and J. Hoch, *J. Biomol. NMR*, 1993, **3**, 569.
10. P. Schmieder, A. S. Stern, G. Wagner and J. C. Hoch, *J. Biomol. NMR*, 1994, **4**, 483.
11. D. Rovnyak, J. C. Hoch, A. S. Stern and G. Wagner, *J. Biomol. NMR*, 2004, **30**, 1.
12. D. Rovnyak, D. P. Frueh, M. Sastry, Z.-Y. J. Sun, A. S. Stern, J. C. Hoch and G. Wagner, *J. Magn. Reson.*, 2004, **170**, 15.
13. K. Kazimierczuk, J. Stanek, A. Zawadzka-Kazimierczuk and W. Koźmiński, *Prog. Nucl. Magn. Reson. Spectrosc.*, 2010, **57**, 420.
14. B. F. Logan, *Properties of High-Pass Signals*, PhD Thesis, Columbia University, New York, 1965.
15. D. L. Donoho, *IEEE Trans. Inf. Theory*, 2006, **52**, 1289.
16. E. J. Candès, J. Romberg and T. Tao, *IEEE Trans. Inf. Theory*, 2006, **52**, 489.
17. D. J. Holland, D. M. Malioutov, A. Blake, A. J. Sederman and L. F. Gladden, *J. Magn. Reson.*, 2010, **203**, 236.
18. S. Hu, M. Lustig, A. P. Chen, J. Crane, A. Kerr, D. A. C. Kelley, R. Hurd, J. Kurhanewicz, S. J. Nelson, J. M. Pauly and D. B. Vigneron, *J. Magn. Reson.*, 2008, **192**, 258.
19. M. Lustig, D. L. Donoho and J. M. Pauly, *Magn. Reson. Med.*, 2007, **58**, 1182.
20. R. Otazo, D. Kim, L. Axel and D. K. Sodickson, *Magn. Reson. Med.*, 2010, **64**, 767.
21. D. J. Holland, M. J. Bostock, L. F. Gladden and D. Nietlispach, *Angew. Chem., Int. Ed.*, 2011, **50**, 6548.
22. K. Kazimierczuk and V. Y. Orekhov, *Angew. Chem. Int. Ed.*, 2011, **50**, 5556.
23. S. G. Hyberts, A. G. Milbradt, A. B. Wagner, H. Arthanari and G. Wagner, *J. Biomol. NMR*, 2012, **52**, 315.
24. B. K. Natarajan, *SIAM J. Comput.*, 1995, **24**, 227.
25. D. L. Donoho, *Commu. Pure Appl. Math.*, 2006, **59**, 907.
26. I. Drori, *EURASIP J. Adv. Signal Process.*, 2007, **2007**, 20248.
27. B. Adcock, A. Hansen, C. Poon and B. Roman, 2013, arXiv:1302.0561 [cs.IT].
28. S. G. Hyberts, H. Arthanari and G. Wagner, *Top. Curr. Chem.*, 2012, **316**, 125.
29. S. G. Hyberts, K. Takeuchi and G. Wagner, *J. Am. Chem. Soc.*, 2010, **132**, 2145.

30. K. T. Block, M. Uecker and J. Frahm, *Magn. Reson. Med.*, 2007, **57**, 1086.
31. J. A. Tropp and S. J. Wright, *Proc. IEEE*, 2010, **98**, 948.
32. G. Davis, S. Mallat and M. Avellaneda, *Constr. Approx.*, 1997, **13**, 57.
33. D. Needell and J. A. Tropp, *Appl. Comput. Harmon. Anal.*, 2009, **26**, 301.
34. K. Bredies and D. A. Lorenz, *SIAM J. Sci. Comput.*, 2008, **30**, 657.
35. K. Kazimierczuk and V. Y. Orekhov, *J. Magn. Reson.*, 2012, **223**, 1.
36. I. Daubechies, M. Defrise and C. De Mol, *Commun. Pure Appl. Math.*, 2004, **57**, 1413.
37. A. S. Stern, D. L. Donoho and J. C. Hoch, *J. Magn. Reson.*, 2007, **188**, 295.
38. J. C. Hoch and A. S. Stern, *NMR Data Processing*, Wiley-Liss, 1996.
39. M. J. Bostock, D. J. Holland and D. Nietlispach, *J. Biomol. NMR*, 2012, **54**, 15.
40. S. Foucart and H. Rauhut, A Mathematical Introduction to Compressive Sensing, *Birkhäuser*, Basel, 2013.
41. J. Yang and Y. Zhang, *SIAM J. Sci. Comput.*, 2011, **33**, 250.
42. M. M. Bronstein, A. M. Bronstein, M. Zibulevsky and H. Azhari, *IEEE Trans. Med. Imaging*, 2002, **21**, 1395.
43. R. Chartrand, *IEEE Signal Process. Lett.*, 2007, **14**, 707.
44. E. J. Candès, M. B. Wakin and S. P. Boyd, *J. Fourier Anal. Appl.*, 2008, **14**, 877.
45. A. E. Yagle, 2009. http://web.eecs.umich.edu/~aey/sparse/sparse11.pdf.
46. S. Ji, Y. Xue and L. Carin, *IEEE Trans. Signal Process.*, 2008, **56**, 2346.
47. P. Schniter, L. C. Potter and J. Ziniel, *IEEE Trans. Signal Process.*, 2009. http://www2.ece.ohio-state.edu/~schniter/pdf/tsp09_fbmp.pdf.
48. V. A. Morozov, *Sov. Math. Dokl.*, 1966, **7**, 414.
49. R. C. Puetter, T. R. Gosnell and A. Yahil, *Annu. Rev. Astron. Astrophys.*, 2005, **43**, 139.
50. F. Bauer and M. A. Lukas, *Math. Comput. Simul.*, 2011, **81**, 1795.
51. D. L. Donoho and B. F. Logan, *SIAM J. Appl. Math.*, 1992, **52**, 577.
52. S. G. Hyberts, D. P. Frueh, H. Arthanari and G. Wagner, *J. Biomol. NMR*, 2009, **45**, 283.
53. S. G. Hyberts, G. J. Heffron, N. G. Tarragona, K. Solanky, K. A. Edmonds, H. Luithardt, J. Fejzo, M. Chorev, H. Aktas, K. Colson, K. H. Falchuk, J. A. Halperin and G. Wagner, *J. Am. Chem. Soc.*, 2007, **129**, 5108.
54. B. E. Coggins and P. Zhou, *J. Biomol. NMR*, 2008, **42**, 225.
55. E. Kupče and R. Freeman, *J. Magn. Reson.*, 2005, **173**, 317.
56. J. Stanek and W. Koźmiński, *J. Biomol. NMR*, 2010, **47**, 65.
57. K. Kazimierczuk, A. Zawadzka, W. Koźmiński and I. Zhukov, *J. Magn. Reson.*, 2007, **188**, 344.
58. J. A. Högbom, *Astron. Astrophys. Suppl.*, 1974, **15**, 417.
59. T. Blumensath and M. E. Davies, *Appl. Comput. Harmon. Anal.*, 2009, **27**, 265.
60. B. E. Coggins, J. W. Werner-Allen, A. K. Yan and P. Zhou, *J. Am. Chem. Soc.*, 2012, **134**, 18619.
61. Y. Matsuki, M. T. Eddy and J. Herzfeld, *J. Am. Chem. Soc.*, 2009, **131**, 4648.

62. R. W. Gerchberg, *Opt. Acta Int. J. Opt.*, 1974, **21**, 709.
63. A. Papoulis, *IEEE Trans. Circuits Syst.*, 1975, **22**, 735.
64. S. Paramasivam, C. L. Suiter, G. Hou, S. Sun, M. Palmer, J. C. Hoch, D. Rovnyak and T. Polenova, *J. Phys. Chem. B*, 2012, **116**, 7416.
65. C. L. Lawson, *Contributions to the Theory of Linear Least Maximum Approximation*, PhD Thesis, University of California, Los Angeles, California, 1961.
66. C. M. Thiele and W. Bermel, *J. Magn. Reson.*, 2012, **216**, 134.
67. M. Misiak, W. Koźmiński, K. Chmurski and K. Kazimierczuk, *Magn. Reson. Chem.*, 2013, **51**, 110.
68. M. Urbańczyk, D. Bernin, W. Koźmiński and K. Kazimierczuk, *Anal. Chem.*, 2013, **85**, 1828.
69. S. W. Provencher, *Comput. Phys. Commun.*, 1982, **27**, 229.
70. M. A. Delsuc and T. E. Malliavin, *Anal. Chem.*, 1998, **70**, 2146.
71. C. L. Lawson and R. J. Hanson, *Solving Least Squares Problems*, Prentice-Hall, Englewood Cliffs, NJ, 3rd edn, 1995.
72. A. N. Tikhonov, *Sov. Math. Dokl.*, 1963, **4**, 1035.
73. A. Beck and M. Teboulle, *SIAM J. Imaging Sci.*, 2009, **2**, 183.
74. M. Doneva, P. Börnert, H. Eggers, C. Stehning, J. Sénégas and A. Mertins, *Magn. Reson. Med.*, 2010, **64**, 1114.
75. Y. Shrot and L. Frydman, *J. Magn. Reson.*, 2011, **209**, 352.
76. X. Qu, D. Guo, X. Cao, S. Cai and Z. Chen, *Sensors*, 2011, **11**, 8888.
77. Y. Kwak, S. Nam, K. V. Kissinger, B. Goddu, L. A. Goepfert, W. J. Manning, V. Tarokh and R. Nezafat, *J. Cardiovasc. Magn. Reson.*, 2012, **14**, W24.
78. S. Hiller, I. Ibraghimov, G. Wagner and V. Y. Orekhov, *J. Am. Chem. Soc.*, 2009, **131**, 12970.
79. M. Mobli, A. S. Stern, W. Bermel, G. F. King and J. C. Hoch, *J. Magn. Reson.*, 2010, **204**, 160.
80. A. Zawadzka-Kazimierczuk, K. Kazimierczuk and W. Koźmiński, *J. Magn. Reson.*, 2010, **202**, 109.
81. D. Rovnyak, M. Sarcone and Z. Jiang, *Magn. Reson. Chem.*, 2011, **49**, 483.
82. A. Gautier, H. R. Mott, M. J. Bostock, J. P. Kirkpatrick and D. Nietlispach, *Nat. Struct. Mol. Biol.*, 2010, **17**, 768.
83. R. Koradi, M. Billeter and K. Wüthrich, *J. Mol. Graph.*, 1996, **14**, 51.
84. W. F. Vranken, W. Boucher, T. J. Stevens, R. H. Fogh, A. Pajon, M. Llinas, E. L. Ulrich, J. L. Markley, J. Ionides and E. D. Laue, *Proteins*, 2005, **59**, 687.
85. J. Nováček, A. Zawadzka-Kazimierczuk, V. Papoušková, L. Žídek, H. Šanderová, L. Krásný, W. Koźmiński and V. Sklenář, *J. Biomol. NMR*, 2011, **50**, 1.
86. K. Kazimierczuk, A. Zawadzka and W. Koźmiński, *J. Magn. Reson.*, 2009, **197**, 219.
87. T. S. Sørensen, T. Schaeffter, K. O. Noe and M. S. Hansen, *IEEE Trans. Med. Imaging*, 2008, **27**, 538.

Subject Index

algorithms
    fast maximum likelihood
        reconstruction (FMLR),
        112–113
    high-resolution iterative
        frequency identification for
        NMR (HIFI-NMR), 103–106
    iterative soft thresholding
        (IST), 283
    lower-value algorithm, 151–153
    multidimensional NMR
        spectroscopy, 271–272
aperture synthesis procedure, 171
automated projection spectroscopy
    (APSY), 107–110

backprojection
    applications of, 158–161
    filtered backprojection,
        156–157
    hybrid backprojection/lower-
        value (HBLV), 154–156
    lower-value algorithm, 151–153
    radial sampling
        projection-slice theorem,
            120–123
        quadrature detection,
            123–124
    from radial to random,
        161–166
    reconstruction
        ambiguity of radially
            sampled data, 149–151
        higher-dimensional
            spectra, 135–137

information content,
        149–151
inverse radon transform,
        134–135
lattice analysis and
        related reconstruction
        methods, 128–130
lattice of possible peak
        positions, 125–127
point response function
        for, 137–149
polar Fourier transform,
        134–135
radon transform,
        130–133
without filtering, 153–154
band-selective excitation short-
    transient (BEST)
    band-selective pulse shapes,
        14–18
    BEST-HSQC *vs.* BEST-TROSY,
        18–22
    ^{13}C-detected experiments,
        22–24
biomolecular NMR, 207–217
Bloch–McConnell equations, 6–7

Cholesky decomposition, 51
CLEAN
    artifacts via decomposition,
        176–177
    in biomolecular NMR,
        207–217
    decomposition, 177–178
    gain parameter, 179–180

historical background,
170–171
implementations of
projection–reconstruction
NMR, 203
randomized sparse
nonuniform sampling,
203–207
uses of, 200–202
mathematical analysis of
and compressed sensing,
193–199
linear equations, 185–192
NUS inverse problem,
181–185
notation, 171–174
problem, 174–176
reconstructing, 180–181
compressed sensing (CS) theory,
102, 268
compressive sampling matching
pursuit (CoSaMP), 273
computer-assisted structure
elucidation (CASE), 247–248
coupling-based flip-back
techniques, 10
covariance NMR
direct covariance NMR, 221–226
indirect covariance NMR
applications of, 236–239,
242–244
artifact detection, 231–235
generalized indirect
covariance (GIC),
226–228
optimizing spectra,
239–242
principle, 226
signal/noise ratio, 228–231
unsymmetrical indirect
covariance (UIC),
226–228

data acquisition approaches, 98–100
DemixC, 245
direct covariance NMR, 221–226

fast maximum likelihood
reconstruction (FMLR), 110
algorithm, 112–113
FDM. *See* filter diagonalization
method (FDM)
filter diagonalization method
(FDM)
4D NMR, 91–93
1D NMR, 78–82
theory
harmonic inversion
problem, 64–68
harmonic inversion
problems, 71–72
hybrid FDM, 70–71
multi-D FDM,
regularization of,
76–78
multi-D FDM, spectral
estimation by,
72–76
multi-D spectral
estimation, 71–72
spectral estimation
problem and
regularized resolvent
transform, 68–70
3D NMR, 86–91
2D NMR, 82–86
4D NMR, 91–93
FMLR. *See* fast maximum likelihood
reconstruction (FMLR)
Fourier transform (FT) NMR,
34–35
free induction decay (FID),
34, 62

generalized indirect covariance
(GIC), 226–228
gradient-echo principles, 38

harmonic inversion problem (HIP),
61, 64–68
^1H–^1H dipolar interactions, 6
higher nominal signal-to-noise
values, 228

high-resolution iterative frequency
   identification for NMR
   (HIFI-NMR), 102
     algorithm, 103–106
hybrid backprojection/lower-value
   (HBLV), 154–156
hybrid FDM, 70–71

indirect covariance NMR
   applications of, 236–239,
     242–244
   artifact detection, 231–235
   generalized indirect covariance
     (GIC), 226–228
   optimizing spectra, 239–242
   principle, 226
   signal/noise ratio, 228–231
   unsymmetrical indirect
     covariance (UIC),
     226–228
interpretation, NMR spectroscopy,
   100–101
iteratively re-weighted least squares
   (IRLS), 277–278
iterative soft thresholding (IST), 283

linear prediction (LP) extrapolation
   application, 54–57
   best practices, 57–58
   coefficients, 51–52
   history of, 50–51
   signals of known phase,
     53–54
   stability requirement, 52–53
linear prediction (LP) method, 63
longitudinal relaxation enhanced
   (LRE)
     amide proton, 11–13
     paramagnetic relaxation
      agents, 7–8
     protons other than amides, 14
     selective spin manipulation,
      8–11
     Solomon and Bloch–
      McConnell equations, 6–7

matrix-pencil method (MPM), 65, 66
maximum entropy (MaxEnt)
   reconstruction
     applications, 263–265
     linearity of, 258–259
     non-uniform sampling,
      259–260
     parameter selection, 257–258
     random phase sampling,
      260–262
     reconstruction and
      deconvolution
       J-coupling, 262–263
       linewidths, 263
     theory, 254–257
maximum likelihood (MLE) estimate,
   111–112
mono-exponential function, 11
multi-D FDM
   regularization of, 76–78
   spectral estimation by, 72–76
multidimensional decomposition
   techniques, 246
multidimensional NMR
   spectroscopy
     algorithms, 271–272
       iterative hard thresholding
        (IHT), 273–274
       iterative soft thresholding
        (IST), 274–275
     applications, 283–291
     convex relaxation methods
      conjugate gradient
       descent, 275–276
      YALL1, 276
     higher dimensional
      spectroscopy, 291–299
     and implementation, 278–282
     iteratively re-weighted least
      squares (IRLS), 277–278
     non-convex minimisation,
      277–278
     terminology, 282–283
     theory, 269–271
multi-D spectral estimation, 71–72

NMR Consensus Trace Clustering, 245
NMR sensitivity, 2–3
non-negative matrix
  factorization, 245
^{15}N polarization enhancement
  pathway, 20–21
null transverse equilibrium
  magnetization, 34
NUS inverse problem, 181–185

one-dimensional Fourier transform
  (FT) NMR, 34–35
1D NMR, 78–82
orthogonal matching pursuit
  (OMP), 273

polar Fourier transform,
  134–135
polarization-enhanced fast-pulsing
  techniques
    BEST
      band-selective pulse
        shapes, 14–18
      BEST-HSQC *vs.* BEST-
        TROSY, 18–22
      ^{13}C-detected experiments,
        22–24
    experimental sensitivity,
      3–5
    experimental time, 2–3
    inter-scan delay, 3–5
    longitudinal relaxation, 3–5
    NMR sensitivity, 2–3
    proton longitudinal relaxation
      enhancement
        amide proton, 11–13
        paramagnetic relaxation
          agents, 7–8
        protons other than
          amides, 14
        selective spin
          manipulation, 8–11
        Solomon and Bloch–
          McConnell equations,
          6–7

SOFAST-HMQC
  Ernst-angle excitation,
    24–26
  implementations of,
    26–28
  UltraSOFAST-HMQC,
    28–29
post-processing, NMR spectroscopy,
  100–101
projection–reconstruction NMR, 203
proton longitudinal relaxation
  enhancement
    amide proton, 11–13
    paramagnetic relaxation
      agents, 7–8
    protons other than amides,
      14
    selective spin manipulation,
      8–11
    Solomon and Bloch–McConnell
      equations, 6–7

QR decomposition (QRD), 51
quantum dynamics simulation, 61

radiofrequency (RF) electromagnetic
  emissions, 170

sensitivity curve, 5
signal-to-noise ratio (SNR), 2, 268
single-scan two-dimensional NMR
  experiment, 38
singular value decomposition (SVD),
  51–52
SOFAST-HMQC
  Ernst-angle excitation, 24–26
  implementations of, 26–28
  UltraSOFAST-HMQC, 28–29
Solomon equations, 6–7
statistical post-processing, NMR
  spectroscopy, 102–103
statistical total correlation
  spectroscopy (STOCSY), 245
SVD. *See* singular value
  decomposition (SVD)

3D NMR, 86–91
two-dimensional Fourier transform
    (FT) NMR, 34–35
2D NMR, 82–86

ultrafast (UF) NMR spectroscopy
    principles of
        direct-domain acquisition,
            42–43
        generic scheme of,
            38–39

indirect domain
    information, 40–42
    magnetic field gradients,
        35–38
    spatial encoding, 39–40
processing
    basic procedure, 43–45
    SNR considerations in,
        45–46
unsymmetrical indirect covariance
    (UIC), 226–228, 227